ANIMAL WASTES

This book contains revised material based upon papers presented at the Seminar on Animal Wastes organized by the Regional Office for Europe of the World Health Organization and the Government of the Czechoslovak Socialist Republic in cooperation with the United Nations Development Programme and held at the Czechoslovak Research and Development Centre for Environmental Pollution Control, Bratislava.

ANIMAL WASTES

edited by

E. PAUL TAIGANIDES, P.E., Ph.D.

Professor of Environmental Engineering,
Department of Agricultural Engineering, Ohio State University,
Columbus, Ohio, USA

APPLIED SCIENCE PUBLISHERS LTD
LONDON

APPLIED SCIENCE PUBLISHERS LTD
RIPPLE ROAD, BARKING, ESSEX, ENGLAND

ISBN: 0 85334 721 2

WITH 127 TABLES AND 98 ILLUSTRATIONS

Printed in Great Britain by Galliard (Printers) Ltd, Great Yarmouth

Foreword

There is growing emphasis in the Regional Office for Europe of the World Health Organization (WHO) on public health problems related to pollution of the environment, and many of the activities now being undertaken require multi-disciplinary participation by physicians, engineers, biologists, chemists, legislators and economists. The inter-country programme of WHO concerning the promotion of environmental health includes, for example, work concerning planning methods in environmental pollution control, the study and development of information systems, the development of multi-lingual glossaries on air pollution, water pollution, solid wastes and ecology, the study of basic sanitation problems in Europe due to international migration and tourism, studies of the health hazards of sea bathing and of non-ionizing radiation, the production of a manual covering analytical methods used in water pollution control, and various aspects of occupational health and industrial hygiene.

These activities are designed to assist member countries in the solution of environmental health problems and the development of strategies in environmental management. The European Office also has a considerable number of country programmes, in cooperation with the United Nations Development Programme (UNDP) and individual governments, concerned with the development of water supplies and basic sanitation, with sectoral studies and other pre-investment projects concerning the development of sanitary services, and also a broad spectrum of environmental pollution control projects, sometimes concerned with the strengthening of institutions, such as the development of an Environmental Pollution Abatement Centre in Katowice, Poland, to serve the heavily industrialized regions of Silesia, and sometimes with comprehensive studies of pollution in particular geographical areas, such as the project for environmental pollution control in the metropolitan area of Athens, Greece.

Since 1971 a large-scale project has been in progress in cooperation with the Government of the Czechoslovak Socialist Republic with the object of developing a Federal Centre for Environmental Pollution Control. Located in Bratislava, the Centre has initiated extensive activities to stimulate and carry out research and development work on air, water and solid waste pollution control. It is significant that, when one considers the wide-ranging problems which are being experienced in the field of environmental pollution abatement in a country such as Czechoslovakia with a rapidly expanding industrial economy, the Centre chose as a subject for an important seminar held under the auspices of the project, with considerable international participation, the subject of animal wastes.

Such is the importance of this problem and its interests, not only to Czechoslovakia but to virtually all other European countries, that the European Office warmly responded to the suggestion of holding such a meeting. I wish to convey my congratulations to the Government of the Czechoslovak Socialist Republic and to the Centre on the successful outcome of the meeting. The hard work put in by so many people is justified by the response of participants in providing constructive and practical discussions to the various contributions by a distinguished international group of authors.

The increasing industrialization of agriculture has meant the concentration of large numbers of animals in single production units and the consequent creation of extremely complex pollution problems, the environmental effects of which, and the costs involved in their solution, sometimes being of a magnitude which has led either to the abandonment of plans for their construction, or to the curtailment of expansion or even the enforced closure of existing enterprises. The stage has been reached where it has been seriously suggested that large-scale animal production units should be sited in city areas rather than in the countryside, so that a municipal sewerage system would be available for disposal of wastes. Such a solution would, of course, have serious socio-economic effects and it is now commonly accepted that the problem of disposal of animal wastes must be solved by the agricultural industry itself. There has therefore been intensive research and development activity concerned with the design of stock units, the treatment of liquid and solid wastes, their ultimate disposal, their effects on the environment and, more positively, investigations into the possibilities of by-product recovery and re-use.

The public health aspects of the problem influence the siting decision for new developments, the type of construction, the way in which the unit is managed and the means of disposal of the wastes produced. In addition to public health nuisance due principally to atmospheric pollution, contamination of water can occur due to pathogenic bacteria, viruses and parasites, to organic pollution causing, for example, deoxygenization and the deposition of suspended matter, and enrichment due to nutrients and toxic matters including heavy metals, sterilizing agents and pesticides. If the animal wastes are applied to land, there are also dangers of direct contamination of crops intended for human consumption. Animal wastes present a difficult environmental problem but also present intriguing possibilities as a resource.

LEO A. KAPRIO
Regional Director
Regional Office for Europe
World Health Organization
Copenhagen
Denmark

Preface

This book contains revised material based upon papers presented at the Seminar on Animal Wastes (SEMAW), organized jointly by the Regional Office for Europe of the World Health Organization and the Government of the Czechoslovak Socialist Republic, in cooperation with the United Nations Development Programme, held in Bratislava, Czechoslovakia (ČSSR).

Unlike previous conferences on the same topic, held regularly over the past decade in the USA and elsewhere, SEMAW had several unique features. Only topics relevant to large modern feedlots were discussed. Modern large feedlots require and can afford considerable technology inputs. Abstracts of the papers were circulated for comment to all the other authors, government officials and feedlot specialists to ensure that the papers covered assigned topics. Technical presentations were arranged into small, coherent sessions of two to three papers followed by long periods of discussion, so as to stimulate effective participation by the audience. Workshops were organized to develop guidelines for the practical application of current technology and to identify gaps in knowledge. Modern large feedlots with unique waste management systems were studied during two days of organized excursions in Slovakia.

Seminar papers were written as chapters in a book, coordinated and edited to avoid non-essential duplication. Furthermore, the papers were revised after the Seminar to incorporate comments by other authors, by Seminar participants, and research findings reported during the Seminar or in the published literature. Papers were authored by leading authorities on animal waste management, well known for their expertise, and by government officials who are in charge of developing and/or administering agricultural pollution control policies.

This book is not a literature review of research or of available technology, nor are the papers reports of research or work done by the author alone. The papers contain the essential information and data needed by engineers, planners or scientists to evaluate alternative systems for feedlot waste and wastewater handling, treatment, disposal and utilization on the basis of their technical, economic and ecological feasibility.

This book is intended to serve as a practical reference book for planners, engineers, scientists, administrators and policy-makers working on all aspects of large feedlot development and feedlot operation, waste management and environmental health. It may also serve as a textbook for college and university undergraduate courses in agricultural waste management and utilization.

I wish to thank the authors of the papers for their cooperation, patience and willingness to rewrite their papers several times.

I am deeply grateful to my co-workers in the WHO Regional Office for Europe and in the Czechoslovak Research and Development Centre for Environmental Pollution Control for giving me the opportunity to coordinate this Seminar and edit its Proceedings and for their support and cooperation.

I am deeply indebted to the Ohio State University Officials, Professor G. L. Nelson (Chairman of the Agricultural Engineering Department), Professor R. M. Kottman (Dean of the College of Agriculture), and Professor H. Bolz (Dean of the College of Engineering), who allowed me to accept this demanding assignment. I wish also to thank my secretary, Mrs Linda Baker, for superb secretarial services, and my family, Maro, Paul, Tasos and Katerina, for their understanding and love.

E. PAUL TAIGANIDES

Contents

Part II: Technologies for Processing and Treatment of Animal Wastes

Part III: Utilization and Disposal of Animal Wastes

Acknowledgements

This Seminar was organized at the Czechoslovak Research and Development Centre for Environmental Pollution Control, under the sponsorship of the Czechoslovak Federal Ministry of Agriculture and Food, the Federal Ministry for Technical and Investment Development, and in cooperation with the Regional Office for Europe of the World Health Organization and the United Nations Development Programme.

The Seminar was attended by over 200 persons including several government ministers, 6 representatives of international agencies, engineers and scientists from 15 countries of Europe and North America, and public health officials, veterinary hygienists, animal nutritionists, researchers, economists and farm production managers from Czechoslovakia. Experts from 13 countries presented papers on selected topics. These topics were selected to serve the growing need for practical design information on planning and operating waste management systems for large animal feedlots.

Some 30 people from our own Centre were intimately involved in organizing this Seminar and making local arrangements for the participants. Their contribution to the success of the Seminar is hereby acknowledged.

The fine cooperation and contribution to the success of the Seminar by members of the sponsoring ministries and the WHO Regional Office for Europe and of the Technical Programme Coordinator and Editor are acknowledged with our deep gratitude.

We hope that the exchange of ideas during the Seminar and the publishing of the Proceedings in the form of a reference book for planners, engineers, scientists, administrators and policy-makers will contribute to the development of good management systems for the wastes from modern large animal feedlots and improve environmental health.

J. Fratrič
Director
Czechoslovak Research and Development Centre
for Environmental Pollution Control
Bratislava, ČSSR

PART I

The Animal Industry and the Quality of Environment

1

Agriculture in Czechoslovakia

Emil Čakajda

First Deputy-Minister of Agriculture and Food of the Czechoslovak Socialist Republic, Prague, ČSSR

The socialist reconstruction of Czechoslovakia carried out under the leadership of the Communist Party has brought about considerable changes in all aspects of life. At present, the Czechoslovak Socialist Republic is a country with advanced industry and agriculture. Even with a limited land fund (0·5 ha of farmland, 0·36 ha of arable land per head of population), the intensity of agricultural production provides a high level of nutrition for the inhabitants of the country.

Under previous five-year plans, the dynamic development of agriculture provided for a gradual reduction of imports of grain without increasing imports of protein components. Hence in the consumption of foodstuffs, particularly those of animal origin, Czechoslovakia is self-sufficient with her own produce. As to the consumption of basic foodstuffs, Czechoslovakia is among the leading countries of the world. For instance, in 1973 annual consumption reached the following levels:

Meat, total (on bone)	77·3 kg
Fats (value of pure fat, net value)	20·0 kg
Milk and dairy products	205·6 kg
Eggs	291·0
Sugar	36·7 kg
Cereals, total (value of flour)	109·1 kg
Fruit (value of fresh fruit)	40·0 kg
Vegetables (value of fresh vegetables)	78·5 kg
Beer	146·4 kg

Czechoslovak agriculture represents a large-scale production branch of the national economy fully utilizing not only the natural and climatic conditions (average temperature 7·8 °C, average rainfall 717 mm) but also the social, economic and political conditions for its dynamic development. Agriculture is undergoing a gradual transition to the industrial type of production. The foundations of this development were laid in 1949 when efforts were started to organize state-owned agricultural enterprises and to found Unified Agricultural Cooperatives on a large scale. The result of this socialist process is that at present more than 92 % of all farmland is owned by the Unified Agricultural Cooperatives and state agricultural organizations. (Total farmland area is 7 060 000 ha, arable land area 4 943 000 ha; the rest, 538 000 ha, belongs to small landowners and to a small number of individual private farmers.) Today there are 2746 Unified Agricultural Cooperatives with an average area of 1500 ha of farmland, and 284 state farms with an average area of more than 5500 ha of farmland.

The present structure of agricultural production is based mainly on the natural and climatic conditions. Particular importance is attached to the considerable irregularities of terrain and to the peculiarities of climate representing a transition from maritime to continental climatic conditions. The structure also fully respects the requirements of the national economy.

Cereals (mainly wheat and barley) are grown on a prevailing part of the arable land (56 %); fodder crops are grown on 29·6 %, potatoes, fruit and vegetables on 5·6 % of arable land, and the rest is represented by technical crops of which 4·2 % is formed by arable land under sugar beet. The structure of plant production corresponds with that of animal production. The main emphasis is laid on meat production, the production of pork still representing a prevailing part. All the remaining products of animal production share percentages proportionate to requirements.

Since 1948, gross farm output has increased to approximately a twofold level and market production has reached a level more than 3·3 times as high. At present the degree of the intensity of Czechoslovak agriculture can be characterized by the following indices (results of 1974):

Cereal yields, total	38·3 metric centners/ha
of which: Wheat	39·9 metric centners/ha
Barley	39·2 metric centners/ha
Sugar beet yields	397·3 metric centners/ha
Fodder crops on arable land, total, in dry condition	45·0 metric centners/ha
Potatoes	163·2 metric centners/ha
Meat produced/ha of farmland	237·0 kg/ha
Milk produced/ha of farmland	763·0 litres/ha
Eggs produced/ha of farmland	904/ha

Czechoslovak agriculture has a large material and technical base including the means of mechanization, chemization, large-capacity premises, agricultural building industry, etc.

Science and research, special services and agricultural education are highly developed. Advanced industry, particularly agricultural engineering and chemical industry, serves as an important potential for the further development of agriculture. In the development of Czechoslovak agriculture, an increasingly high importance is attached to advancing international cooperation.

Today, Czechoslovak agriculture has 142 000 modern tractors of all existing power classes. There are 49·5 ha of farmland per physical tractor and 139 ha under cereals per one grain harvester-thresher. Farms have 189 015 trailers and 19 743 trucks. Considerable progress has been made in the use of large-scale technologies in the production of bulk feeds (60 % is ensiled and stored as low-moisture haylage, 15 % is pressed, and only 14 % is dried in the traditional way). For this purpose, there are 416 drying plants producing 624 300 tons of dried feeds.

In recent years, successful results have been achieved in the mechanization of sugar beet and potato growing and harvesting. Full mechanization is available for 75·1 % of the area under sugar beet. However, due to the difficult terrain and soil conditions, the harvesting of potatoes is, at the present time, fully mechanized in only 60 % of the area under this crop (in the socialist sector).

Chemistry is an important factor in the development of plant production. The quantity of commercial fertilizers used at present represents 220 kg P and N per 1 ha of farmland. Chemical plant protection has been raised to a high level.

In animal production, the existing and, particularly, the intended and newly introduced concentration of animals enables a complex utilization of the most modern technology and modern means for the main technological processes.

The large-scale introduction of scientific and technical progress in Czechoslovak agriculture has been brought about mainly by the fact that our engineering, chemical and building industries fully meet the main material and technical requirements of agriculture; there is close cooperation with other countries, particularly within the CMEA, in this field. The Government of the Czechoslovak Socialist Republic is making planned efforts to create favourable conditions for this process.

From the viewpoint of the specialization and concentration process now taking place in agricultural production (which is considered as an objective and regular consequence of development), an increasingly important role is played by specialized state organizations rendering their services to enterprises in primary production. Czechoslovak agriculture has well-developed plant-breeding and seed-producing organizations fully meeting our requirements for high performance and high quality seed and planting material. The continuous development of animal-breeding work is the task of the State Animal-Breeding Administration, which at present is mainly engaged in solving the problems of the improvement of Czech and Slovak cattle breeds both for meat and milk efficiency. The main task of this organization in pig breeding is the preparation of a hybridization programme for the selection of a pork type of pig. Highly valuable services for the development of large-scale agricultural production are rendered by the State Veterinary Administration which has fully sanitated our farm animal herds in a short time in recent years, having eradicated diseases such as brucellosis in cattle and bovine tuberculosis; today, the veterinary service fully controls the health and veterinary conditions of large-scale animal breeding.

The development of the state feedstuff industry is an important intensifying factor in the development of agriculture in Czechoslovakia. At present this industry produces more than 6·5 million tons of feed mixtures, fully sufficient to meet the requirements for these mixtures in the main categories of farm animals in large-scale breeding.

As to the chemization of plant production, successful development has been started in the agro-chemical enterprises using scientifically justified methods of plant nutrition and protection; these enterprises are established as joint agricultural enterprises by cooperative farms and state farms.

In farm mechanization, valuable services are rendered to cooperative as well as state farms by Machine and Tractor Stations; projecting and building organizations provide services in the field of capital construction.

The system of large agricultural enterprises and advanced special services directed from one centre through a system of interlinked central, regional and district state directing authorities fully guarantees a planned development of Czechoslovak agriculture relying on the latest scientific findings and their introduction in practice.

The main source of scientific findings is a broad scientific and research base for the field of agriculture enriched by a whole range of research institutes for various branches of industry and by numerous basic research institutes of the Czechoslovak Academy of Sciences and universities. A valuable contribution to our research efforts is provided by the active international cooperation of Czechoslovak research institutes with similar institutions of the member countries of the CMEA as well as other countries. The research base directed by the Ministries of Agriculture and Food employs more than 8600 highly qualified workers.

There are five specialized agricultural and veterinary universities which have existed for many years in Czechoslovakia and which today have more than 10 000 students. Furthermore, we have a broad network of secondary agricultural schools and apprentice schools completely covering the demand for medium-level technical cadres and the requirements for workers in the main farm-worker professions.

Today, Czechoslovak agriculture employs fewer than 1 million persons, as distinct from the 2·4 million working in agriculture in 1948. More than 100 000 of this total number are university graduates and workers with secondary-level education. This is a basis creating good prerequisites for further dynamic development of our agriculture in the years to come.

The achieved intensity of agriculture and labour productivity are the main factors underlying the formation of a good economic situation in Unified Agricultural Cooperatives and on state farms. In keeping with the purposeful economic policy of the State based on guaranteed stable purchase prices of farm produce and on additional price and grant instruments, Czechoslovak agriculture creates favourable economic conditions for a rapid planned technical development and modernization of its base as well as for the meeting of all the personal and social needs of agricultural workers. The average personal incomes of workers in agriculture reach almost the same level as those of workers in the whole national economy.

The conditions of work and life are being very quickly balanced between agriculture and industry. The Czechoslovak village, the home of about 40 % of all inhabitants (only about 30 % of country families depending on work in agriculture), provides the same, or even better, conditions for the mode of life and standards of living, in comparison with conditions prevailing in towns.

The achieved political, social, economic and technical results of Czechoslovak agriculture make it possible to set before us further objectives in all fields.

In the field of production this means taking into account the expected growth of population and the increase of foodstuff consumption, that self-sufficiency in the main foodstuffs from the country's own resources should be secured, even despite the continuous decrease of the available land fund.

Introducing scientific and technical progress on a large scale, we are making efforts to secure a high organizational and technical level of agricultural enterprises and to develop them as large enterprises with an industrial character of production and organization.

A continuous growth of the standards of living for workers in agriculture and an all-round development of Czechoslovak villages should be secured in keeping with the development of the whole national economy.

Any impairment of the biological character of the landscape and deterioration of the environment of our socialist society by a further rapid development of agriculture should be prevented.

In the Marxist-Leninist conception, environment is not a separate part of social life. It is a result of the interaction of man, society and nature.

The results of this care for the environment in the Czechoslovak Socialist Republic demonstrate that the social approach to its requirements is based on the unity of the interests of the whole society; the practical implications of this care are the result of the efforts of all the components of the society.

For the development of the environment, the socialist State uses the results of scientific and technical progress and secures their utilization to the benefit of the society according to the plans of national economic development. The State solves and overcomes problems entailed by economic advance and by the changes in the mode of life which have assumed a revolutionary character in the rural population of Czechoslovakia in the past decade.

Unfortunately, this cannot be said about damages that may be caused to the environment in the period of an advanced 'agricultural revolution' which may affect the ecological basis of life with consequences beyond the present state of scientific knowledge.

Hence it is an important task of science to present proposals and to show ways of reaching an optimum harmony between the measures increasing the intensity of production growth on the one hand and the effort to maintain a good environment on the other.

If this objective is to be achieved, the latest findings of science and technology should be utilized on a mass scale in all fields. Hence we believe that, for instance in plant production, the varietal composition of crops will change more quickly than at present, the existing material being replaced by higher-yielding varieties; there will be a continuous process of quantitative and qualitative improvement of the system of plant production and protection, and soil fertility will be increased by ameliorating practices. These factors will play a decisive role in providing a generally higher degree of utilization of the land potential of Czechoslovakia.

In animal production, animal nutrition will be considered as the decisive field, besides improvement-breeding and veterinary work. Using the latest findings of science, we choose radical approaches to the solution of the problems of the production, harvesting, preservation and use of bulk feeds in order to set in effective motion the vast sources which are still frequently accounted as losses. Similarly, in the production of feed mixtures we make efforts to achieve an optimum quality in order to provide a high effectiveness of animal production as a whole.

Naturally, the assertion of these objectives, both in plant production and animal production, requires the necessary conditions to be created also in the field of farm organization in order to enable the maximum utilization of high-performance machines and equipment and to provide for the complete mechanization or automation of all operations. The main purpose is to introduce industrial modes of production in agriculture.

Czechoslovak agriculture has already started such efforts successfully. The gradual implementation of the programme of specialization creates, at the same time, favourable conditions for increased concentration in all fields of production. The main criterion is an optimum utilization of natural conditions and the technical and technological parameters of the existing and future machines and equipment. As to the main crops and the main categories of cattle, the situation under our conditions is that optimum parameters, judged from the viewpoint of specialization and concentration, can be obtained in an area ranging between 4000 and 7000 ha of farmland. These considerations are based on the assumption that one specialized crop, for instance sugar beet, will be grown in this area besides cereals and fodder crops, and that at least one category of cattle will be kept within this unit (for instance dairy cows). The production of pigs, poultry and eggs, requiring no direct connection with land, is organized separately on the basis of different principles. Two approaches are used in the solution of the problems of creating the spatial and organizational conditions to make these intentions reality: first, the achievement of the necessary farm size through a gradual voluntary fusion of smaller units (a majority of state farms have reached such a size since their origin): and second, the voluntary and mutually beneficial cooperation of enterprises, either in all activities or in some selected activities. The second form, *i.e.* cooperation, precedes the merging of enterprises in a majority of cases.

Cooperating activities are practised both by cooperatives and by state organizations. In addition, even in these forms of cooperation, there are some activities in which the optimum organization exceeds the limits of the given scope (for instance the fields of plant production, plant protection, open herd turnover, high-performance drying facilities, repairing and building activities, etc.). Higher forms of the cooperation of agricultural enterprises are practised in such cases; Unified Agricultural Cooperatives and state farms combine their efforts to create a specialized joint agricultural enterprise for such an activity. Joint agricultural enterprises are also designed to cover a major part of the production of pigs, poultry and eggs. At present there are about 360 enterprises of this type. There is a considerable variation in the area for which these special activities are carried out (ranging from 15 000 to 30 000 ha, as a rule). However, in some activities (*e.g.* egg production) this area is much larger. As mentioned above, the degree of concentration in different activities mostly depends on the achieved technical and technological parameters, on the requirements for transport, raw materials and consumers' demand, as well as on the requirement for environmental conservation.

Facilities with the following animal concentrations are at present being built for animal production:

Cows	600–1200 heads
Calves	1200 heads
Beef cattle	1000 heads
Pigs	up to 10 000 heads
Hens	150 000 birds

Under Czechoslovak conditions (population density 140/km^2), the possibility of a safe system of excrement disposal is a limiting factor on further concentration, particularly in pig production and also, to some degree, in cattle breeding. Unfortunately the existing systems do not fully guarantee the protection of the environment.

I believe that the exchange of findings and experience concerning excrement disposal will contribute actively to a gradual solution of these complex problems to the benefit of all involved. I also believe that available methods will be found, suitable from the viewpoint of water management, hygiene, technology, environmental conservation, etc., as well as from the economic point of view.

These problems are related to the processing and utilization of excrements from large-capacity animal production facilities. The solution of this range of problems is of key importance for further development determined by overall societal needs. Expert opinion was demanded through the Czechoslovak Research and Development Centre for Water Protection from Pollution, under the UNO/WHO Programme, in order to facilitate the use of foreign experience. The result was the recommendation to organize a special seminar on this subject, because even worldwide experience concerning this problem was found insufficient for the expected developmental trends.

Hence by organizing this Seminar, Czechoslovakia becomes one of the pioneer countries in the solution of the problems of environmental protection through the effective disposal and use of excrements in highly concentrated animal production. We consider the decision of WHO to join the organizing effort for the Seminar as an expression of confidence and as an appreciation of the work of our research institutions and of our practical efforts. Although the major form of international cooperation by the ČSSR in the field of research is the coordination of the scientific and technical research plan mainly with the plans of other socialist countries, I should like to express my gratitude to all the institutions of the United Nations participating in the preparation of the Seminar, for their initiative and for the means expended; my thanks also go to individual workers for the effort with which the Seminar has been prepared.

At the same time, I should like to assure all participants that we shall do our best to make this event an example of the effectiveness of international cooperation to the benefit of all countries involved, we intend to make the Seminar a practical expression of the Czechoslovak principles of international cooperation.

The path of the development of Czechoslovak agriculture, as described above, sets high requirements in all aspects. It will demand high material and technical investments, much political, organizing and technical effort and, naturally, much time. However, all the prerequisites for the practical implementation of the principles of this path are being created in the Czechoslovak Socialist Republic. It will bring about a considerable increase in the intensity of agricultural production, a further rapid growth of social labour productivity, and a continuous increase in the standards of living of farm workers. A valuable contribution will be brought to the all-round enhancement of the Czechoslovak economy and to the improvement of the conditions of our citizens' lives.

2

Animal Feedlots: Development, Trends, Problems

B. A. Runov

Vice-Minister, USSR Ministry of Agriculture and Food, Moscow, USSR

INTRODUCTION

Certainly no greater issue arising from the growth in world population is so pressing in its demand for solution as the problem of creating an adequate food supply. Among the food sources which are considered optimal for human nutrition and health are meat, milk and eggs. Animal products contain high concentrations of protein, most or all of the amino acids and some of the vitamins essential for human growth, a balanced content of hydrocarbons, fats and minerals, and in addition they taste good. The low calorie content of animal products makes them superior to most other human food sources.[1]

In animal production operations we are beginning to see an assembly-line type of mass manufacture of eggs, milk and meats. Poultry, cattle, swine and sheep can be raised in complete confinement with an automatically regulated environment, automatic feeders, automatic waterers and mechanized removal of eggs and milk. A significant innovation in animal production has been the change from pasture to raising livestock on an industrial basis in animal feedlots.

FEEDLOT DEFINITION

The term 'feedlot' actually implies many things, but all of them basically involve increased reliance upon the use of modern technology to introduce a systems approach to livestock management in order to ensure the highest level of productivity. It may involve the mechanization of many farm processes into flow lines using maximum automation, the combined construction of farm buildings with related process areas optimally joined, and the efficient use of centralized control methods in the overall management of the complex.

More specifically, the term 'feedlot' is defined as a facility which meets the following three criteria:[2]

1. High concentration of animals held in a small area for periods of time ranging from weeks to years for the purpose of production of meat, milk, eggs, breeding stock or the stabling of horses.
2. Majority of the feed and water is transported to the animals.
3. Land area occupied by the animals will neither sustain vegetation nor be available for crop production.

Pasture and range-grown animals may meet the first criterion and, on occasions, the second, but do not meet the third criterion because by definition a pasture or range produces hay and crops for the animals to pasture on.

From the standpoint of pollution control, the major difference between feedlots and pasture is that the first is a 'point source' of pollution, while the latter is a non-point or diffused source of pollution. Under pure pasture operations there should be no pollution development because theoretically the wastes from the animals provide the plant nutrients required to grow hay or the necessary grass to sustain the animals. In pasture production, animal wastes are recycled *in situ* and do not become a burden to the environment.

In animal feedlots, wastes are generated at such rates and quantities that they cannot possibly be assimilated by the land area within the feedlot facility. Wastes therefore accumulate and must be transported away for disposal within time-intervals ranging from daily to several months. In some instances wastes may accumulate or be stored for one year, but eventually they would have to be disposed outside the feedlot facility boundaries.

Feedlots may be complete confinement units with the animals being housed in covered structures, with their environment being mechanically controlled and without the animals being exposed to the direct effects of rain, wind or solar rays. Open feedlots may be partially covered to provide shade and some protection from extreme heat, rain or wind.

Concentration per animal may vary depending on the severity of the climate, but is always less in open feedlots than in housed feedlots. For beef cattle, for example, space per animal decreases from 200 m^2/animal in good pasture land to 20 m^2/head in open feedlots, and to 2 m^2/head in housed feedlots with slotted floors. As space per animal is decreased, requirements for skilled management, higher technology and energy increase. Cost per animal, however, is decreased, making feedlots an economical method of raising high-quality animal food under controlled conditions. Production of animals in confinement feedlots results not only in high-quality product but also in uniform quality.

In developing feedlot facilities, the designer should keep the farmstead flexible so as to be able to incorporate into the facility new innovations and technologies. Also, management systems and approaches change every few years. These changes should be adopted with a minimum of cost and effort. Good planning is time and money well spent.

The first step in feedlot planning is selecting the proper site. Major facilities for which plans must be made are all-weather access roads, feed grain and roughage storage, feed mill if feed needs to be prepared daily, housing and/or corral structures for the animals, feeding and watering systems including feed bunks, waterers, feeding roads and alleys, hospital area and working chutes to immunize or deliver medicine to animals, manure handling facilities, and offices and sanitation facilities for the people working in the feedlot.

Important considerations in planning for these facilities are drainage, orientation of the facility in terms of access by animals to solar light and energy, in terms of wind protection, in terms of wind direction so as to avoid dispersal of odours towards human settlements, ventilation of each part of the feedlot with fresh air, exposure to the drying rays of the sun, etc. Also, feed processing and feeding facilities should be oriented for efficient handling of materials from storage without interfering with waste handling or other daily operations on the feedlot.

PRODUCTION FEEDLOTS

Table I presents typical input–output figures for cattle, swine, poultry, sheep and horse feedlots. The table gives ranges on the main inputs and major outputs of a typical animal feedlot. The main inputs to a feedlot are the animals themselves, feed, water and bedding material. Although straw and other bedding materials are still used extensively, they are used mainly in smaller-size feedlots and their use is decreasing every year. There are several reasons for this, the main ones being the scarcity and cost of bedding materials, the high labour requirements, and the recent trend towards animal feedlots being separated from crop production.

Manure is the largest single item which has to be handled on the feedlot. The large quantities of manure to be handled, the distaste exhibited by workers towards handling waste, the inherent properties of manure, and the potential for disease and pollution make waste handling and management the biggest single problem in feedlot management.

For a 30 000 beef cattle feedlot, the daily feed grain requirement would be in the range of 240–360 tons per day, or on average 110 000 tons of feed grain per year, according to Table I. One can then estimate from such figures mill capacity and storage facilities if grain is not to be raised locally. If surrounding cropland is to provide the feed grains for the feedlot, then the amount of cropland area within an economic transport area of the feedlot site can be estimated on the basis of average annual yields. For example, if

TABLE I

TYPICAL INPUT–OUTPUT DATA IN MODERN ANIMAL FEEDLOTS
(data compiled from refs. 3–6)

Animal type	*Feedlot type*	*Animal size range (kg/head)*	*Time in feedlot (days)*	*Materials input*		*Materials output*	
				Feed (kg/head/day)	*Water (litre/head/day)*	*Product (kg/head)*	*Waste (manure) (kg/head/day)*
CATTLE							
Beef	Open	250–500	100–180	8–16	40–120	500	2–20
	Housed (slotted floor)	250–500	100–180	8–16	40–120	500	10–30
	Housed (solid floor)	250–500	100–180	8–16	40–120	500	10–30
Dairy	Stalls	500–650	—	15–25	60–320	10–40[a]	40–60
	Free stalls	500–650	—	15–25	100–130	10–40[a]	40–60
	Cow yard	500–650	—	15–25	120–320	10–40[a]	40–60
SWINE							
Pork pigs	Open (dirt)	20–100	150–180	1–3	4–20	100	1–3
	Housed (slotted floor)	20–100	150–180	1–3	4–20	100	1–5
	Open (solid floor)	20–100	150–180	1–3	4–20	100	1–5

POULTRY							
Broilers	Housed (litter)	0–2	40–60	0·05–0·1	0·1–0·2	1·5–2	0·05–0·06
Layers	Housed (cages)	1·5–2	400	0·1–0·12	0·15–0·2	0·6–0·9[b]	0·1–0·2
	Housed (litter)	1·5–2	400	0·1–0·12	0·15–0·2	0·6–0·9[b]	0·1–0·2
Turkeys	Open	2–14	120–170	0·2–0·3	0·3–0·5	8–14	0·3–0·6
Ducks	Open, wet	0·5–4	40–60	0·2–0·3	40–130	3–4	—
SHEEP							
Lambs	Housed	30–60	40–150	2–3	4–7	40–60	1·5–3
Sheep	Housed	50–100	40–150	2–4	7–13	Wool and mutton	2–4
	Open	50–100	40–150	2–4	7–13	Wool and mutton	2–4
HORSES	Stable	300–600	—	9–14	30–40	Recreation, work	20–60

[a] Milk in kg/cow/day.
[b] Eggs/hen/day.

maize feed grain production averages 5 tons/ha/year, then 22 000 ha of cropland would be required.

Beef Cattle Feedlots

Higher efficiencies in beef feedlots are sought through the development of early-maturing breeds, improved feed rations and greater applications of technology. Because feed cost accounts for 60–80 % of the total production costs, it is important that beef feedlots develop their own feed sources. Such methods as the use of mixed feeds, premixes and supplements, pellets, cubes, etc., in combination with high-quality hay and silage produced by the feedlot site can reduce total energy intake by 20–25 % per unit of gain.

Beef calves are raised on rangeland or in pasture. They are brought into feedlots when they are one to two years old weighing 200–300 kg. Feed rations are programmed to contain higher proportions of hay at the beginning of the feeding period in the feedlot. Fat content of the feed increases at the end of the feeding period. The exact proportion of each feed constituent depends on several factors, including availability and cost of each ingredient, weight of animal and the goal for the quality of the end-product.

Over 90 % of beef cattle are raised in open lots. Only a small percentage is grown in housed feedlots, but this number is steadily increasing, particularly in northern climates which are subject to extremes of temperature.

Dairy Cattle Feedlots

In planning a dairy cattle facility, several decisions need to be made about the type of housing, the animal breed, the milking parlour and the rearing of calves within the feedlot premises or at another location.

The major types of housing are: stall barn with milk room, free-stall barn with milking centre, and cow yard with milking centre. Over 50 % of cows are raised in stall barns. In this system, milk cows and replacements are retained at a fixed location where they are fed and cows are milked. Barns are insulated and mechanically ventilated. Cows spend 100 % of their time in the barn where they are fed and milked manually or by portable milkers with the milk being piped to a milk room where it is cooled and stored. In some cases, cows spend 80 % of their time outside the barn pasturing in nearby fields. Bedding or rubber mats may be used in such barns. Manure is usually collected over straw or mechanical gutter cleaners and moved out of the building on a daily basis. Storage facilities for solid and/or liquid manure are therefore needed.

Free-stall barns are only used by less than 20 % of the dairy cattle operations, but this is a rapidly expanding method, particularly on large feedlots. Where pasture is not available, milk cows and replacement heifers are kept under roof in barns but are allowed free movement between resting stalls and feeding areas. Where pasture is available, the animals spend as much as 80 % of the time on pasture. The barns are generally not insulated and are naturally ventilated. In very severe climates, insulation and mechanical ventilation can be found.

The milking centre includes a milking parlour and milk room. Cows are milked in the parlour twice daily and the milk is mechanically transferred to the milk room where it is cooled and stored. Milking equipment is cleaned daily in both rooms. Over 90 % of the free-stall barns still use bedding in the resting area. Manure is usually collected in alleys and mechanically scraped out of the barns. Some used bedding may be added to the manure to improve its handleability. For these systems, semi-solid wastes from the barn and milking centre are generally the same as those from the stall barn system. The remaining 10 % of free-stall barns utilize liquid manure systems split equally among three types: solid floors with liquid storage, slotted floors, and liquid flush.

In cold regions, solid-floor barns with separate liquid manure storage are becoming more popular than slotted-floor barns with sublevel storage. In warm regions, liquid flushing systems predominate with collection of the diluted wastes and daily or other periodic irrigation of fluid wastes to the land. The milking centre wastes for these liquid manure systems are generally added to the manure storage to reduce the total solids concentration for ease of pumping.

Cow yards with milking centre are used with approximately 25 % of the dairies. They are predominant in regions with warm climates. Where the climate is hot and dry, milk cows and replacements are maintained in open-sided shelters which provide shade, and in cooler climates the shelters are partially enclosed and include bedded packs or free stalls.

Bedding is seldom used in large yards in dry climates but often used in shelters in cooler climates. Manure may be removed for field spreading weekly from paved yards, or after the winter season from bedded shelters and partially paved yards. With large earthen yards, manure may be mounded periodically and removed annually for field spreading. The liquid waste discharge from this type of production system consists of milking centre wastes and runoff resulting from precipitation on the exposed surfaces of the cow yard.

Swine Feedlots

Three major types of swine feedlots are used: (a) open lots or pasture lots; (b) fully roofed buildings with completely or partially slotted floors; and (c) solid concrete floors with partial or full roofs.

Open dirt or pasture lots account for the majority of the world's pig production. However, due to the large space requirements (60–500 pigs/ha) and several other reasons including lack of odour and pollution potential control, the tendency is towards housed facilities. Runoff is the primary waste output from these lots. Animal density, rainfall/runoff relationships and other hydrological factors (land slope, infiltration rate, rain intensity and duration, etc.) are therefore the primary factors to consider with respect to pollution control facilities. Lots with more than 250 pigs/ha will not support vegetation and are considered feedlots.

Buildings with complete roofing and slotted floors are a recent development in the swine producing industry and presently account for 20% of the total production capacity. These buildings generally consist of two types, those with only a portion of the floor space slotted and those with the entire floor space slotted. Slotted floors with temporary storage pits underneath reduce the hand labour required to clean pens and are thus responsible for the continuing trend towards fully enclosed houses with totally slotted floors.

Many units for market hogs were developed on the basis of 0·07 m^2/animal with one-third slotted. Many of these buildings serve as a combination nursery and finishing unit. A common pen size is 3·0 m by 7·3 m for three litters or approximately 30 pigs. Storage capacity in the pit is 0·1–0·3 m^3/pig for a depth of 0·6–1·2 m.

Buildings with partially or totally slotted floors may have storage pits, oxidation ditches or under-house lagoons incorporated as an internal component of the production unit. The latter two systems are used in less than 10% of swine feedlots.

Systems with manure storage pits predominate and account for over 90% of all slotted-floor operations. Pits are generally filled with a minimum of 0·15 m to a maximum of 0·6 m of water before pigs are placed in slotted-floor pens, or after complete wastewater discharge. This allows for better cleaning when pits are emptied as well as reducing odours initially. Spillage and overflow from waterers and mist from fogging for summer cooling add to the quantity of liquid which must be handled. The amount of washwater employed to clean different partially slotted units represents the major difference in the volume and concentration of wastes stored in partially or totally slotted production units. Additionally, the manner in which pit waste is discharged will affect concentration. Many storage pits are completely emptied only every three to six months to reduce labour and water precharge requirements. Pits may be equipped with an overflow pipe to control water level and supernatant may thus be continuously released or partially discharged as necessary. The concentration of a supernatant overflow or partial discharge will be less than the average concentration of the total waste load for a complete pit emptying.

Production units with solid concrete floors may be partially or totally roofed. About 25% of the swine production capacity is of this type and units can vary from those which are partially open, having 2 m^2 of floor space per market animal, to those which are completely roofed with only

1 m^2 of floor space. The most prevalent practice is to have 1–1·5 m^2 per animal with one-half to two-thirds under roof.

Bedding of wood shavings, straw or sawdust is used in some farrowing houses and nurseries because of its insulation and absorptive characteristics; however, this represents an insignificant portion of the wastes from the swine industry.

Wastes on the concrete pen floors are periodically washed or scraped into a collection gutter. High-pressure, low-volume hose systems allow more rapid and efficient cleaning and thus large reductions in washwater quantities. Drinking-cup spillage, fogging water and urine continuously flow into the collection gutter. Rainfall and roof runoff that have access to the concrete floors or drainage that enters the waste collection and conveyance system contribute to the amount of wastewater that must be handled. The amount of rainfall that must be handled is small, amounting to an average of about 1–2 litres/day/animal due to the high animal stocking rate. These liquid wastes leaving the gutter may enter a concrete tank or some other temporary storage, a lagoon, or be discharged to adjacent land.

Poultry Feedlots

Poultry feedlots are of four major types: broiler, layer, turkey and duck feedlots.

Three basic operations are involved in broiler feedlots: breeding stock, production of broiler chicks, growth and slaughter of broilers. These operations may be separate or integrated into one facility. Breeder stock are kept in houses which include nesting, litter-covered floor pits, and a slotted or wire-covered perching area over pits. The wastes consist of a mixture of manure and litter plus manure scraped from the pit. The weight of the breeder stock birds is about 2·7 kg for hens and 3·6 kg for roosters, with an average of one rooster for every ten hens. At the end of their useful life ($1\frac{1}{2}$ years) these birds are usually sold as roasting chickens.

Broilers are usually raised in houses using a floor litter system. The birds are grown to a weight of about 1·8 kg in approximately eight weeks. The waste is in the form of mixed litter and manure. The litter is replaced periodically and may remain in the house for as long as a year. It may be turned with a plough between batches of chicks, and various chemicals (to aid in composting) and more litter may be added each time.

The egg-laying industry is comprised of two different operations: laying hen production and egg production. Breeding birds are maintained similarly to broilers, and the wastes are similar. Roosters and hens weigh about 3·6 kg and 2·7 kg respectively. The growing pullets are maintained in cages over pits (20 %) or in floor litter systems (80 %). In a very few cases floor litter systems are used in the first few weeks followed by cage systems until laying age.

Laying hens are maintained in several types of housing systems. These are: cages over dry pits (approximately 70% of all operations), floor litter/pit perch (20%), slat-wire/litter pit (5%), cages over dry pits (ventilated) (3%), and cages over wet pits (2%). The cage systems use several set-ups for the cages which are only different in geometry with no effect on the waste load. Both dry and wet pit systems utilize mechanical removal of wastes. Fans are used in the ventilated system for drying the wastes. The pit in this type of system is generally deeper than pits used in the other system. In the wet pit system, water is added to facilitate pumping of the waste in the form of a slurry.

The floor litter/pit perch system utilizes a litter-covered floor area with nests for laying and perches mounted over pits. Food and water are available in the floor area. The slat-wire/litter pit system is essentially the same except that slatted floors or wire meshes are used over the pits and food and water are placed in this area.

The breeding stock operations for turkey feedlots are separate from the growing of turkeys for slaughter. Breeding stock are maintained on litter. Litter absorbs the manure and is mechanically removed periodically.

Over 70% of the turkey feedlots are open lots with the remainder being mainly housed lots. Open lots consist of a brooder house where the turkey poults are kept for the first eight weeks after hatching and outdoor confinement areas where the birds are fed to a finished market weight. The latter period averages nine weeks for hens and 15 weeks for toms. Normally open lots grow one flock of birds per year. Some open lots utilize the brooder house to produce a second flock, in which case these birds are finished in confinement. The wastes produced in the brooder house consist of the manure excreted by the birds and the litter material which is used to cover the floor; both are removed from the house by mechanical means between groups of birds.

The normal bird density is 600–1300 turkeys per hectare of open feedlots. Confinement or range areas are rotated or waterers and feeders are relocated periodically to prevent excessive accumulation of wastes in one area. This procedure is also helpful in disease and parasite control. These areas are ploughed under and planted in cover crops during idle periods. Pollution potential develops only during periods of heavy rainfall.

In housed feedlots, production of slaughter birds consists of poult growing for the first eight weeks after hatching in a brood house and then feeding to finish market weight in confinement rearing houses. The finishing houses are similar to brood houses with the exception that more space is allowed for each bird. Both the brood house and finishing house utilize litter on the floor and wastes are removed periodically by mechanical means.

Duck raising facilities may be considered as being in two major groupings, wet and dry lots. The primary difference between the two is the amount of water used. In feedlots, the primary purpose for allowing ducks free access to swimming water is for improvement in the quality of the feathers (used as down). However, the quality of down is apparently not materially affected by growing in total dry lot facilities. Many of the larger producers have integrated facilities, that is, breeding, hatching, growing and slaughtering facilities are located on the same complex with the waste treatment plant usually designed to handle the entire waste load of the facility.

The largest group (80 % of the population) are the 'wet' lots in which the ducks have access to water runs. Local surface and groundwater is channelled to supply the birds with swimming areas ('runs') and to facilitate a controlled discharge of wastewater for treatment and disposal. The slopes leading to the water's edge collect faecal material which is washed into the water during rainstorms. These runs may be combined with some shelter facilities. For the last few weeks of the growing period, the birds are completely raised on the run and adjacent land. The amount of water provided in these 'wet' lots ranges from 40 to 130 litres/duck/day.

A second category of feedlot is the 'dry' lot, peculiar mainly to continental climate areas. The main difference from a wet lot is the reduced amount of water used in the raising of the duck. Dry lot facilities are usually constructed with flushing troughs placed under a wire floor portion of the building. Feeders and waterers are also in this area providing for collection of some of the manure. The remainder of the floor is solid covered with litter. Flushing results in the dilution of the manure and movement of the slurry into the processing plant. Water usage for a dry lot generally ranges up to 15 litres/duck/day.

Sheep Feedlots

Cereal grains make up the bulk of lamb fattening rations. However, roughage is also mixed with the feed. Bedding is used only in housed feedlots.

Open lots include completely exposed dirt lot operations as well as partial confinement (*e.g.* sheltered feeding areas) where the open area is a corral. The wastes from these operations include both manure and runoff. Breeding flocks in open lots are stocked at a rate of from 2 to 20 m^2/lamb for fattening; 2–10 m^2/lamb are allocated.

Housed lamb production represents the most modern and concentrated lamb feeding operation. Wastes from such an operation are either solid (scraped) or liquid (pumped) depending on the chosen waste management scheme.

Horse Feedlots

Considerable amounts of bedding, ranging from 4 to 20 kg/horse/day, are used both for manure absorption and for cushioning the horse when it lies down. Straw, sawdust or wood shavings may be used as bedding material.

Stalls should be cleaned frequently, if not daily. The amount of manure produced per horse varies from 15 to 22 kg/horse/day. Mechanical gutters have been used to remove manure, but bedding must be scraped to the gutter cleaner. Usually horses are housed one horse per stall.

In open lots where sufficient land is available for horses to roam around, several horses may be confined to a pen area.

CONCLUSION

As the population of the world grows, the need for more food production also grows. We in agriculture, as the food and fibre providers of the world, have a greater responsibility for human survival today than ever before. At the same time, however, we have the challenge and the opportunity to serve society in a noble task.

Animal production for human food will continue to be important in the foreseeable future. Humans like the taste of meat, poultry, milk and eggs. Animal products are not only among the most nutritious foods, but also have the unique capability of satisfying the palate.

The challenge to animal scientists, public health officials, government policy-makers and engineers is to continue to strive for more food from animals with less feed grains and roughage, energy and land resources. Higher efficiencies and good-quality products have been achieved in animal feedlots, but economic waste management under environmentally sound principles could limit progress in achieving even higher plateaux of productivity in animal feedlots.

REFERENCES

1. Kottman, R. M. and Geyer, R., Future prospects of animal agriculture. In: *Livestock Waste Management and Pollution Abatement*, pp. 9–18, American Society of Agricultural Engineers (ASAE), St Joseph, Mich., 1971.
2. Loehr, R. C. and Denit, J. D., Effluent regulations for animal feedlots in the USA (this volume, pp. 77–89).
3. Runov, B., *Animal Feedlots in Canada and in the United States of America*. Agricultural Publishing House, Moscow, 1970.
4. Environmental Protection Agency, *Effluent Limitation Guidelines for Feedlots*, EPA 440/1–73/004. Washington, DC, 1973.
5. American Society of Agricultural Engineers, *Livestock Waste Management and Pollution Abatement* (Proceedings of the 1971 International Symposium on Livestock Wastes at Ohio State University). ASAE, St Joseph, Mich., 1971.
6. Taiganides, E. P. and Stroshine, R. L., Impact of farm animal production and processing on the total environment. In: *Livestock Waste Management and Pollution Abatement*, pp. 95–8. ASAE, St Joseph, Mich., 1971.

3

World Demand for Animal Products for Human Food, 1970–2000

W. H. BARREVELD

Agricultural Industries Officer, Food and Agriculture Organization, Rome, Italy

The World Food Conference held in November 1974 has made it abundantly clear that 'increasing the rate of food production substantially, especially in developing countries' is the number one priority to mankind to meet the challenge of an explosive population growth and the urgent need to increase the quality of life. Out of about 4000 million people, 460 million are actually starving, while 2000 million people are badly nourished.[1] During the last 20 years, food production in the developing countries, inhabited by 74 % of mankind, hardly kept pace with population increase and this has led to a stagnation of food production per head as is illustrated in Figure 1.[2, 19]

The outlook is further overshadowed by the prognosis that over the next ten years the largest population increase will take place in the developing countries by an amount of 765 million, who are least equipped to cope with the situation due to lack of skilled human manpower and technological and financial resources. Looking to the future,[3] a preliminary forward

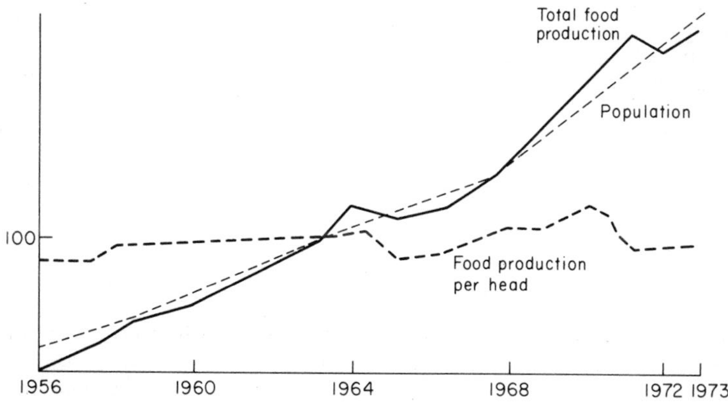

FIG. 1. Food and population balance in developing countries (indexes 1961–5 = 100).

assessment of food demand and supply has been made for the period up to 1985 on the basis of assumptions concerning growth of population and gross national product, and excluding any major changes in government policies or relative prices. On this basis, world food demand in the 1970s and 1980s is calculated to grow at a rate of 2·5 % per annum, of which 2 % represents population increase and 0·5 % increased purchasing power. This, however, masks major differences between country groups: thus in the developed countries taken as a group the growth rate of demand is projected at 1·6 % per annum, but in the developing countries at 3·7 % in terms of farm value.

The larger portion of the world population living in developing countries will have the utmost difficulty in keeping food production in line with increased demand and will therefore have to revert mainly to the cheapest resources such as cereals and other vegetable protein supplies to fulfil its basic calorie and protein needs. This prediction is further supported by an analysis of the present consumption pattern of animal and vegetable protein in different parts of the world. With increased affluence, people will spend more of their income on protein foods, in particular proteins derived from animal origin, as is illustrated in Table I.[4]

TABLE I

PROTEIN CONSUMPTION PER HEAD IN GRAMS PER DAY (from ref. 4)

Region	*Cereals*	*Animal protein*	*Other sources*	*Total*
North America	15·9	70·7	11·6	98·2
Australia and New Zealand	22·6	63·4	8·1	94·1
Western Europe	27·0	48·5	12·4	87·9
Eastern Europe	42·3	35·8	12·5	90·6
USSR	43·6	35·6	12·4	91·6
Argentina, Paraguay and Uruguay	26·5	57·4	10·0	93·9
Japan	25·8	31·8	19·3	76·9
Central America	31·6	22·8	14·4	68·8
Caribbean	21·4	22·8	13·8	58·0
Africa	33·3	12·1	15·6	61·0
Near East	45·1	12·2	8·6	65·9
South Asia	32·3	6·3	9·7	48·3
China	31·8	8·8	15·9	56·5

Taking 40 g/head/day as the minimum daily protein intake required,[5] provided sufficient calorie intake is assured, the average total protein intake figures suggest that protein deficiencies are not caused by inadequate total supplies but by bad distribution. To what extent this is the case is shown in Table II.[3]

TABLE II

ESTIMATED NUMBER OF PERSONS RECEIVING LESS THAN THEIR MINIMUM PROTEIN REQUIREMENT IN 1970

Region	*Population (million)*	*Percentage receiving less than minimum*	*Number of persons receiving less than minimum (million)*
Developed countries	1 074	3	28
Developing regions, excluding Asian centrally planned economies	1 747	20	360
Latin America	283	13	37
Far East, excluding China	1 020	22	221
Middle East	171	20	34
Africa	273	25	68
World, excluding Asian centrally planned economies	2 821	14	388

Protein deficiencies are most severe in the developing countries as a whole, and Africa, the Far East and the Near East in particular, areas which correspondingly show the lowest production rate of animal products per head of human population, per person active in agriculture and per animal unit (Table III). From this table we can distinguish three systems of animal production: large areas of land and low labour input (Oceania); little land, high labour input (Europe); and the extensive land use and high labour input of Africa. It is clear that the means and methods of increasing animal production differ between these regions.[7]

From the foregoing the following main conclusions can be drawn:

1. Increasing food production in the world is the number one priority to meet food demand over the next decades.
2. Increased affluence will create a shift towards more protein consumption, and of animal protein in particular.
3. Highest consumption per head and production of animal protein is found in the developed countries.

In addition to these long-term trends, during the last few years the world has had to cope with diminishing food reserves. It started in 1972 when bad weather seriously affected cereal production in several subcontinents simultaneously. World food stocks fell by 38 million tons, and a further decrease of 11% took place in 1973. The concurrent sharp rises in petroleum prices, reflecting themselves in increased costs for processing, transportation and distribution, caused further sharp rises in food and feed

TABLE III

PRODUCTION OF ANIMAL PROTEIN IN KILOGRAMMES, 1970[a] (from refs. 7 and 8)

	Per head of human population	*Per person active in agriculture*	*Per 1 000 ha of agricultural land*	*Per animal unit*[b]
Europe (excluding USSR)	20·1	191	38 085	62·5
North America	26·5	1 056	11 907	50·0
Oceania	62·4	1 625	1 866	17·9
USSR	18·6	115	7 336	44·2
Africa	2·4	8	542	5·6
Latin America	8·7	66	4 113	11·1
Far East	1·8	7	7 125	6·3
Near East	4·5	21	2 990	13·1
China	2·6	9	7 518	16·1
World	8·0	38	6 345	22·3

[a] Protein content:

Beef	15%
Mutton and goat	12%
Pork	11%
Poultry	14%
Eggs	11%
Milk, cow and buffalo	3·5%
Milk, sheep and goat	5%

[b] Animal units:

Cattle	0·8
Sheep and goats	0·1
Pigs	0·2
Poultry	0·01

prices, which resulted in decreased and changing consumption patterns of animal products.

The combination of long-term trends and the recent tumultuous events in the food and feed market make long-range demand projections no easy task. Tables IV–VII are abstracts from FAO's commodity demand projections up to 1980,[9] and world demand projections compiled for the World Food Conference for 1985 and 1990.[3]

It is essential to bear in mind, however, that these are assessments of a future commodity situation which would arise from a continuation of trends in production and demand under given assumptions, which for the 1980 commodity projections have been as follows: (a) constant prices and unchanged national production and consumption policies over the projection period; (b) continuation of recent trends in per capita incomes, with a small acceleration in the rate of increase in the developing countries; and (c) population growth at rates anticipated in the UN demographic projections.

All major animal products for human food have been included, but for this paper the geographical distribution has been restricted to three major groups of countries:

Economic Class I: North America, Western Europe, Oceania and others (Israel and Japan).
Economic Class II: Africa, Latin America, the Near East, Asia and the Far East.
Economic Class III: Asian centrally planned economies, USSR and Eastern Europe.

All tables for meat products refer to carcass weight excluding offals and slaughterhouse fats. For milk and milk products the figures are in terms of whole milk, equivalent on the basis of fat content of milk in producing countries.

TABLE IV

CONSUMPTION AND PROJECTED DEMAND OF BEEF AND VEAL MEAT

	Consumption				*Demand*	
	1964–6		1969–71		1980 *projected*	
	Total ('000 tons)	*Per head (kg/year)*	*Total ('000 tons)*	*Per head (kg/year)*	*Total ('000 tons)*	*Per head (kg/year)*
Economic Class I	18 388	26·6	21 208	29·2	27 483	34·1
Economic Class II	8 237	5·4	9 556	5·5	13 551	5·9
Economic Class III	6 521	5·8	8 590	7·0	11 771	8·0
World	33 146	9·9	39 354	10·6	52 805	11·6

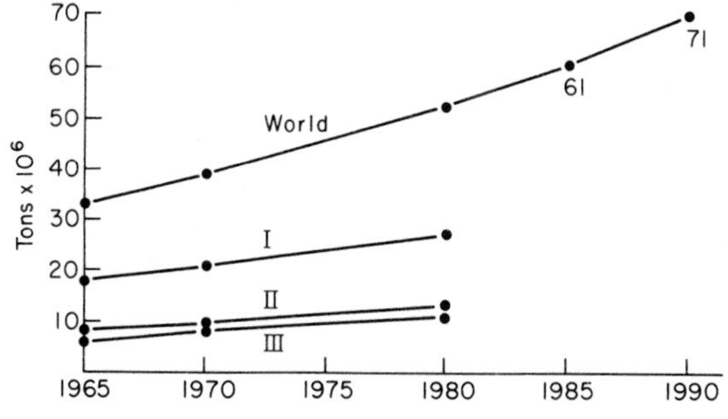

TABLE V

CONSUMPTION AND PROJECTED DEMAND OF MUTTON AND LAMB MEAT

	Consumption				*Demand*	
	1964–6		1969–71		1980 *projected*	
	Total ('000 tons)	*Per head (kg/year)*	*Total ('000 tons)*	*Per head (kg/year)*	*Total ('000 tons)*	*Per head (kg/year)*
Economic Class I	2 349	3·4	2 620	3·6	3 199	4·0
Economic Class II	2 273	1·5	2 537	1·4	3 921	1·7
Economic Class III	1 703	1·5	1 914	1·6	2 633	1·8
World	6 325	1·9	7 071	1·9	9 753	2·1

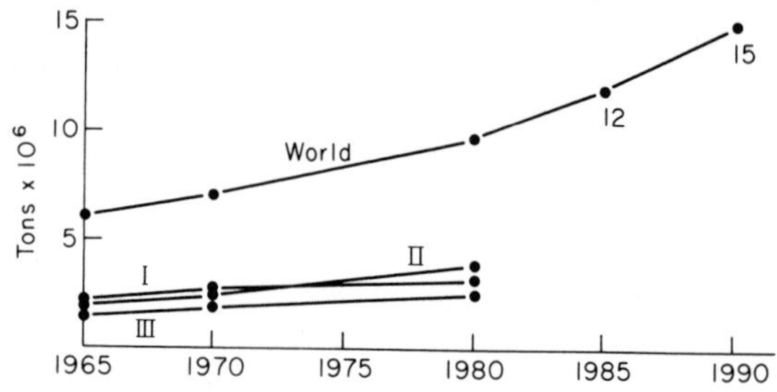

TABLE VI

CONSUMPTION AND PROJECTED DEMAND OF PIG MEAT

	Consumption				*Demand*	
	1964–6		1969–71		1980 *projected*	
	Total ('000 tons)	*Per head (kg/year)*	*Total ('000 tons)*	*Per head (kg/year)*	*Total ('000 tons)*	*Per head (kg/year)*
Economic Class I	14 509	21·9	17 338	23·8	20 295	25·2
Economic Class II	2 907	1·9	3 636	2·1	5 586	2·4
Economic Class III	12 826	11·4	14 800	12	20 502	14
World	30 242	9·0	35 774	9·7	46 383	10·1

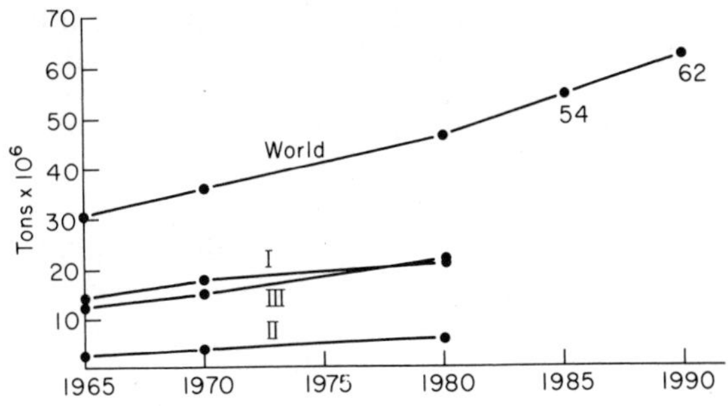

TABLE VII

CONSUMPTION AND PROJECTED DEMAND OF POULTRY MEAT

	Consumption				*Demand*	
	1964–6		1969–71		1980 *projected*	
	Total ('000 tons)	*Per head (kg/year)*	*Total ('000 tons)*	*Per head (kg/year)*	*Total ('000 tons)*	*Per head (kg/year)*
Economic Class I	6 794	9·8	9 156	12·6	13 561	16·8
Economic Class II	1 662	1·1	2 086	1·2	3 578	1·6
Economic Class III	3 304	2·9	4 444	3·6	6 438	4·4
World	11 760	3·5	15 686	4·2	23 577	5·2

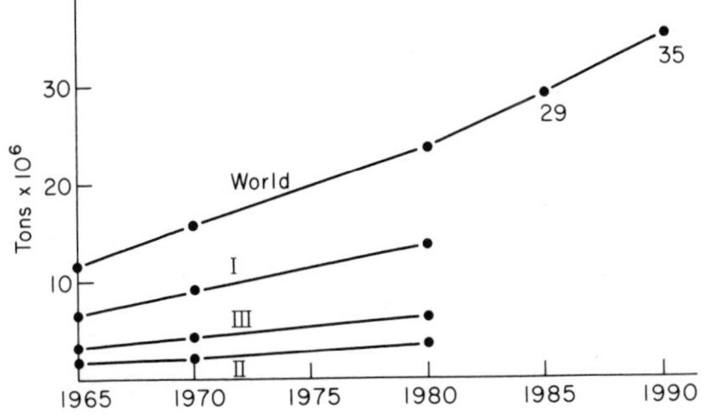

A summary of the above demand projections for meat in 1980 is given in Table VIII. Added to this table are alternative demand figures under recent revised assumptions of population and income growth, but still based on constant prices. The main differences between new and previous gross domestic product assumptions are in assuming that in the majority of developed countries economic growth in the medium term is expected to be depressed by the effects of present energy difficulties and that high rates of growth will be enjoyed by petroleum exporting countries.[9]

TABLE VIII

SUMMARY AND REVISED DEMAND PROJECTIONS FOR 1980 ('000 tons)

	Beef and veal		*Mutton and lamb*		*Pig meat*		*Poultry*		*Total*	
	Dem. proj.[a]	*Alt. dem.*[b]	*Dem. proj.*[a]	*Alt. dem.*[b]	*Dem. proj.*[a]	*Alt. dem.*[b]	*Dem. proj.*[a]	*Alt. dem.*[b]	*Dem. proj.*[a]	*Alt. dem.*[b]
EC I	27 483	26 440	3 199	3 125	20 295	19 470	13 561	13 034	64 538	62 069
EC II	13 551	13 952	3 921	4 133	5 586	5 479	3 578	3 690	26 636	27 254
EC III	11 771	11 907	2 633	2 653	20 502	21 190	6 438	6 657	41 344	42 407
World	52 805	52 299	9 753	9 911	46 383	46 139	23 577	23 381	132 518	131 730

[a] 1980 demand as in the 1970–80 FAO Agricultural Projections, revised.
[b] Demand projections based on constant price, revised GDP and population growth rate assumptions.

According to the new demand projections, the 1980 demand for the main categories of meat would be slightly (0·6 %) lower at the world level than in the revised projections, but substantial shifts between the economic classes would take place. Demand in Economic Class I would be 4 % below and in Economic Classes II and III 2·3 % and 2·5 %, respectively, above the previously projected levels. The new demand projections for meat in Economic Class I would be 2·5 million tons lower, over 40 % of the reduction being due to a decline in demand for beef and veal. The demand for pig meat and poultry meat in Economic Class I would be about 820 000 tons and 520 000 tons lower compared with the quantities shown in the 1980 FAO projections.

World demand for milk and milk products for human consumption is projected to grow from 373 million tons whole milk equivalent in 1970 to 462 million tons by 1980 (see Table IX).[10] This means an average annual increase of 2·2 %, only slightly higher than the expected population growth. In most developed countries per capita consumption would level off or even decrease.[17]

From the projections the general conclusion can be drawn that total as well as per capita demand for animal products for human consumption

TABLE IX

CONSUMPTION AND PROJECTED DEMAND FOR MILK AND MILK PRODUCTS (from ref. 10)

	Consumption						*Demand*		
	1964–6			1970			1980 *projected*		
	Total incl. animal feeding (million tons)	*Total for food*	*Per head (kg)*	*Total incl. animal feeding (million tons)*	*Total for food*	*Per head (kg)*	*Total incl. animal feeding (million tons)*	*Total for food*	*Per head (kg)*
EC I	195·76	181·50	262·5	198·89	185·33	255·0	217·28	205·37	255·0
EC II	75·17	72·37	47·3	83·75	80·70	46·2	123·89	119·97	52·4
EC III	105·91	91·36	81·0	122·94	107·13	87·0	157·47	136·67	93·4
World	376·84	345·23	103·0	405·58	373·16	100·7	498·64	462·01	101·4

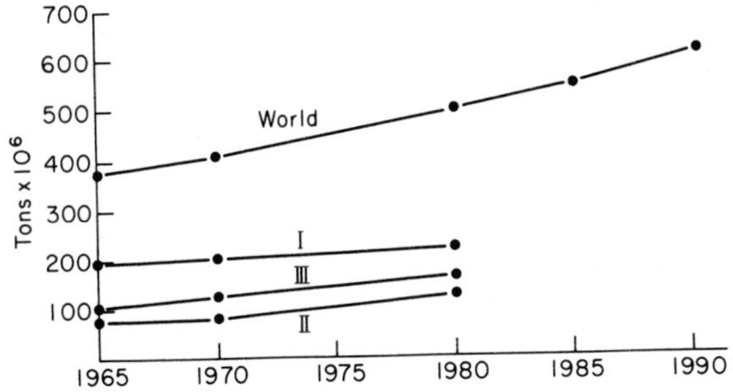

shows an upward trend in the next few decades. It is estimated[3] that under the influence of rising income levels the annual percentage demand for expensive foods would increase more rapidly than for demand for direct consumption of cereals, starchy roots and sugar. Thus the demand projections for the developing market economy countries show annual rates of growth of 3·0% for cereals but 4·8% for fish and 4·7% for meat.

The obvious question will be raised whether the world will be able to meet the estimated demand projections for animal products and what main constraints towards reaching these goals can be expected. Since there has emerged a clear divergence in sophistication and type of production systems in the developed and the developing countries, these two groups of producing countries will be analysed separately.

During the last 25 years the more affluent societies, starting in the USA, have been able to combine their agricultural production potential with animal genetic upgrading, advances in technology and feeding efficiency into an animal products industry, which has brought large-scale animal protein consumption within reach of the majority of their populations. The system is a valid one, but remains vulnerable as has been demonstrated by the recent period of increases in cereal prices, corresponding animal protein costs and decrease in demand.

Reliance on cereal resources for feed use has reached proportions which are stirring up mounting criticism at the international level and have led to appeals[11] to the high-income countries for restraint in affluent consumption of animal protein. This concern is not without justification, if we compare the uses to which cereals are put. Per capita consumption of cereals in developing countries is about 220 kg/year, nearly all for human consumption. In Canada and the USA consumption runs at about 1000 kg, of which only about 70 kg is used for direct human consumption, the rest being used to feed animals. Together the high-income countries use something like 370 million tons of grain annually for livestock feeding, which is more than the total human consumption of cereals in China and India combined.

The argument is less valid on the national level because here we must remember the economic fact that a large part of the grain supplies is available precisely because of the demand for livestock feed; if this were not there, less would be produced.[4] When, however, large quantities of grain are taken from the world market for animal feed purposes in periods of relative scarcity, so as to cause abrupt price increases, countries dependent on grain imports for human consumption will find it increasingly difficult to finance basic food requirements for their populations.

A second criticism heard against a grain-fed intensive animal industry is that the accumulated energy inputs for growing the feed and the animals cause a disproportionate drain on non-renewable energy resources.[12] For instance, the example is given that to put one glass of milk on the table an added energy input by man equivalent to half a glass of diesel fuel is required.[12] This would imply that a food system with this degree of energy expenditure cannot be made global as it would require 80 % of the world's current energy account.

It can therefore be expected that the aspects of food/feed competition and high energy demands in modern animal husbandry will be of increasing international concern which may influence livestock production technology and policy thinking in the future. In fact, under pressure of the recent period of high cereal and protein feed prices, a noticeable shift has already taken place in the USA and Europe leading to less grain feeding for shorter periods, and a search for cheaper ingredients such as manioc and banana flour and recycled waste products from the food-processing industries.[13, 14]

The fastest response to the rising demand for animal products will undoubtedly still come from factory-type feedlots in the consuming countries and from established exporting countries. A constraint in beef production may be the inability to obtain sufficient calves.[4] A further constraint towards full expansion of animal protein production in the developed countries could be an expanding use of meat extenders and analogues derived from plant protein. As meat extenders, soy protein extracts can improve the taste, chewability and water absorption of the final product. Nearly all the technical problems concerning meat extenders have been solved. Their use in sausages, hamburgers, TV dinners, etc., is therefore limited only by legislation. If these restrictions are relaxed, the extent of the use of extenders will depend primarily on the relationship of their price to that of meat.[15] Use of meat extenders is gradually gaining ground in the USA, Europe and the USSR. Some 25 000 tons were produced in 1973 in the USA, which represents some 125 000 tons of meat-like products. This figure was only 0·7 % of total US red meat consumption, but equalled 21·5 % of its total fresh, chilled and frozen meat imports.[18]

The FAO forecast for the extent to which meat substitutes will cut into the total meat market by 1980 ranges from 1 % to 6 % depending on restrictive policies, promotional success and the margin in price below meat.[4] The success of another contender, the single-cell proteins or bio-proteins, will again depend on economic factors, technological improvements, acceptability and legislation.

This group of products could either develop into a bonus to the animal products industry by providing a high-protein animal feed for monogastric animals, or become a challenge when products for direct human consumption would be developed from this source. The present investigations and actual single-cell protein production capacity envisaged by 1980 are given in Table X.[6] It is important to note that the major efforts are concentrated in Europe, obviously based on a reluctance to be dependent on the import of soybean. The abundance of soybean in the USA is also the probable reason why single-cell protein development has not progressed very far in that country.

The majority of substrates are based on non-renewable hydrocarbon sources. Mention should, however, be made of research efforts in the partial microbial upgrading of renewable carbon resources such as sugar, starch and cellulose-containing waste materials, including animal wastes to be used in liquid and wet feeding systems.

In summing up, therefore, in the developed countries constraints towards animal protein production along present lines will be a continuing and perhaps increasing concern on the international level of cereal use for feed use and depletion of non-renewable energy sources. The possible effect will be accentuated if world grain and energy prices remain in relatively short supply, increasing their costs. A second major constraint may be the accelerated use of meat extenders, though this will be mainly governed by the introduction of permissible legislation by local governments and by price relationships.

TABLE X

POTENTIAL WORLDWIDE PRODUCTION AND PILOT FACILITIES FOR SINGLE-CELL PROTEIN (SCP) BY 1980

Country	*Organism*	*Substrate*	*Estimated capacity (tons/year)*
Czechoslovakia			
Slovnaft Kojetin	Yeast	Ethanol	60 000
Slovnaft Kojetin	Yeast	N-paraffins	100 000
Finland			
United Paper Mills	Fungi	Sulphite liquor	10 000
France			
British Petroleum, Lavera	Yeast	Gasoil	100 000
Group Français des Protéines	Yeast	N-paraffins	40 000
Germany			
Biological carbon research	Algae	Carbon dioxide	Pilot
Italy			
Italproteine, Sardinia	Yeast	N-paraffins	100 000
Liquichimica, Calabria	Yeast	N-paraffins	200 000
Mexico			
Sosa Texcoco	Algae	Carbon dioxide	Pilot
Romania			
Roniprot	Yeast	N-paraffins	60 000
Republic of China	Yeast	N-paraffins	1 000
Switzerland			
Nestlé	Bacteria	Ethanol	1 000
United Kingdom			
British Petroleum	Yeast	N-paraffins	100 000
ICI	Bacteria	Methanol	100 000
Shell Chemical	Bacteria	Methane	1 000
RHM Foods	Fungi	Carbohydrates	10 000
Tate & Lyle	Fungi	Carbohydrates	4 000
USSR	Yeast	N-paraffins	200 000
United States			
Amoco Foods Co.	Yeast	Ethanol	5 000
LSU/Bechtel	Bacteria	Cellulose	10 000
University of California	Fungi	Cellulose (newsprint)	Pilot
Japan			
Mitsubishi Gas	Yeast	Methanol	100 000
Sumimoto	Bacteria	Methanol	100 000
Kyowa Hakko	Yeast	N-paraffins	100 000

Turning our attention to prevailing constraints in the developing countries, a different picture emerges. For a better understanding of the problems and potential of animal protein production in the developing countries, the following statistics are relevant.[7]

TABLE XI

RELEVANT STATISTICS ON LIVESTOCK NUMBERS AND TYPES IN DEVELOPING COUNTRIES

	Percentage of total world
Human population	74
Agricultural land	58
Livestock (numbers):	
Cattle and buffalo	70
Sheep and goats	63
Pigs	60
Livestock (production):	
Cattle and buffalo: beef	34
milk	21
Sheep and goat	50
Pigs	37

From Table XI one can conclude that although the majority of livestock is located in the developing countries, its total productive capacity is never over 50 % but more often only one-third of total world production. This is further illustrated in Table XII, which shows a lower meat production per animal in the developing nations (Class II) compared with the developed (Class I) and the centrally planned economies (Class III). Furthermore, it is shown that over a 20-year period, increases in output per animal and per head have been minimal, indicating that animal production in developing

TABLE XII

AVERAGE ANNUAL MEAT PRODUCTION BY ECONOMIC CLASS (kg/animal)

Economic Class:	*I*		*II*		*III*	
	1948–52	1968–72	1948–52	1968–72	1948–52	1968–72
Beef	48·9	74·3	12·6	13·9	25·5	40·5
Sheep and goats	5·1	6·5	3·3	3·4	4·9	5·6
Pork	79·6	95·2	21·2	22·7	44·7	50·3
Chicken	2·6	6·8	0·8	1·1	1·2	2·1
Total meat (kg/head)	41·7	67·4	9·9	10·4	13·9	28·0

countries has been due mainly to increase in livestock numbers which have more or less kept pace with the increase in population.

The major constraints that hamper a quick improvement in animal protein production from resources that are potentially available in the developing countries are summarized as follows:

1. A fast population growth, and marginal average per capita production of food products. In consequence, a keen competition between humans and animals for food resources has arisen, the animal being dependent mainly on natural vegetative growth unfit for direct human consumption.
2. Low producing animals and inefficient conversion of imbalanced feed resources as is shown in Tables XI and XII.
3. Extreme cyclical variations in climatological conditions and seasonal fluctuations in the availability and quality of feedstuffs from grazing. The recent decimation of flocks and herds in the Sahel zone caused by a prolonged drought will be only too well remembered. But seasonal fluctuations too can lead to serious economic losses. Data reported from a livestock centre in a Near Eastern country showed that calves were no heavier at 15 months of age than they were at 8 months after having passed through the dry season. This loss of 7 months of growing time and the extra feed needed to regain lost weight is doubly expensive, and increases the amount of feed for the same weight gains.[16]
4. Animal diseases, which lower production or prohibit the introduction of viable animal production schemes in potentially suitable areas. A typical example referring to the latter constraint is the occurrence of animal trypanosomiasis which affects an area of 7 million km^2 in tropical Africa.
5. The comparative lack of expertise in animal feeding and management, the lack sometimes of infrastructure and distribution systems, and perhaps more than anything else the lack of financial resources to take steps to improve the situation, should also be mentioned as constraints to increased animal protein production.

Under these circumstances, how will the developing countries, and in particular the poorest, cope with procuring more animal products for their people? With no appreciable changes expected in population growth rates over the next decades and a continuing competition between supplies for food and feed use, the preferred choice of species to be used are ruminants, which can thrive on more fibre and less protein than other species of farm animals.[20] A second major thrust should be directed towards upgrading genetic stock not only by importation of new genes but by testing local breeds of livestock under improved management conditions.

For more and better feed supplies, much more emphasis should be put on the better utilization of permanent pastures and meadows. This area is twice as big in developing countries as the area of arable land (Table XIII).[7,8] Opening up this vast potential would, however, in certain cases require large pre-investment; the best example is the eradication of tsetse fly, which it has been estimated would cost US$100 million per year for 20 years.[3] But by the control of animal trypanosomiasis an additional 120 million cattle could be raised with annual meat production worth about US$750 million.

TABLE XIII

PERMANENT MEADOWS AND PASTURES (from ref. 8)

	Total area ($ha \times 10^6$)	*Arable* ($ha \times 10^6$)	*Arable area* (%)	*Pastures and meadows* ($ha \times 10^6$)	(%)
Developing countries:	6 625	663	10	1 378	20·8
Africa	2 382	189	8	701	29·4
Latin America	2 056	119	5·8	455	22
Near East	1 205	85	7	182	15
Far East	892	269	30	39	4·4
Other	90	1	1	1	1
Centrally planned Asian countries	1 144	116	10	319	28

Further promising areas for increasing and improving feed supplies are: the introduction of plants with the highest photosynthetic capacity; the efficient use of farm wastes and by-products from the agro-industries; increasing the digestibility of feedstuff by pretreatment; and the use of non-protein nitrogen sources in ruminant feeding.[21] Seasonal fluctuations in feed availability from grazing could at least in part be compensated for by the building up of stocks for emergency feeding and improved timing of slaughtering. Investigations and pilot schemes with promising results have been initiated but much more is required to make a substantial impact.

As far as FAO is concerned, its policy with regard to increasing animal protein production was clearly spelt out by the member governments during the Third Session of the Committee on Agriculture in April 1975:

> 'A priority compatible with ecological considerations should be given to the development of ruminant production using marginal land and land unsuitable for crop production. Also the integration of livestock with crop production should be stressed. In view of the food/feed competition, the whole issue of a feed base for ruminants in relation to land use including forage/livestock production relationships should be given more emphasis in FAO's work.'

If people's preference for animal protein products persists in the next few decades, and there are reasons to believe that it will, it is expected that under the pressure of increasing world demand for food grains and fossil fuels, there will be a shift to greater reliance on forages and agricultural by-products for ruminant feeding on a worldwide scale. This will not necessarily imply a diminished economic feasibility but will lead to a more rational use of available world resources which are of no direct food use to man.

The combination of grazing and feedlotting only to improve the quality and palatability of the meat with a minimum amount of grain is expected to be the trend in animal agriculture in the developed and developing countries. Thus, instead of the divergence in production systems that has emerged in the world with its inherent problematic issues and consequences, there will be an increasing tendency towards a more uniform approach towards providing the world with its needed animal protein. The pooling of potential feeding resources, manpower, technical skills and investment resources from both the developing and the developed world is needed to achieve this in the quickest and most effective way.

REFERENCES

1. Development Forum, 2–8 Nov. 1974. Centre for Economic and Social Information/OPI, United Nations.
2. El Shazly, K. and Jasiorowski, H. A., *The Potential for Using Unconventional Feed Resources in Animal Nutrition*. FAO, 1975.
3. *World Food Conference: Preliminary Assessment of the World Food Situation, Present and Future*. UN, E/Conf/65/Prep/6, 1974.
4. Abbott, J. C., *Proteins: Progress and Prospects*. FAO, 1973.
5. *Energy and Protein Requirements*. Report of a Joint FAO/WHO Ad Hoc Expert Committee, 1973.
6. Wells, J., Analysis of potential markets for single cell protein. Paper presented at Symposium on Single Cell Protein, American Chemical Society National Meeting, Philadelphia, 9 Apr. 1975.
7. Jasiorowski, H. A. and El Shazly, K., *World Production of Animal Protein and the Need for a New Approach*. FAO: AGA: AAAP/75/a, Mar. 1975.
8. *FAO Production Yearbook*, 1972.
9. *Review of the FAO Meat Production and Demand Projections to* 1980. FAO: CCP ME 74/3, July 1974.
10. *Agricultural Commodity Projections*, 1970–1980*: Milk and Milk Products*. FAO: CCP 71/20, Sept. 1971.
11. *Declaration of the Rome Forum and World Food Problems*. Institute for Environment and Development, Nov. 1974.
12. Borgstrom, G. The price of a tractor. *Ceres, FAO Review on Development*, Nov./Dec. 1974.

13. Jasiorowski, H. A., *The Current Status and Trends in World Animal Production*. FAO, June 1975.
14. *Some Observations on the Impact of Structural and Other Changes in the Raw Materials Used for Compound Feed Manufacturing*. International Trade Centre, ITC/DTS/MDS/SMR/45(a), Mar. 1975.
15. Beloglavec, D. M., Meat-like products and their possible impact on the demand for meat. *Monthly Bulletin of Agricultural Economics and Statistics*, July/Aug. 1974.
16. Robertson, E. I., *Animal Feeds*. Report to the Government of Libya, FAO, No. 1938, 1965.
17. Winkelmann, F. *Imitation Milk and Imitation Milk Products*. FAO AGA/misc./74/2, 1974.
18. Duda, Z., *Vegetable Protein Meat Extenders and Analogues*. FAO AGA/misc./74/7, 1974.
19. *Provisional Indicative World Plan for Agricultural Development*, Vol. I. FAO, 1970.
20. *Ruminants as Food Producers*. Special Publication No. 4, Council for Agricultural Science and Technology, Ames, Iowa, 1975.
21. Kottman, R. M. and Geyer, R. E., Future prospects for animal agriculture. In: *Livestock Waste Management and Pollution Abatement*, pp. 9–18. ASAE, St. Joseph, Mich., 1971.

4

Criteria and Guidelines for the Selection of Animal Feedlot Sites

PHILIP H. JONES

Professor of Civil Engineering, Institute of Environmental Studies, University of Toronto, Toronto, Ontario, Canada

INTRODUCTION

During the period when feedlots were just coming into prominence, the selection of a feedlot site was a very simple task. There were many alternative sites and the main criterion for selection was to minimize the transportation cost of bringing in the necessary factors of production. However, with the advent of environmental awareness, many new constraints have entered the decision. Some of the new considerations include waste disposal, environmental regulations, climate, topography and the use of surrounding land. The importance of all the factors, both old and new, must be evaluated by qualified professional people and must then be considered by the feedlot managers in the selection of animal feedlot sites.

For purposes of discussion, it is assumed that decisions have already been made regarding the type of animal to be fed and the size of operation. It is also assumed that a commercial-size operation is under consideration.

FACTORS OF PRODUCTION

The most important factor in feedlot site selection is still nearness to the necessary factors of production.[1] The term 'nearness' is intended to imply both in close proximity and easily accessible. The major factors of production are feeder animals, feed, water and labour.

Ample sources of feeder animals, feed grains and roughage must be available in close proximity. It is best to have many alternative sources in case of a crop failure. These factors can always be brought in from further away, but only at a much greater cost.

Also, in areas where inclement weather is to be expected, potential periods of isolation should be considered when planning feed storage. Thus, the availability of these factors is certainly the key to the successful operation of an animal feedlot.

The adequacy and reliability of a water supply is important both from a quantitative and a qualitative point of view. The quantity of water consumed varies according to the type of equipment employed and the type of livestock housed. For example, swine consume 7–15 litres/day, beef consume 35–45 litres/day, and poultry consume 0·2–0·5 litres/day. Unless the operation is connected to a municipal supply, storage is important. Usually the rate of consumption at specific times (*i.e.* when pens are being washed) will exceed the capacity of the well pumps. Also, well production will fall during times of high demand such as hot summer days. Thus, a storage tank with a second pump or pressure system is desirable. It is also important to ensure that the water source used by the livestock operation is well protected from pollution by others.

The availability and type of labour force required is also an important consideration. What professional needs exist and how frequently will they be called upon to serve the operation? As the size and complexity of an operation increases, so does the need for professional assistance. Veterinary surgeons and geneticists may be required on a consistent basis to service large finishing programmes. On the business side, accountants and other management professionals may be needed. Based on the answers to these questions, the decision may be taken as to how 'remote' an operation may be established from an urban centre having these services.

MARKET OUTLETS

Though not as important as the factors of production, it is advisable to have processing plants as near to the feedlot site as possible and that these processing plants in turn be as near to the consumer as possible. This factor is important because it is less expensive to ship a processed carcass than a live animal. Thus, by locating near the processing plants, savings may be gained in transportation costs. Fortunately, it is also advantageous to the processing plant to locate near the feedlot.

WASTE DISPOSAL

Another important consideration in the selection of an animal feedlot site is the disposal of manure. The most economical method of disposing of this manure is to spread it on nearby cropland. Thus, the availability of land for waste disposal is an important consideration in site selection.

Nitrogen is the single most significant component in manure, in terms of crop production on land used for disposal of animal wastes. Excessive

TABLE I

LAND REQUIREMENTS FOR UTILIZATION OF MANURES IN AN INTEGRATED LAND–LIVESTOCK SYSTEM
(modified from ref. 2)

Size of operation	*Minimum corn land requirements*		
	N excreted (*kg*)	*Crop utilization*[a] (*ha*)	*Pollution control*[b] (*ha*)
2 million broilers, 10 weeks	14 060	800	400
200 000 layers, 365 days	11 340	800	400
20 000 hogs, 175 days	10 440	800	400
4 000 feeders, 365 days	12 700	800	400
2 000 dairy, 365 days	12 700	800	400

[a] Minimum land areas in continuous corn for the complete use of the N-component.
[b] Minimum land areas in continuous corn or the maximum application of N which will not reduce corn yield or cause water pollution.

nitrogen can cause certain cereal crops to grow very quickly and collapse prior to germination or ripening. Thus, the rate of application of manure to land is controlled by the concentration of nitrogen in the manure and the crop to be grown.

Table I is based upon the nitrogen requirements for growing corn on the land used for manure disposal in cold northern climates.[2] Corn has a higher nitrogen requirement than most other grain crops and is therefore a most appropriate crop for sites irrigated with manure. The table also indicates the absolute minimum land requirement for disposal of manure so that the nitrogen applied does not reduce the yield of corn or cause water pollution.

In addition to finding available land for waste disposal, consideration must be given to the climate. If there are prolonged periods when the ground is too wet to drive a tractor on or if it is frozen, storage will prove necessary with enough capacity to hold all manure generated during the period. Naturally, this will affect the operating costs.

TOPOGRAPHY AND RUNOFF CONTROL

The topography of the land must be considered in order to gain an idea of what expense will be involved in controlling runoff. If there is an adjacent natural water body which the operation may pollute, then control measures must be taken to prevent runoff. Normally, this involves the construction of a drainage ditch around the feedlot to divert runoff into a holding pond. The contents of the holding pond are then periodically pumped on to nearby cropland. Runoff control is important for two reasons in site selection: firstly, there must be room available on the site for an appropriate size holding pond, and secondly, there must be nearby land available to dispose of the contents of the holding pond. The usual requirement is emptying the holding pond within two weeks after a big storm. Size of pumps and other equipment is thus calculated from the volume of wastewater to be removed against the hours of operation within the two-week period.

Holding pond size is normally designed on the basis of a particular high-intensity/short-duration storm that one might experience in the area of the site under consideration. The usual storm design is rainfall of 10 or 25 years' recurrence interval and 24-hour duration. Furthermore, the size of the pond will vary according to the permeability of the soil and the expected frequency with which the pond would be emptied.

A second consideration in feedlot site topography is the possible danger of groundwater contamination. Leachates from manure contain high concentrations of soluble organic carbon, nitrogen in various forms, and phosphorus. If the soil is sandy it will have little capacity to adsorb phosphorus and its high porosity will allow rapid percolation of the other soluble components. Clay soils with limited permeability are more suitable, allowing the polluted leachates to be drained away to a collection point where appropriate disposal steps can be taken.

USE OF SURROUNDING LAND

A newer and very important consideration in feedlot site locations is the land use patterns of the area.[1] The development plans for the area must be examined to see if there are incompatible land uses being proposed such as intensive or expensive residential areas.

Even if the neighbours are farmers, it is important to evaluate the land, for if it is so valuable as to preclude it from legitimate farm operations and the farmer neighbours own their land, it may be only a matter of time (perhaps one or two years) before they decide to sell the land for some non-farm purpose. This may then cause the feedlot to be moved prematurely. Under such conditions, it may be possible to lease a site, anticipating a move in a very short time.

A new class of problem is emerging in certain parts of the world where the city dweller is moving into the country to seek refuge from urban pollution. This class of person is neither used to, nor is he likely readily to tolerate, what he would consider to be further pollution problems and the invasion of his home and surroundings with perfectly normal agricultural odours or noises. For this reason, it is important to consider the direction of prevailing winds. This can be accomplished by developing a figure showing wind direction and percentage of time the wind might blow from a given direction. These figures may be produced monthly, quarterly or annually. It is certainly necessary to know the percentage wind direction during the summer months of any potential site for livestock production.

ENVIRONMENTAL REGULATIONS

Many countries in the world have developed a set of regulations for the control of nuisance and pollution caused by indiscriminate development of livestock industries and thoughtless management practices of the waste of these industries. Together with this federal legislation must be considered the various state laws and associated regulations concerned with environmental quality, water and air pollution, health, nuisance and trespass.

Regulations generally control and restrict the locations of livestock breeding and feeding establishments with respect to minimum distances from surface water, residential dwellings, municipalities, recreational areas and arterial highways. They also control such things as geological formations where animal industries may be located, available areas for manure disposal, and control of transport of liquid manure in open vehicles on public highways. Further regulations may relate to storm water runoff control facilities, groundwater contamination, and odour and fly control. Frequently these regulations are controlled by some kind of permit arrangement issued through appropriate agencies within the level of government in question. Each jurisdiction will usually tend to regulate these matters through its own particular administrative procedures, but generally the regulation is based on protecting the rights of individuals and societies from nuisance, odour or damage.

Thus, in selecting an animal feedlot site, consideration must be given to existing regulations controlling feedlots in that particular area.[3] One might also question what legal protection exists to prevent subsequent legislation requiring the operation to take prohibitively expensive control measures after it has been established.

For those countries without specific regulations on animal feedlots, the most significant laws concerning site selection would be those common laws concerning nuisance and trespass. The law of nuisance subdivides into private and public nuisance. Under this common law, the person or persons whose person or property has been injured by a harmful substance discharged into water or air has a cause for action against the person or firm responsible.[4] A public nuisance is where a large number of people are affected or where a public resource is harmed.

The common law of trespass concerns the invasion of the right of an owner or occupier of a property to its exclusive possession or enjoyment. Odours, gases or dusts passing from one property to another have been considered as a form of trespass. Also, water-transported manure solids washed upon another's land have been interpreted as trespass.[4]

Thus, in selecting a feedlot site, consideration must be given to all types of regulations, both specific regulations affecting animal feedlots and common law practices regarding nuisance and trespassing.

INNOVATION TO RESOLVE SITING PROBLEMS

The reasons for locating intensive livestock operations in the rural countryside are largely traditional. The reasons do not apply to the same extent with corporate agribusiness as they do to family operations where these matters represent a 'way of life'. The need to have livestock in the country was originally related to the need for locally grown feed and also the provision of pasture land. These needs do not exist in intensive specialized livestock operations as the feed is usually shipped in and thus its need to be located near a shipping centre is much greater than its need to be in the rural countryside. The following are two innovative arrangements which might reduce to a minimum some of the problems associated with site selection for intensive livestock operations.

Industrial City or Park Setting

In an industrial setting such as one might find in the heart of many large metropolitan centres, it might be considered that an intensive livestock operation could easily locate.

Advantages:

Access to substantial water supply to maintain clean, odour-free operation.

Accessibility to high-flow city sewer systems to transport and dilute the animal wastes prior to treatment in a municipal plant (if possible).

Easy access for feed and market.

Large labour pool.

Access to skilled professional services and capital.

Disadvantages:

Danger of noxious industries harming the livestock.

Loss of nutrients in waste which should, from a conservation viewpoint, be recycled to the land.

Possible problems of odours with neighbouring industries.

If city waste flow is low, possible expense of sewer surcharge causing the cost of waste handling to increase rapidly.

Clearly this is an alternative that has not been thoroughly examined and perhaps warrants further thought.

Agribusiness Park

This represents an entirely new suggestion for locating intensive livestock operations together. It is basically like an industrial park but located in the rural countryside. It presupposes the managed location of mutually supportive agribusiness activities. Conceptually, there would be a core of intensive livestock operations with different livestock as neighbours. Adjacent to this central area would be support services such as tractor sales and service, farm equipment, pumps, feeders, etc. Government extension offices might also be located somewhere in the centre, as would professional services.

Surrounding the centre and buffering the activity from the urban centres would be large tracts of arable land suitable both to receive the wastes of the livestock and also provide cereal grain and roughage. In the arable land might be located a farm residential area where all the personnel might live with perhaps some schools and/or colleges of agriculture and veterinary medicine.

Advantages:

All people would be agriculturally oriented and farm odours might be less offensive to them.

The surrounding land would be available for waste disposal.

Potential for growing feed at minimum cost.

Cooperative use of expensive capital equipment rendering it more efficient.

Disadvantages:

Difficult to assemble such a large area of land without long-range planning and government support.

If the agribusiness park is then surrounded by a large area of crop-producing land, not only would buffering be achieved but animal wastes would provide the necessary fertilizer to produce the feed necessary for the livestock. In addition, by instituting a centrally managed handling system, costs could be substantially reduced and benefits substantially increased.

REFERENCES

1. Byrkett, D. L., Modelling the optimal location of the cattle feeding industry with particular emphasis on environmental considerations. Ph.D. dissertation, Ohio State University, 1974.

2. Webber, L. R. and Lane, T. H., The nitrogen problem in the land disposal of liquid manure. In: *Proceedings of Agricultural Waste Management Conference*, Cornell University, Ithaca, NY, 1969.
3. Byrkett, D. L., Taiganides, E. P. and Miller, R. A., Effects of environmental legislation on cattle feedlot location. In: *Proceedings of International Symposium on Livestock Wastes*. ASAE, St Joseph, Mich., 1975.
4. Willrich, T. L. and Miner, J. R., Litigation experiences of five livestock and poultry producers. In: *Livestock Waste Management and Pollution Abatement*, pp. 99–101. ASAE, St Joseph, Mich., 1971.

BIBLIOGRAPHY

Agricultural Code of Practice for Ontario. Ontario Ministry of Agriculture and Food, Toronto, Ontario, Canada, Apr. 1973.

Animal Waste Management. Council of State Governments, Washington, DC, Sept. 1971, 205 pp.

Badger, D. D. and Cross, G. R., Economic implications of environmental quality legislation for confined animal feeding operations. In: *Livestock Waste Management and Pollution Abatement*, pp. 204–7. ASAE, St Joseph, Mich., 1971.

Canada Animal Waste Management Guide. Canada Dept. of Agriculture, Ottawa, 1972.

Cath, W. S., Summary of existing state laws. *Animal Waste Management J.*, Sept. 1971, pp. 17–18.

Doig, P. A. and Willoughby, R. A., Response of swine to atmospheric ammonia and organic dust. *J. American Veterinary Medical Assoc.*, **159**(11) (1971) 1353–61.

Dominick, D. D., Animal waste management and the environment. *Animal Waste Management J.*, Sept. 1971, pp. 11–14.

Environmental Protection Agency, *Beef Cattle Feedlot Site Selection*, EPA-R2-72-129. USGPO Stock No. 5501–00455, Washington, DC, 1972.

Loehr, R. C., Animal wastes—a national problem. *ASCE J. Sanit. Eng. Div.*, **95**(SA2) (1970) 189–221.

Taiganides, E. P. and Stroshine, R. L., Impact of farm animal production and processing on the total environment. In: *Livestock Waste Management and Pollution Abatement*, pp. 95–8. ASAE, St Joseph, Mich., 1971.

Webber, L. R., Animal wastes. *J. Soil Water Conserv.*, **26**(2) (1971) 47–50.

5

Layout and Design of Animal Feedlot Structures and Equipment

MERLE L. ESMAY

Professor of Agricultural Engineering, Michigan State University, East Lansing, Michigan, USA

INTRODUCTION

The next step, after a site for a feedlot has been selected and developed, is to plan for the structures and the equipment needed for the type of animals to be raised at the site. In developing plans for the buildings and associated structures and equipment, both the objectives of the animal production unit and the constraints to production must be seriously considered.

The main objective in planning the production unit is to develop a structure which will provide the type of environment needed for optimum economic animal production and at the same time provide good working conditions for humans. The management and engineering decisions that need to be made are:

(a) type of confinement facilities—sheltered or unsheltered, warm or cold, and associated animal space requirements;
(b) type of shelter construction—wood, stone, metal, or other materials;
(c) feed handling system—type of storage bins, silo (vertical or bunker), mechanized conveyance of feed to animals or do animals move to feed;
(d) orientation of facilities with respect to climatic factors and access to roads;
(e) layout to optimize traffic patterns—animal traffic, feed, water and materials transport, vehicle traffic in and out of feedlot, and traffic pattern of workers; and
(f) waste disposal—storage, treatment, final disposition, mode and handling.

In making decisions for the development of feedlot structures, the planner is constrained by:

(a) human health and safety considerations, not only of workers but also of the consumers of the animal products;

(b) economic advantages and disadvantages of each alternative;
(c) animal health and relative productivity under alternative systems of production;
(d) environmental pollution limits plus odour, dust and insect nuisance considerations.

Obviously all plans for the production facilities will have to fit the feedlot site after considering such things as soil conditions, climate and proximity of human dwellings.

FEEDLOT FACILITIES

There are two basic types of feedlot facilities: unsheltered and sheltered confinement units. Open feedlots are mainly unsheltered and suited for beef cattle operations in moderately cold and semi-arid climates. The trend in temperate climates is towards sheltered confinement for large dairy, poultry and pig fattening feedlots. (See the schematic outline of Fig. 1 for beef cattle.) Sheltered feedlots require considerably more engineering and thus need to be discussed in greater detail.

No matter what the physical facilities are, there are some basic requirements which must be made. These requirements are for water, feed, air, animal space and layout in relation to climate, materials handling and other considerations.

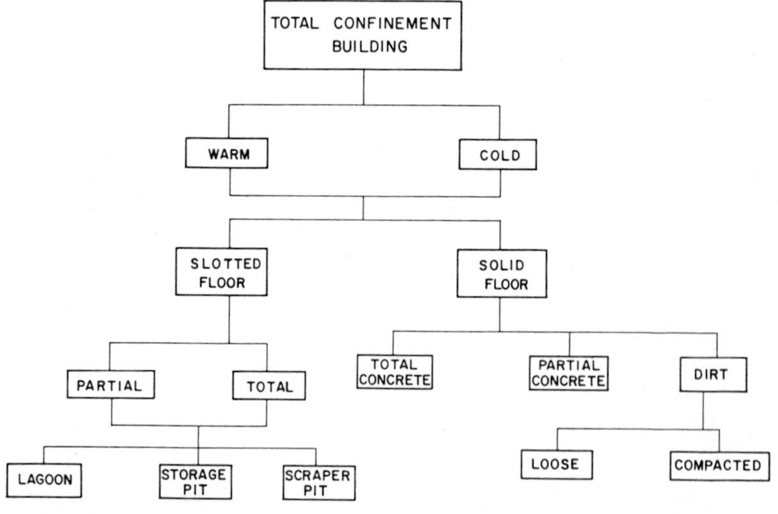

FIG. 1. Total confinement building variations for beef cattle shelters (from ref. 1).

Animal Space Requirements

Space requirements per animal, in general, decrease as the degree of confinement and environmental control increases. Air volume per animal is provided by mechanical ventilation and is discussed later. Table I gives general guidelines as to feedlot and floor area requirements for various animals in warm or cold confinement. Space per animal in feedlots for beef cattle is highly dependent on the climatic conditions. The minimum feedlot space per animal of 20 m^2 is possible only in arid climates having 25 cm or more moisture deficit per year (the deficit is annual lake evaporation minus rainfall). In more humid climates where the feedlots become muddy and the manure does not dry out from the sun and wind, up to four times as much space may be necessary.

In fact, it is in the more humid and colder climates that partially or all-surfaced feedlots and partially or totally sheltered feedlots become economically feasible. Muddy feedlots can reduce beef cattle gains as much

TABLE I
FEEDLOT AND SHELTER REQUIREMENTS FOR ANIMALS
(from refs. 1, 2, 4 and 6)

Animal type	*Feeder space (cm/head)*	*Floor space (m^2/animal)*
Beef cattle:		
Open feedlots		
Arid climates	30	18–20
Cold humid climates	40	50–80
Partially surfaced, sheltered feeding	30	10–15
Sheltered confinement		
Solid floors	25	3
Slotted floors	25	2
Warm-type confinement	25	2
Dairy cattle:		
Free-stall	40	6–7
Stanchion-type	100	6–8
Hogs (fattening to 100 kg):		
Nursery		0·4
Growing, slotted floor		0·8
Growing, concrete floor		1·2
Poultry-egg production:		
Floor birds		
Cold		0·2
Warm		0·1
Caged birds		
Warm		0·04–0·08

as 30 %. Under adverse climatic conditions, open feedlots may increase feed requirements and lengthen feeding periods. For example, in an area that has a moisture surplus and average coldest month temperature of 5 °C, open feedlots increase feed requirements by 15 % and lengthen the feeding time by 15 % as compared to sheltered housing.

The decision as to whether or not to provide sheltered confinement for beef cattle must be based on economics. Important factors are rate of gain, feed conversion efficiency, cost of feed and value of marketable animal. The sheltered confinement space per animal may be as low as one-tenth to one-fortieth of that for open feedlots. The costs are, however, considerably higher per unit of area. Construction costs vary from country to country and from time to time, and so cannot be estimated. It is estimated, however, that the cost per animal for covered space with solid concrete floor is twice that for open lots, and covered space with slotted floor is three times as high. These increased costs per animal must be made up by increased gains, shorter feeding periods, better management of animals, better pollution control and better working conditions.

The space requirements for dairy cattle (see Table I) are about the same in free-stall and stanchion barn housing. Management of the cattle and feed handling is, however, quite different for the two systems. In stanchion barns, feed and water must be distributed to each stall as each animal is confined to that one position, other than for milking. Originally in stanchion-type barns the cows were also milked in place. Today most

TABLE II

FEED REQUIREMENTS FOR CONFINED ANIMALS

(from refs. 2–6)

Beef cattle in dry lot:

Type and weight of animal	*Feeding period (months)*	*Total gain (kg)*	*Feed/animal*			
			Corn	*Protein*	*Silage*	*Hay*
Steer calves (200–450 kg)						
Liberal grain	9	300	1 600	140	—	800
Liberal roughage	11	250	925	200	2 800	400
Heifer calves (200–400 kg)						
Liberal grain	9	200	1 200	90	—	800
Liberal roughage	10	200	925	135	1 400	450
Yearling steers (300–525 kg)						
Liberal grain	7	225	1 600	115	—	850
Liberal roughage	8	225	925	135	3 200	725
Heavy steers (435–525 kg)						
Liberal grain	3·5	90	740	85	910	200
Liberal roughage	4·0	115	130	135	3 200	90

TABLE II—*contd*

Dairy cattle feed per year:

Type and weight of animal	*Hay equivalent (kg)*	*Hay (kg)*	*Silage (kg)*
Cows, 450 kg (4 500 kg of 4% milk)	3 000	1 500	4 500
For each 45 kg additional body weight	225	100	450
Dairy heifers (1–2 years)	1 500	1 000	1 800
Dairy calves (under 1 year)	900	700	700

Daily feed for swine:

Type and weight of animal (kg)	*Daily gain (kg)*	*Total air-dry feed (kg)*
Growing hogs		
5–10	0·3	0·5
10–20	0·4	1·0
35–55	0·8	2·4
55–80	0·8	3·0
80–100	0·9	3·5
Bred gilts and sows		
135	0·4	2·5
225	0·3	2·9
Lactating gilts and sows		
160	—	5·0
200	—	5·7
Young and adult boars		
135	0·4	2·7
225	—	3·4

Poultry (total feed requirement per year):

Body weight (kg)	*0 eggs/year (kg)*	*200 eggs/year (kg)*	*300 eggs/year (kg)*
1·4	22	34	40
1·6	24	37	43
1·8	26	39	45
2·0	28	40	47
2·3	30	42	49
2·5	32	44	51
2·7	34	46	52

TABLE III

WATER INTAKE REQUIREMENTS FOR CONFINED ANIMALS
(from refs. 2–6)

Type and weight of animal (kg)	*Water/animal (kg/day)*
Beef cattle:	
Calves	
200–300	30
300–400	35
Dry cows	
400–600	35
Heifers and steers	
400–500	40
Fattening calves	
200–300	30
Fattening yearlings	
300–450	35
450–550	50
Dairy cattle:	
Calves—250	20
Heifers—300–400	30
Small cows—400	45
Large cows—500	
Milk 10–25 kg/day	75
Milk up to 40 kg/day	90
Dry cows	45
Hogs:	
15	3
40	4
60	8
100	12
Dry sows	18
Lactating sows	25
Poultry (100's of birds):	
Growing chicks	15
Pullets	20
Laying hens (moderate temperature)	30
Laying hens (32 °C)	35

modern stanchion barns also have separate milking parlours. The free-stall arrangements have appreciable feed and waste handling advantages for dairy herds larger than 100 or 200 milking cows. Stanchion barns are traditionally warm-type insulated structures with mechanical ventilation. Free-stall shelters may be either warm or cold. If the free-stall shelters are of the warm type, then the feeding facilities must also be enclosed in the warm environment. Dairy cows must either by confined completely in a warm environment or they must be left completely in the cold environment so that they can become acclimatized.

The space requirements for hogs and poultry indicated in Table I are for environmentally controlled housing.

Feed and Water Requirements

Feed and water requirements for livestock are given in Tables II and III. Storage requirements for the various feed materials can be planned from the data provided. The length of feeding period and the length of storage periods for various feeds must be determined. Many feed materials must be stored on the livestock farm from one harvest season to the next unless there are available outside sources of supply. Generally, to be assured of continuous year-round feed supplies, it is best to have total on-farm storage. This is particularly true for the roughages of hay and silage. Hay is bulky and not easily transported and silage is not really marketable as it is not a stable product out of the storage silo.

Hay and silage storage structures make up the major storage structures for livestock farms due to the bulk of the material and the need for year-round supplies. Silage storage units may be either of the upright or horizontal bunker type. The upright silos are more adaptable to completely mechanized distribution of silage for smaller herds of less than 500 animals. For larger beef and dairy herds where more than 1000 tons of silage storage are required, the horizontal silos are more economical than the upright silos. The entire livestock confinement facilities must then be planned for fenceline feeding from a mechanical unloading wagon.

SHELTERED CONFINEMENT

Confinement shelters may be either of the warm or cold type. Cold confinement shelters provide protection from the natural climatic factors of wind, rain and snow, but do not appreciably modify the air temperature from the outside climatic conditions. The buildings are then uninsulated and provided with sufficient openings to allow for natural air movement throughout. The warm confinement structures are insulated, closed and mechanically ventilated. Inside environmental temperatures in the warm structures are maintained in the range of 20 °C while the temperature in the cold structures fluctuates within 2 or 3 °C of the outside climatic temperature. The economic factors of production, feed utilization and management dictate as to whether to provide a warm or cold shelter. Young stock must have the warmer temperatures so warm housing is necessary. Cold housing is quite satisfactory for adult cattle, both beef and dairy, in regions where the coldest month average temperature is not below 6–8 °C.

Confinement housing may also be classified according to the type of flooring in the shelter. Solid floors are quite common and may range from compacted earth to total concrete. The other type of floors are slotted, either total or partial slats. Slotted floors are used to reduce animal space requirements and to provide for mechanized handling of the manure as a liquid or semi-liquid. Slotted floors are used the most for hog facilities, and are being used for beef, and some partially slotted floors are used for dairy.

TABLE IV

RECOMMENDED SLAT SIZE AND SPACING

(from ref. 4)

Animal	*Slat spacing*	
	Narrow slats (3–8 *cm*)	*Wide slats* (10–12 *cm*)
Beef	Not recommended	4–4·5 cm
Dairy:		
Cows	Not recommended	4–4·5 cm
Calves	2 cm between 2·5 × 5 cm slats on edge in an elevated stall	3 cm
Swine:		
Farrowing	1 cm (2·5 cm behind sow)	1 cm (2·5 cm behind sow)
Pigs		
10–20 kg	1·25 cm	2–2·5 cm
20–100 kg	Not recommended	2·5–3 cm
Sows	Not recommended	Not recommended
Sheep:		
Ewes	2·5 cm	2·5 cm
Lambs	1·5–2 cm	2–2·5 cm
Feeders	2–2·5 cm	2·5 cm

Concrete slats are the most common and durable, but are the heaviest and thus require the strongest supports. Wood wears and warps and is chewed by hogs, leaving irregular slot openings. Manufactured slotted-floor systems of steel, aluminium and plastic are more uniform than wood or concrete. They may also be easier to handle, install and replace than concrete, but generally are more expensive. When selecting slats, consider initial cost, predicted life, intensity of use (wear rate), strength, corrosion, fire, noise and replacement cost. Slats should be stable; twisting, flexing or shifting may cause cracks in the slat material or slat coating, or may catch or pinch animals. Table IV provides information on slat spacing for various animals.

Warm Confinement Shelter

No animal shelters would be necessary in a continuously perfect climate. Such a climate is prevalent in few places in the world. In most temperate climates of the world, some protection from both cold and hot weather is necessary. Warm shelters then serve the purpose of climate modification. For intensive production of many animals bred for high output and fed scientifically balanced rations, a warm-type shelter with a controlled environment is economically possible and desirable. An optimized environment will develop the genetic potential of the confined animals.

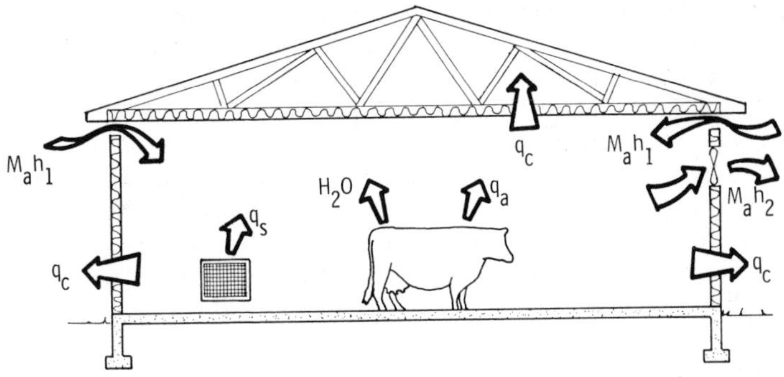

FIG. 2. An energy balance must be maintained for the environmentally controlled structure. q_a = heat from animals; q_s = head added supplementally; q_c = heat conducted through the insulated walls and ceiling; M_ah_1 = mass of incoming air × the enthalpy per unit of air mass; M_ah_2 = mass of outgoing air × the enthalpy per unit of air mass.

A closed, insulated, mechanically ventilated warm-type animal shelter must be designed and maintained with a stable heat and moisture balance between the inside and outside. Animals through their metabolic processes produce both heat and moisture, as well as excreta and waste gases such as carbon dioxide and methane. The animal heat production can be used beneficially in the winter time to keep the shelter warm and to heat incoming ventilation air, while in the summer time it can be detrimental. The moisture produced by the respiratory system as vapour can be a problem in the winter as it must be removed from the shelter to prevent the relative humidity from going above the 75–80% maximum desired level. In the summer time, vaporization can be used as a means to convert sensible heat to latent heat for removal from the shelter without increasing the air temperature. Figure 2 illustrates the energy balance that must be maintained within a warm-type animal shelter.

Energy Balance

The energy balance for the environmental control system may be stated as:

$$q_a + q_s + M_a h_1 = q_c + M_a h_2$$

where q_a = heat from animals or birds (kcal/hr); q_s = heat added supplementally (kcal/hr); $M_a h_1$ = mass (kg) of incoming air multiplied by enthalpy (kcal/kg/hr); q_c = heat conducted through the exterior surfaces of the building (kcal/hr); $M_a h_2$ = mass (kg) of ventilation air exchange multiplied by the outgoing enthalpy h_2 (kcal/kg/hr).

All these factors are manipulatable within limits. The animal heat production q_a is controllable by varying the number, size and kind of animals or birds confined in a given shelter space. The supplemental heat q_s is by definition artificially controlled by man. The mass M_a of ventilation air exchange into and out of the building can be mechanically controlled with fans. The incoming air enthalpy h_1 is, however, dependent on the outside climatic conditions which are not controllable. The product $M_a h_1$ is controlled by varying the mass of air exchange. The outgoing ventilation air enthalpy h_2 depends on the inside environment which can be controlled. The heat conduction q_c through the exterior surfaces can be controlled within limits by the amount of insulation material used in constructing the walls, ceiling, foundation and floor.

Structural Insulation

How much insulation to provide is a design function which depends on the economics of insulation costs versus supplemental heat costs. Heat flow through walls, ceilings and other exterior materials can be calculated by the equation:

$$q_c = UA(t_1 - t_0)$$

where q_c = overall heat flow (kcal/hr); U = the overall structural average coefficient of heat transmission from the air on one side to the air on the other side of walls, ceilings, etc. (kcal/m^2-°C-hr); A = area (m^2); t_1 and t_0 = inside and outside air temperatures in °C.

The overall heat transmission coefficient U for a given wall or ceiling can be calculated by the equation:

$$U = \frac{1}{1/f_1 + L_x/k_1 + 1/f_0}$$

where f_1 and f_0 = the film or surface conductance inside and outside of the wall or ceiling respectively (it represents the time rate of heat exchange by conduction, convection and radiation for a unit area of surface air film; L_x = the thickness of any homogeneous wall or ceiling material; k_1 = the coefficient of conductivity per unit of thickness of any homogeneous wall or ceiling material.

Ventilation Air Exchange Heat Loss

The heat exchange for livestock buildings due to ventilation air exchange is:

$$q_v = M_a(h_2 - h_1)$$

where q_v = ventilation air heat exchange (kcal/hr); M_a = mass of ventilation air exchange (kg/hr); h_1 = enthalpy of the ventilation air entering the building (kcal/kg), h_2 = enthalpy of the ventilation air leaving the building (kcal/kg).

The mass of ventilation air exchange through a livestock building must be sufficient during the coldest weather to remove the water vapour directly produced by the metabolic systems of the animals (see Table V). If the ventilation air exchange rate is not sufficient to do this, one of two things, or both, may occur. Firstly, the inside building environmental air relative humidity may be caused to rise above the 75–80 % level which is detrimental to animal health; and/or secondly, condensation of moisture from the environmental air on to the cooler surfaces may occur.

In most temperate climates, adult animals may be housed in well-insulated houses without the need for supplemental heat during the cold season. In these cases, the amount of cold weather air exchange is limited to that possible with the metabolic heat from the animals after the conduction heat losses have been subtracted. A test of whether it is possible to get by without supplemental heat is to calculate the air exchange possible at the average cold weather temperatures that occur 2·5 % of the time in the location of concern and determine if this air exchange will in fact remove the water vapour produced directly by the metabolic systems of the animals without increasing the relative humidity of the inside environmental air above the 75–80 % level. The column entitled 'minimum exchange ventilation rates' shown in Table V has been calculated to remove the water vapour produced directly by the animals.

At the average monthly cold weather climatic temperature the air exchange rates shown in the centre column of Table V should be possible. If these average temperature air exchange rates are not possible with the metabolic heat produced by the animals, then some changes in design must be made. Either supplemental heat is necessary, more insulation must be applied in the building side walls and ceiling, or a higher density of animals must be placed in the given building space. The average cold weather air exchange rates are sufficient to bring about some additional dehydration of the faecal matter and/or litter besides being sufficient to remove the water vapour produced directly by the animal or bird metabolic systems. Both the minimum and average cold weather air exchange rates are sufficient to provide ample fresh air (oxygen) and to remove undesirable gases (CO_2 and methane).

Hot weather or summer ventilation air exchange rates must be adequate to remove the excess heat produced by the animals without allowing the

TABLE V

RECOMMENDED VENTILATION AIR EXCHANGE RATES FOR CLOSED, INSULATED CONFINEMENT HOUSING IN TEMPERATE CLIMATES IN CUBIC METRES PER HOUR PER ANIMAL

(from ref. 2)

Type of animal	*Minimum exchange during winter*[a]	*Average exchange during winter*[b]	*Summer exchange minimum for hot days*
Layers (light)	0·5	0·9	10
Layers (heavy)	0·6	1·1	14
Broilers, 1·5 kg	0·2	0·5	8
Pullets, 1·5 kg	0·2	0·5	8
Turkeys (breeders)	1·3	3·5	45
Turkeys (hens)	1·0	3·0	35
Turkeys (toms)	1·3	3·5	40
Sow and litter	15	50	500
Hogs, 90–100 kg	10	35	100
Hogs, 20–25 kg	2	5	25
Dairy cows, 455 kg	35	75	1 000
Dairy calves, 45 kg	5	10	100
Beef animals, 455 kg	35	75	1 000
Beef animals, 227 kg	25	50	600
Sheep, 90 kg	10	20	100
Sheep, 23 kg	2	10	40

[a] The minimum winter ventilation air exchange is designed for the minimum winter temperatures that occur only 2·5% of the time.
[b] The average winter ventilation air exchange is based upon the average temperature for the coldest month of the season.

inside building environmental air to increase significantly above the outside high climatic temperatures. A 1–2°C increase above outside climatic temperatures should be the maximum, even in humid climates. In low relative humidity climates (10–30% relative humidity), evaporative cooling should be utilized to the extent that the inside environmental temperature is the same as or lower than the outside climatic temperature. This means that all the sensible heat produced by the animals or birds must be converted to latent heat by using it to evaporate available water in the building (from the waterers, excreta or litter). This amount of evaporation can normally be accomplished with a good standard ventilation system in which there is adequate air exchange and distribution to evaporate that much water. In the dry climates (10–30% relative humidity levels prevailing), specifically designed evaporative coolers may be used in which the incoming climatic air is drawn through a continuously wetted porous material in order to lower the dry-bulb temperature nearer to the wet-bulb temperature.

The summer minimum ventilation rates given in the right-hand column of Table V provide ample summer ventilation air exchange in high-density buildings with fairly high outside climatic relative humidities of 30–70 %.

Air Distribution

Adequate air distribution is an essential part of providing an optimum environment in livestock and poultry structures. Correct calculations may determine how much ventilation air exchange is necessary and adequate mechanical equipment may be provided to move the specified amount of air, but unsatisfactory conditions may still result if the ventilation air is poorly distributed throughout the housing structure. Air movement and distribution is one of the critical factors that must be controlled properly in order for the environment to provide the desired physiological responses from the animals and birds. A high-velocity draught, for example, can be detrimental. A draught is defined as air motion that causes undue localized feeling of coolness or warmth to any part of the body of an animal or bird. Higher air velocities can, however, be advantageous for cooling during hot weather. As long as the ambient air temperature is less than the body temperature, heat loss from the animal to the air will be increased as the air velocity increases. The equation for this heat transfer is:

$$q_{cv} = hA_{cv}V^n(t_s - t_a)$$

where q_{cv} = convective heat transfer (kcal/hr); h = heat transfer coefficient (kcal/m^2-°C-hr); A_{cv} = effective convective surface area (m^2); V^n = air velocity (m/hr) to some exponent—the $\frac{1}{2}$ power is assumed for normal conditions; $t_s - t_a$ = difference between the animal's surface temperature and the environmental air temperature in °C.

The design of good ventilation air distribution within a building is based on sound logic. Air is an invisible gaseous fluid that will flow the easiest route provided—as does water. Therefore, if ventilation air outlets are located close to the inlets, the air will take the shortest route between these two points. Thus, the air will bypass the main part of the structure and not provide any effective ventilation. There is no magic in just installing a few fans in a building unless the air flow patterns are controlled. If exhaust fans are used, then inlets for the air must be provided and *vice versa*.

Ventilation System Design

An exhaust fan ventilation system is preferable to the pressure-type system under most conditions for the following reason: pressure-type or fans mounted in the structure to blow air into the building will dump excessive quantities of cold air intolocalised spots in the building causing undue stress on the housed animals or birds. The air can be distributed through properly designed air duct systems, but these are costly to install and add a continuous friction load to the movement of air. During hot weather there may be some advantage in the higher air velocities that might be attained with blow-in type fans, but this does not offset the cold weather disadvantage. Reversible fans might be used, but they have found little acceptance to date.

The simplest and most economical ventilation system is the cross-flow type (as illustrated in Fig. 3) obtained by locating the exhaust fans along one wall of long, narrow buildings and the air intake openings in the opposite wall. With this arrangement there is no alternative but for the ventilation air to move across the building throughout its full length from the inlet to the exhaust fans. The maximum width of buildings for the cross-flow exhaust system is about 15 m, depending on density of animals in the house and the installation of equipment that might interfere with air flow, such as cages, pen partitions, feeders, etc.

The typical gable-roof type livestock or poultry structure should be provided with a horizontal, insulated ceiling at the wall plate level. This elimination of the gable space from inside the building enhances air flow control and thus the provision of an optimum environment. Also, it is simpler and more economical to insulate the ceiling adequately as compared with attempting to insulate the roof framing. A low-cost but effective fill-type, fibre-glass or rock wool insulation material can be used above the ceiling material. Figure 3 illustrates the insulated ceiling as well as the location of inlets and exhaust fans.

The ventilation air inlet design and location is important to the successful performance of a ventilation system. The location of the inlets is more critical than the location of the exhaust fans. For example, the equal spacing of single exhaust fans along one wall of the building is not too important. They may in fact be located two or three together if that provides some economic saving for installation, wiring, maintenance and care. The inlet, however, must provide for the admission of ventilation air equally the full length of the building without causing undesirable draughts. Thus, a continuous narrow slot located at the top of the wall is ideal. The inlet can then be provided with a directional baffle that causes the air to flow in a given direction, as for example horizontally adjacent to the ceiling or vertically adjacent to the side wall. The optimum direction depends on the confinement location or animals and equipment within the house and the outside temperature conditions. Summer warm air should always be directed downward to maximize its cooling benefit. Winter cold air might be directed either horizontally or vertically to maximize its diffusion before it strikes any animals. This is particularly critical if the animals are confined to a given location by cages or pens and cannot move from a draught.

The air inlet slot width should be designed and constructed to allow the specified amount of ventilation air to enter the building with a 150 m/min slot velocity. This means that the slot must be a different width for ventilation during summer and winter seasons as some 10–15 times as much air must be exchanged during hot as compared with cold weather (see Table V). Ideally, an automatically self-adjusting-width slot inlet will provide the ventilation at a constant velocity. This can be achieved automatically with a device that changes the slot width by sensing the static air pressure difference between the inside and the outside of the building. Many successful slot inlets for exhaust ventilation systems are, however, functioning completely satisfactorily with only two opening width possibilities, one for winter and one for summer. Other slot inlet openings are hand adjustable for all widths and generally only need be changed every few weeks.

During cold climatic conditions it may be that only one out of ten, or fewer, fans will operate; thus air is not exhausted very continuously along the length of the building. As long as the air inlets are designed to let air into the building evenly along its length this is not critical. However, for this reason a narrow slot inlet is sometimes also located on the fan side of the building in between the fan locations. This fan side inlet should only have one-fourth the maximum width of the slot inlet in the wall opposite the fans (see Fig. 3).

POULTRY HOUSING

An example of a poultry housing system for 200 000 laying hens is discussed in so far as the design of buildings and equipment is concerned for a fully mechanized and controlled environmental system. The following management-type assumptions and design decisions are made for the egg production system:

1. Started pullets will be brought in at 20 weeks of age from a separate enterprise.
2. The laying hens will be confined in a stair-step cage system consisting of four rows of double-decked 30 cm × 40 cm cages with three hens per cage.
3. All materials necessary for the life support of the laying hens will be mechanically conveyed:
   (a) feed;
   (b) water;
   (c) ventilation;
   (d) egg removal;
   (e) excreta removal.
4. The 200 000 laying hens will be housed in ten houses, each housing 20 000 hens and approximately 12 m by 123 m in size (see Fig. 3). They should be located not closer than 30 m apart for good air circulation and fire protection.
5. The ventilation air exchange will be provided with twelve

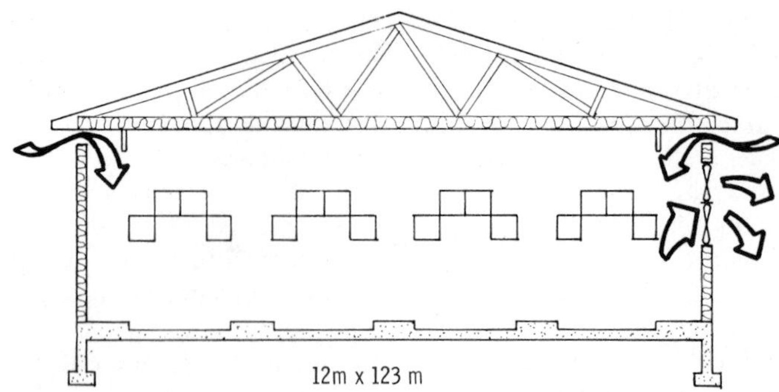

FIG. 3. Cross-section of an egg production house for 20 000 laying hens. Double-deck, stair-step cages, each 30 cm × 40 cm for three hens. Provide twelve 17 000 m^3/hr propeller-type fans, each thermostatically controlled. All fans to be on one side of the building. Incoming air comes through slots in each wall with the ones in the fan wall only in between fans and one-fourth as wide as the opposite wall slot.

17 000 m^3/hr propeller-type fans. All fans operating will provide the 10 m^3/hr required for hot weather conditions. Two fans will provide the average cold weather ventilation air exchange during the coldest month and one fan should operate at all times to provide the minimum cold weather ventilation air exchange. Each fan will be thermostatically controlled.

6. The air inlet system should be designed for three-fourths of the hot weather ventilation air to enter on the side opposite the fans and not more than one-fourth on the fan side. The inlet slots should have openings sufficient to allow air to enter at not more than 150 m/min. There are 48 hens housed per 30 cm of house length. Thus, in hot weather with all fans, $10 \times 48 = 480\ m^3$/hr of air must enter for each 30 cm of house length at a velocity not greater than 150 m/min or 8 m^3/min. The inlet slot in the opposite wall from the fans should be not less than 15 cm wide and direct the air vertically downward during hot weather. The slot inlet on the fan side of the building should not be more than 5 cm wide. These slot inlets must be adjustable down to about one-fourth of the width opening for cold weather operation.
7. Feed storage for one week of feeding should be provided for each house. A minimum feed requirement of 100 g/hen-day would require 2 metric tons per day or 14 metric tons of storage for one week.
8. Mechanical dropping pit cleaners will be provided for daily cleaning. This system necessitates daily application of the excreta to the land and assumes that no air or water pollution problems result.

If this presents a problem, then in-house dehydration should be maximized with in-pit fans and stirring, so cleaning may only need to be every month. Also, final heated air drying to 15% moisture content may be desirable for long-time storage and convenience of handling and utilization.

9 The ceiling and walls must be provided with insulation equivalent to an R-value of 1·75 m^2-hr-°C/W.

DAIRY HOUSING

An example of a dairy housing system for 2400 milking cows is discussed in so far as the design of buildings and equipment is concerned for a fully mechanized system. The replacement herd will not be included in this discussion. To keep 2400 continuously milking cows, approximately another 3000 dry cows and young stock must be maintained—a ratio of 1·25 additional animals for every one milking. Of this maintenance group, housing for young calves is most critical, particularly during cold weather. An insulated, closed, environmentally controlled building is required for young stock which must then also be provided with a supplemental heating system. Raising healthy, disease-free young stock is necessary for a successful dairy enterprise.

The following management-type assumptions and design decisions are made for the milk cow housing system:

1. The 2400 milking cows will be housed in two separate housing systems, each designed for 1200 milk cows in a free-stall system. Each 1200 cow unit will have one milking parlour and two 600 cow free-stall housing units similar in layout to Fig. 4. Each 600 cow housing unit will then be divided into four groups of 150 cows each.
2. A polygon milking parlour with four eight-herringbone stall sides is recommended. The milking parlour is operated as a unit by two persons with cows coming into two herringbone stall sides from one barn and two from the other. Each holding area is designed for a single group of 150 cows from each barn. The milking parlour must have a milking rate of near 150 cows/hr in order that the maximum waiting time for cows in the holding area be not more than 2 hr per milking.
3. Required milk storage will depend on the production average of the cows and the timing of milk pick-up. If an annual production rate of 7000 kg/year/cow is assumed, then daily production will be near 20 kg/day/cow or 24 000 kg/day for the 1200 cow herd. Then storage and cooling capacity for at least 24 000 kg of milk is required for once a day pick-up, or half that for twice a day pick-up.

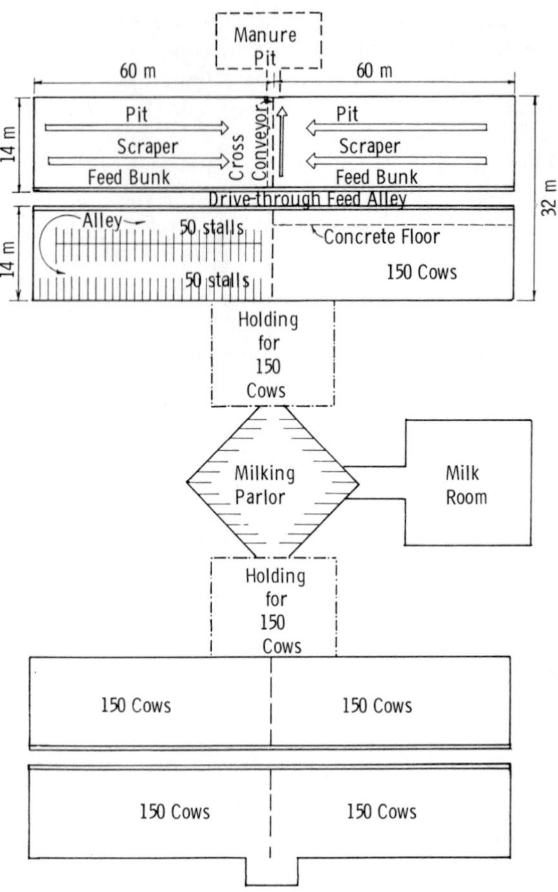

FIG. 4. Milk cow facility for 1200 cows.

4. Roughage and most of the feed ration requirements for the cows will be fed in bunks extending through the centre of each 600 cow barn and on either side of a drive-through alley-way. Each 150 cow group will have about 60 m of feed bunk space, or 0·4 m/cow. The 2400 cow milking herd should be divided into 150 cow groups according to level of milking. A minimum amount of grain (about 1 kg/cow) is fed to the cows in the milking parlour in order to attract them into the parlour. The rest of the required grain ration is fed to each of the 150 cow groups based upon milk production. The hay–silage–grain mixture of feed is distributed in the feed bunks from a self-unloading wagon. The storage silos need not be

immediately adjacent to each barn when wagon distribution is used. The silage storage units can be bunker-type silos which for over 1000 ton capacity are more economical than upright silos.

5. Annual feed storage capacity per milking cow needs to be as follows for a 25 % hay/75 % silage ration: 3600 kg of hay, 9000 kg of silage, 2300 kg of high-moisture corn and 360 kg of bedding. These storage requirements can then be projected for either the 1200 cow unit or for the total 2400 cows and whether storage for the total year's feed is required.
6. The manure is handled as a semi-liquid. Mechanical alley-way scrapers pull the manure towards the centre of each 600 cow barn from four alleys on each end. Each scraper thus must move the manure for an alley length of 60 m. If equipment for this capability is not available, the barn design and scraping length must be shortened accordingly. At the centre of each barn, mechanical cross-conveyors are used to pull the manure into the manure collection tank. From there it must be pumped out and distributed on the land without pollution.
7. The 600 cow housing barns may be designed and constructed for a warm environment or for a cold environment. The cold barn means that inside barn temperatures fluctuate approximately with outside climatic temperatures and the barn is uninsulated and quite open for a free flow of natural air movement without undesirable draughts. The cold structure can be constructed for less money than the environmentally controlled one, but its effects on milk production must determine the economics of the total system. The warm environmentally controlled structure is insulated depending on climatic conditions, closed and dependent on fans for ventilation air exchange. Temperature can be controlled so as not to fall below 12–16 °C. Table V must be referred to for design of insulation and ventilation. The inlet for the ventilation air should be in the centre of the buildings and exhaust fans located in both side walls or the inlet and outlet positions could be reversed.

HOG HOUSING

An example of a hog farrowing and production system will be presented and discussed for marketing 100 000 hogs per year. An enterprise of this size will require ten 10 000 hog production units. Each 10 000 hog production unit will consist of one farrowing house, one nursery house and two growing houses as shown in Figs. 5 and 6. The farrowing houses and nursery houses must be insulated, closed and environmentally controlled with supplemental heat for cold weather. The growing houses may be designed as cold or warm housing. It is recommended, however, that the growing houses be warm, insulated, closed, environmentally controlled structures, as the winter-time temperature in the house can be maintained at a minimum of 16 °C without supplemental heat in temperate climates where the coldest month averages not less than −7 °C. The following management decisions and design criteria are suggested.

1. The management plan for the farrowing operation is for four farrowings per year with an average of 8 pigs per litter. The 320 sow farrowing barn (see Fig. 5) will then produce 2560 pigs each quarter year and slightly more than 10 000 per year. Each sow and litter will be held in the farrowing house until the pigs are from 30 to 35 days old. The pigs will then be weaned and moved to the nursery building and the sows moved to dry sow housing.
2. The farrowing house is equipped with farrowing crates in which the sows are individually confined. A pig creep area is provided on either side of the sow confinement crate to prevent injury to the small pigs. Localized supplemental heat can also be provided for the small pigs in the creep area without overheating the sow. This might be done with overhead heaters or with heating elements in the floor of the creep area. Each farrowing crate is provided with an automatic watering device and a feeder. The sow stalls should be

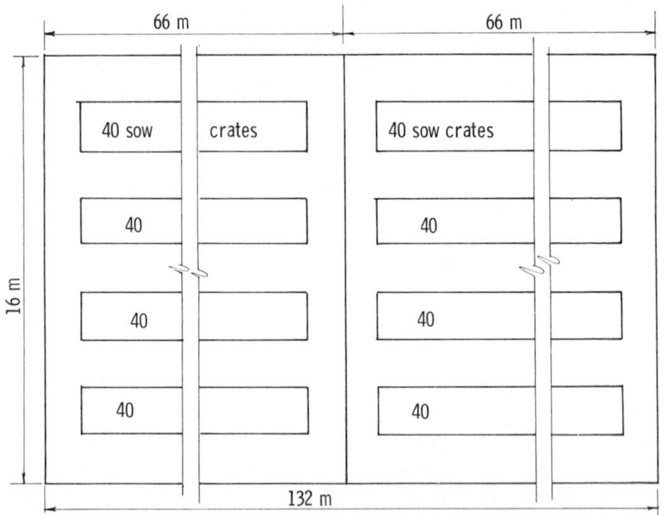

FIG. 5. Plan view of a 320 sow farrowing building for production of 10 000 pigs/year with four farrowings and 8 pigs/litter. The insulated, environmentally controlled building is divided into two 160 sow pens for management convenience. Sow crates are 1·5 m × 2·2 m including width of pig creep areas.

oriented so that there are two feeding alleys. The outside two rows will be faced towards each other for this purpose.

3. A liquid waste handling system is suggested for the hog enterprise. The farrowing house should be provided with a concrete slotted floor. The concrete slats should be approximately 2·5 m long with a top width and a depth each of 10 cm. The sides of the slats should taper inward so that there is a bottom width of 7·5 cm. There should be a 1 cm steel bar near the top of the slat and a 1·5 cm steel bar near the bottom. The slot width provided between slats should be 1 cm. The liquid manure pit should be not less than 1 m in depth below the slotted floor. Supports located lengthwise of the building must be provided for placement of the ends of the 2·5 m long concrete slats.
4. The inside environmental temperature in the farrowing house should be maintained between 18 and 24 °C during cold weather. If the lower temperature is maintained, then some localized supplemental heat will need to be provided during the first few days of each litter. For climatic conditions where the coldest month averages no less than −7 °C, about 400–500 kcal of heat must be provided per hour per sow and litter. Thus, for the 320 sow barn, supplemental heat of about 130 000 kcal/hr must be provided to maintain 18 °C during the coldest climatic periods and up to 160 000 kcal/hr for the higher temperature of 24 °C. The supplemental heat is necessary to compensate for heat loss through the walls and ceiling, and to warm the prescribed ventilation air exchange of 50 m^3/hr per sow and litter. Enough ventilation air exchange must be provided to remove the excess water vapour from the building in order to maintain an acceptable relative humidity.
5. The nursery and growing buildings would be about the same size of 16 m by 75 m (see Fig. 6). All the pigs (approximately 2500) from one farrowing house can be accommodated by the one nursery building for the first two months after weaning. After that they are moved to the growing buildings and provided with twice the space

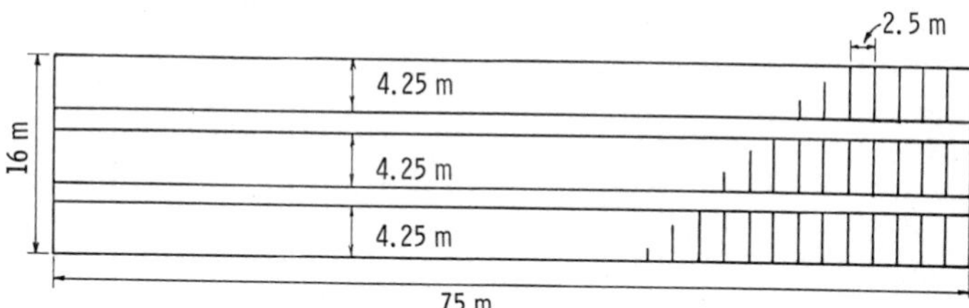

FIG. 6. Basic plan for the nursery and growing buildings. Twice as many small pigs are put in the nursery building as in the growing building. In the nursery there are 28 pigs/pen at 0·38 m^2 each. In the growing building there are 14 pigs/pen at 0·76 m^2 each.

per pig, so two 16 m by 75 m buildings are needed for the 2500 pigs. The nursery and growing houses are divided into pens that are 4·25 m by 2·5 m in size. Approximately 30 of the one-month old pigs may be placed in each pen of the nursery at approximately 0·35 m^2 per pig. After growing for two months in the nursery, the pigs are moved to the growing house and placed 15 to the pen at 0·7 m^2, which is twice the space. The pigs remain in the growing house for approximately three months at which time, at six months of age, they should be ready for market at about 100 kg each.

6. Feed should be delivered to a feeder in each pen of the nursery and growing houses through overhead mechanical conveyors. Water should be available continually in each pen.
7. The nursery and growing houses should have slotted floors similar to the farrowing house except that the slats may be separated a bit more for the older pigs. For the nursery 2·0 cm and for the growing house 3·0 cm slots should be provided between slats. The liquid manure is pumped from the pits when necessary and applied to the land without air or water pollution.
8. The nursery building must be provided with some supplemental heat during the coldest periods, thus the building must be well insulated, closed and mechanically ventilated. According to Table V a minimum volume of air of from 2 to 5 m^3/hr-pig must be provided during cold weather and up to 25 m^3/hr-pig during hot weather.
9. The growing houses only differ from the nursery buildings in that supplemental heat is normally not necessary and more ventilation air exchange per animal is necessary because of the size increase. The range (see Table V) is from 10 to 35 m^3/hr-pig during cold weather and up to 500 m^3/hr-pig in the hottest weather.

BEEF FEEDING FACILITIES

An example of a beef feeding facility for from 100 000 to 200 000 feeder cattle in an arid climate having 25 cm or more moisture deficit per year (the deficit is annual lake evaporation minus rainfall) should be planned basically in a rectangular pattern as illustrated in Fig. 7. The cattle are grouped 200 cattle each in 60 m × 60 m pens. Ten to twenty pens may be lined up in a row on either side of a drive-through feed alley with fenceline feed bunks on each side of the feed alley. Numerous double line drive-through sets of pens may be located side by side. The following management decisions and design criteria are suggested:

1. The 200 cattle in each pen have an equivalent area of 18 m^2 and feed bunk space of 30 cm per animal. The feedlots are unpaved except

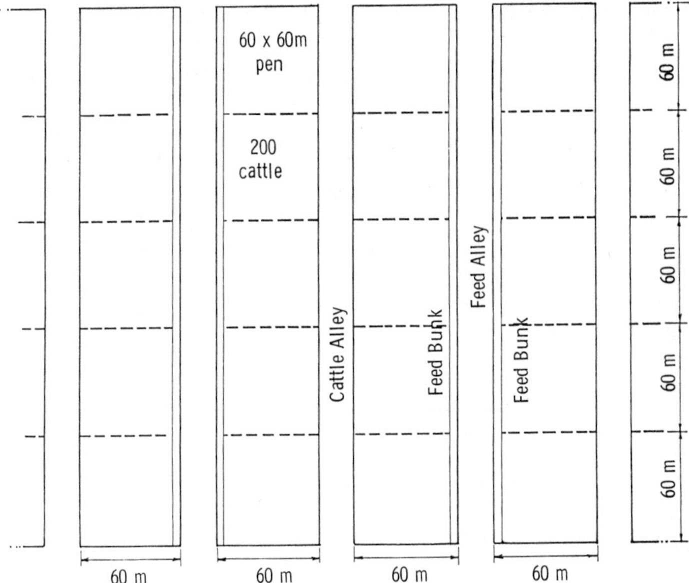

FIG. 7. Open cattle feedlot in arid areas with more than 25 cm of moisture deficit per year. Area per animal is 18 m^2 and feed bunk space per animal is 30 cm. Up to 75 m^2/animal may be needed in more humid areas.

for a strip of concrete about 3 m wide adjacent to the feed bunk on one side of the pen. A 5–8 % slope should drain water away from the feed bunk. An alley-way should also be provided on the opposite side of the pens from the feed bunk for purposes of moving cattle and for drainage of excess water from the pens.

2. Feeding is done from mechanical self-unloading wagons. Feed storage is in centrally located upright silos for grain and concentrates and bunker (horizontal) silos for silage. Fattening cattle should consume at least 2 % of their body weight in dry matter daily. Anything below indicates a low feed conversion efficiency. Cattle for fattening are normally brought into the feedlot at about 300 kg and fattened to around 525 kg. Average daily gain in open feedlots should be somewhat above 1 kg/day/animal.
3. A hospital area should be provided for from 3 % to 5 % of the cattle in the feedlot. This would normally be a covered confined area where some 3 m^2 of floor area is provided per animal. At each end of the cattle handling area there should be facilities for catching and confining groups of cattle.
4. Large, open confinement lots function satisfactorily for arid regions where the cold weather and snow is not too severe. In more

humid temperate climates, with annual moisture deficits of less than 25 cm and where snow and ice are a significant portion of the annual precipitation, open feedlots are not satisfactory. The snow and ice cause bad conditions in the feedlots. Excess water causes mud and converts the manure into unmanageable conditions which may cause both air and water pollution. Each animal must be provided with from $2\frac{1}{2}$ to 4 times the area in these more humid areas of up to 75 m^2/animal. Muddy confinement lots will reduce daily gains by 20–30%. In general, open lots in the more humid temperate climates will increase the feeding time by 15% and require 15% more feed than is required in covered confinement areas with concrete floor or slats.

5. The facilities for covered confinement areas with solid concrete floors will cost twice as much per animal as open lot facilities ($100 as compared with $50/animal). Concrete slotted floors will add another $50/animal. These additional capital costs must be compensated for by increased gains and feed conversions. On slotted floors 1·85 m^2 are provided per 455 kg of animal and on solid floors 2·8 m^2/animal.

REFERENCES

1. Shuyler, L. R., Farmer, D. M., Kreis, R. D. and Hula, M. E., *Environmental Protecting Concepts of Beef Cattle Feedlot Wastes Management*. US Environmental Protection Agency, Corvallis, Oreg., 1973.
2. Lytle, R. J., Esmay, M. L., Muehling, A. J., Van Fossen, L. D. and Brunner, G. E., *Farm Builders Handbook*. Structures Publishing Co., Farmington, Mich., 1973.
3. Esmay, M. L., *Principles of Animal Environment*. AVI Publishing Co. Inc., Westport, Conn., 1969.
4. Midwest Plan Service, *Structures and Environment Handbook*. Agricultural Engineering Dept, Iowa State University, Ames, Iowa, 1975.
5. American Society of Agricultural Engineers, *Agricultural Engineering Yearbook*. ASAE, St Joseph, Mich., 1975.
6. Midwest Plan Service, *Professional Design Handbook*. Agricultural Engineering Dept, Iowa State University, Ames, Iowa, 1975.

6

Management of the Living Environment

Z. Kobos

Deputy Director, Czechoslovak Research and Development Centre for Environmental Pollution Control, Bratislava, ČSSR

The problems of the utilization and disposal of wastes from large animal feedlots are certainly economic in nature, but they are also of legal concern because of their impact on the human and the living environment. The solutions of these problems are extremely important from the point of view of the entire national economy. As such, feedlot waste management is not only the concern of agriculture but almost of all government agencies and operations.

Living environment may be defined as a sum of all components of the world which affect man in a relatively direct manner either as an individual or as the entire human society. The living environment includes man-made structures such as towns, cities, roads, buildings, etc., administrative components such as nations, states, ministries, etc., and natural resources such as soil, water, air and biomass (plants and animals).

The living environment as a whole may be classified into (a) the natural environment, (b) the working environment and (c) the dwelling environment.

Development and protection of the living environment is regulated by political entities at all levels of government (federal government, federal assembly, federal ministries and agencies, national entities, regional, local and global political institutions).

To care for the living environment, it is important that we ascertain scientifically the needs and demands of the human society on the living environment, the effects of changes in the living environment on human society, and also define through scientific and political methods what is the optimum living environment.

The main tools by which we can develop and protect the living environment are planning, economic policies, legal and legislative measures, the utilization of science and technology, public education, and programmes of research.

Many legislative acts of the last 30 years deal directly or indirectly with the care and management of the living environment. These acts are on health, soil conservation, pollution control, water management, noise nuisance control, nature conservation, occupational health and safety, human settlements, etc. In Czechoslovakia at present there exist some 300 legislative acts, ministerial regulations and government decrees bearing directly on the living environment.

In 1970 the Czechoslovak Government undertook an extensive review and update of the entire group of environmental laws and regulations in view of the ever-expanding and complex demands on the environment. This can be done by using computer models to analyse the problem and to synthesize alternative solutions through simulation models.

Modelling and computer simulation of the individual components of the living environment and the utilization of the results thereof in the management of the living environment are becoming common. Two types of models are mainly used to simulate the quality of the living environment: (a) the balance models of production and consumption wastes; (b) the environmental models. The balance model is a mathematical function relating production variables to the resulting sum of wastes. The size of these models ranges from equations with one variable and a constant coefficient, to very complex linear programming models.

Environmental models depict the impact of wastes (solid, liquid, gaseous, heat) on the quality of the individual components of the living environment.

Environmental models may be classified into (a) soil dispersion models, (b) atmospheric dispersion models; (c) waterflow quality models; (d) ecosystem models.

In Czechoslovakia, management of the living environment and its individual components is the responsibility of government organs at federal, republic and local level.

The Federal Assembly, as the supreme legislative body, enacts laws which are valid for the entire country. Programmes on the living environment are included in five-year and long-term economic development plans. The Federal Assembly controls the budget which contains a separate allocation on the living environment.

The Federal Government proposes legislation for consideration by the Federal Assembly, issues decrees to implement laws passed by the Assembly and establishes the network of inspection and monitoring to see that environmental laws are obeyed.

A Commission on the Living Environment, headed by the Deputy Premier, is the advisory body to the Federal Government. The state Planning Commission, responsible for the five-year economic development plans, is also involved in the management of the living environment through programmed improvement in the production technologies and through selection of non-waste technologies.

Federal ministries which are critically important in the management of the living environment are the Ministry of Technical and Investment

Development, the Ministry of Agriculture and Food, the Ministry of Labour and Social Security and the Office of Standards and Measurements.

At the state level, the National Councils of the two Republics plus the executive governments have the overall responsibility of the management of the environment. Each Republic has an advisory council which coordinates and initiates environmental programmes. Ministries with responsibilities and activities relating directly to environmental quality are the Ministries of Health, of Agriculture and Food, of Forestry and Water Management, of Construction and Technology, of Culture and of Education.

Furthermore, National Committees are active in the development and protection of the environment through their public programmes and independence from individual ministries. Members of the National Committees are leading specialists in environmental sciences and technology but mainly citizens who believe in the need to educate their fellow citizens—their neighbours and fellow workers—on matters related to the care for the living environment. They also help organize citizen groups to carry out voluntary work in improving the living environment at their place of work, or in their home district or neighbourhood. Cooperating effectively with the National Committees in this work are the Revolutionary Trade Union Movement and the Socialist Union of Youth.

The problems of the utilization or possible disposal of wastes from large feedlots must be solved not only from the point of view of the needs and interests of agriculture but also from the point of view of broader national interests and aspirations. Even though the involvement of ministries and agencies may at times complicate the solutions to these problems, it is the only way to bring to bear all human concerns and interests on the living environment.

7

Effluent Regulations for Animal Feedlots in the USA

R. C. LOEHR

Director, Environmental Studies Program, Cornell University, Ithaca, New York, USA

and

JEFFERY D. DENIT

Chief, Effluent Guidelines Division, US Environmental Protection Agency, Washington, DC, USA

INTRODUCTION

The use of intensified, mechanized agricultural production techniques has altered the traditional concept of considering pollution from agriculture as 'background' or uncontrollable pollution. An example of such a development is the practice of raising livestock and poultry in concentrated production enterprises, so-called 'feedlots'. As a result, pollution control regulations are being applied increasingly to agricultural operations.

Until recently, in the USA as in many other parts of the world, animal wastes were regulated mainly by local governments. Most states in the United States have had water quality regulations, administrative codes or public health guidelines which are applicable to animal wastes. Even though such wastes may not be explicitly cited as pollution sources in these regulations, the regulations are sufficiently flexible to be applied to animal waste pollution problems.

Many areas in which livestock production is an important factor have specific legislation or guidelines for animal waste management. Individual jurisdictions may have air pollution statutes or guidelines that deal with air quality problems associated with agricultural production such as odours and dust arising from animal production activities.

A number of states have developed registration and permit requirements for control of water pollution from animal feedlots. While the specific requirements vary throughout the United States, they usually include provisions for a feedlot operator (a) to obtain a permit from the appropriate state agency; (b) to correct any pollution hazard that exists; and (c) to ensure that the operation conforms to all applicable federal, state and local laws. Specific minimum pollution control methods or criteria such as runoff retention ponds, dikes, and distance from dwellings may be incorporated into the requirements.

In the United States, recent changes in the federal water pollution control regulations have become constraints which affect agricultural production. These regulations have been developed to control point source pollution from confined animal production facilities, food-processing operations, and all industries in the United States. The purpose of this paper is (a) to describe the federal regulations; (b) to indicate how effluent limitations are developed for point sources; and (c) to describe details of effluent limitations for animal feedlots.

FEDERAL REGULATIONS

Federal water pollution control legislation in the United States, as in other countries, has been developed since the turn of the century but with increasing rapidity in recent decades.

The Water Quality Act of 1965 greatly expanded the scope of federal activities and assigned the federal government a leading role in the control of water pollution. One of the far-reaching effects of the 1965 Act was the provision for establishing water quality standards. Under the supervision of the federal government, each state developed water quality criteria and plans for implementation and enforcement.

The extensive technical, economic and social assessment process necessary to develop the standards fostered a significant change in the basic federal water pollution control philosophy. Emphasis was no longer given to assessing the amount of wastes that could be accepted by waters without causing pollution problems: rather, concern centred on how to keep wastes entirely out of the water.

There also existed a need to centralize the fragmented federal pollution programmes into a single agency. In 1970 the national Environmental Protection Agency (EPA) was established. EPA contains the federal activities dealing with solid waste management, air and water pollution control, water hygiene and other environmental concerns, and has the responsibility to integrate such environmental quality activities in a manner which minimizes all pollution problems. Two years after the establishment of the EPA, the Congress enacted the Federal Water Pollution Control Act, as amended, 1972 (hereafter referred to as the Act). The Act contained the first specific federal requirements for the abatement of agriculturally related sources of pollution. Among many requirements of this highly complex statute, the Act stipulated that each state must develop a waste management plan that was to incorporate identification of agricultural pollution sources and to set forth feasible procedures to control such sources. Similarly, the 'source control' philosophy developed in previous legislation was implemented by requiring technology-based effluent limitations in the control of point sources of waste discharges from all industrial operations including feedlots.

The Act formalized the emphasis on keeping wastes out of surface waters by declaring 'it is the national goal that the discharge of pollutants into the navigable waters of the United States be eliminated by 1985'.[1] To achieve this goal, effluent limitations on point waste sources were to be achieved by the application of best practicable control technology currently available (BPCTCA) by 1977 and of best available technology economically achievable (BATEA) by 1983.

The burden was placed upon industry to implement controls in a manner that represents the maximum use of technology within its economic capability. A number of agricultural operations such as dairy product processing, fruit and vegetable processing, seafood processing, feedlots and fertilizer manufacturing are included among the point sources that are subject to effluent limitations. In determining the effluent reduction made possible by the application of the best practicable or available technology, a 'no discharge' standard could be possible in cases where no discharge is found to be best technology.

Dischargers are required to treat their discharges to the degree indicated by the BPCTCA and BATEA technologies, but they are not required to use any particular method of pollution control. Any method is satisfactory as long as the treated effluent meets the limitations on the parameters set forth for the prescribed levels of technology.

Existing sources are required to achieve effluent limitations based on the assessment of the two levels of technology noted above, BPCTCA and BATEA. New sources are required to meet a third level of treatment which is described as the 'best available demonstrated control technology (BADCT), processes, operating methods, or other alternatives including, where practicable, a standard permitting no discharge of pollutants'.

In summary, the Act contains three different descriptions of treatment technologies, two of which are applicable to existing sources in two roughly five-year intervals, while the third is applicable immediately to new sources. While maintaining the concept of water quality standards as a nationwide baseline, the Act clearly shifted emphasis towards source controls through effluent limitations. This approach was taken because of difficulties in relating waste discharge quality to stream quality and in enforcing previous legislation. The basic assumption, constrained only by the availability of economical control technology, is that the nation will strive towards complete elimination of water pollution.

OVERALL DEVELOPMENT OF LIMITATIONS

The Act directed that the regulation of industrial pollution problems be accomplished upon a basis of technological capability or achievement. Neither receiving water quality, nor effluent dilution before or after discharge, are relevant to the mandate. The crux of this technological basis is in three discrete levels described above.

A great deal of data must be obtained and evaluated by EPA before effluent limitations may be proposed and codified in regulations. Many factors, some specifically stated in the Act, must be evaluated. All these factors, particularly the cost of application of various levels of technology, require considerable investigation. Factors relevant to identifying the technologies which are specified in the Act include:

(a) the total cost of application of the technology in relation to the effluent reduction benefits to be achieved from such application;
(b) the age of the equipment and facilities involved;
(c) the process employed;
(d) the engineering aspects of applying various types of control technologies;
(e) process changes;
(f) the non-water quality environmental impact, including energy requirements.

Utilizing both the technical personnel within EPA and a substantial group of private engineering consultants, the Agency expedited necessary acquisition of data and preparation of reports to obtain an accurate picture of the industry and appropriate technology. As part of determining appropriate effluent limitations, all types of available technology were identified. This extended from the least efficient technology to the most efficient and covered the range of availability of technologies from those in wide use to those available only at the laboratory or pilot stage. For each level of technology, the effluents which result from their application, in terms of the amount of constituents, were identified. Such identification was primarily directed at establishing the performance of actual treatment and control systems in commercial practice at the 'best' or 'exemplary' plants. The limitations focus not only on the technology but also upon the efficient application of the technology.

An analysis of each general level of technology was made to estimate the cost of each technology for individual 'typical' plants and for the entire industry category. Costs were derived to show capital investment, operating and maintenance costs, and energy and power cost. The completed cost analysis was then utilized with aggregate removal efficiency determinations to derive relationships of costs to effluent reduction benefits.

Following the technical and cost investigations, the appropriate

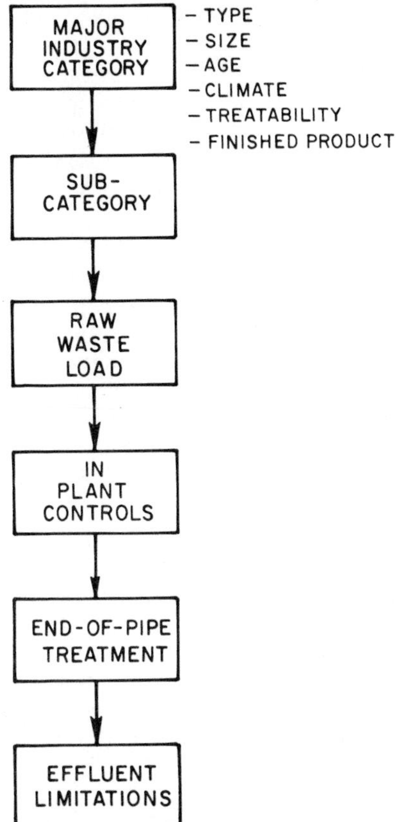

FIG. 1. Schematic of the steps taken to develop effluent limitations based on technology assessment.

limitations were derived and the applicable technologies identified. A schematic of the general methodology is shown in Fig. 1.

FEEDLOT LIMITATIONS

Passage of the Act focused federal regulatory attention upon agriculture in general and livestock operations in particular in a manner previously foreign to this segment of the economy and population. The Act defined concentrated livestock and poultry growing operations (feedlots) as 'point' sources of 'industrial' pollution and required that discharge permits be issued for these operations. While discussion of the permit programme itself is beyond the scope of this paper, the 'National Pollutant Discharge

Elimination System', as it is designated, has resulted in the issuance of nearly 1000 permits to feedlot operations.

The permits provide a mechanism for both administrative and enforcement control of water pollution discharges by EPA and through EPA by each state. Farmers were confronted by regulation of some of the land-oriented aspects of their enterprises. As the primary carrier of pollutants, precipitation runoff, previously considered as 'background' from the land-based feedlot, was now to be regulated. Pollution control historically had been almost entirely a voluntary and cooperative government/citizen programme. In many cases the voluntary approach was effective in minimizing contaminated feedlot runoff from entering surface streams. The element of cooperation remains an important feature of the enforcement measures authorized and implemented under the Act.

From the point of view of the feedlot operator, the permit provides the vehicle for conveying the degree of pollution control expected of him, *e.g.* a guideline for effluent limitations on existing sources of pollution and a standard of performance for new sources.[2] All the detailed considerations underlying this regulation cannot be covered in this discussion. It is possible, however, to summarize the essential philosophy of the limitations and to outline the technology and numerical limitations.

Until 1977, existing feedlot operations are to utilize, upgrade or install pollution abatement facilities. New feedlot installations must utilize the latest techniques. Only by 1983 is enforced adoption of an updated level of technology demanded of the total industry. The regulated feedlot operators must employ either existing or new pollution abatement devices to their maximum efficiency. Anything less would compromise the statutory objective of optimum pollution control.

The feedlot limitations evolved from an intensive thirteen-month undertaking involving a technical study of the feedlots industry, issuance and public review of tentative technical reports and regulations, consideration of all subsequent comments, and modification and final development of the regulations and background information. Feedlots have diverse practices and can include unique circumstances. Nevertheless, there are sufficient consistent similarities to allow reasonable assimilation of the diversity and uniqueness into a manageable framework. In this regard, a definition of a feedlot was developed to serve both as the baseline from which technical analysis proceeded and the primary mode of explaining applicability of the final regulation. This definition reads as follows:

> 'The term feedlot shall mean a concentrated confined animal or poultry growing operation for meat, milk, or egg production, or stabling, in pens or houses wherein the animals are fed at the place of confinement and crop or forage growth or production is not sustained in the area of confinement.'

Within this definition a thorough assessment of the raw materials, production methods, waste characteristics and other factors was conducted to ascertain if indeed the feedlots industry could, or should, be categorized for purposes of establishing effluent limitations. This assessment indicated that the raw wastes from all types of operations were similar on the basis of production units (*e.g.* kg of BOD per 1000 kg of animal weight). The feedlots industry was sub-categorized as shown in Table I to account for

TABLE I
CATEGORIZATION OF THE FEEDLOTS POINT SOURCE CATEGORY

Beef cattle	Open lots
Beef cattle	Housed lots
Dairy cattle	Stall barn
Dairy cattle	Free-stall barn
Dairy cattle	Cowyard
Swine	Open dirt or 'pasture' lots
Swine	Housed, slotted floor
Swine	Solid concrete floor, open or housed
Sheep	Open lots
Sheep	Housed lots
Horses	Stables (racetracks)
Chickens	Broilers housed
Chickens	Layers (egg production) housed
Chickens	Layer breeding or replacement stock, housed
Turkeys	Open lots
Turkeys	Housed lots
Ducks	Wet lots
Ducks	Dry lots

differences in production method, final product, and indirectly for variations in site and location.

Numerous visits to existing feedlots were made and specific site reports were used to verify the feedlot and waste characteristics and the generalized literature data for feedlots. All the existing pollution control systems, except for duck feedlots, consisted of runoff control systems. This is due to the fact that in the United States, adequate land is still available for the disposal and utilization of animal wastes and that a high degree of waste treatment to obtain an effluent quality suitable for discharge to surface waters is not necessary. The one exception is with certain duck feedlots at which biological waste treatment and subsequent discharge to surface waters was being done. Table II indicates the types of pollution control systems that were being used at the time of the effluent limitations study.

TABLE II
SUMMARY OF FEEDLOT POLLUTION CONTROL SYSTEMS CURRENTLY UTILIZED

General basis for control capability	*Production system where used*
Rainfall runoff:	
5 year, 48 hr storm	Open livestock operations
10 year, 24 hr storm	Open livestock operations
25 year, 24 hr storm	Open livestock operations
Long-term storage (60, 90, 120 days)	Open livestock operations
Washdown flushing:	
3·5–11 m^3/head	Swine, open or housed lots
42 m^3/head	Cattle, open or housed
Hydraulic load	Hydraulic flush poultry systems
Production runoff:	
Conventional biological treatment	Duck feedlots

Table II is particularly useful as a display of both the variety of control procedures and the levels of control attained by existing 'exemplary' livestock operations.[7] While not shown in the table, runoff prevention (using totally housed feedlots) and land application of the wastes as the ultimate means of pollution abatement is growing in popularity within economic constraints. Many duck feedlots, particularly those which use open swimming areas, successfully use conventional aerobic biological treatment for pollution control. However, a number of duck feedlots are already accomplishing 'no discharge' by minimizing water use and land disposal of the wastes. Beef, dairy, poultry and swine wastes are routinely disposed of on the land, predominantly as part of a crop production cycle, thus achieving a 'no-discharge' level of pollution control.

Runoff control using classical water conservation and management practices and land disposal are the most important pollution control practices for the feedlot industry in the United States for the foreseeable future, since adequate land for such disposal continues to be available. The existing data indicated that controls based upon runoff volumes as great as or greater than that associated with the 25 year, 24 h storm clearly satisfy requirements as being 'available' and 'achievable' and 'demonstrated'.

Moreover, because of the breadth of operations and locations where this level of control is utilized, it is possible to consider the 25 year event as commensurate with 'practicable'. Certain relevant factors, however, suggested against such a course of action: (a) the 10 year, 24 hr storm is the basis for a number of state and local programmes now under way; (b) a large portion of existing operations which need runoff controls had initiated projects directed at this 10 year storm level; and (c) a significant number of states and regions did not have specific statutory controls and a reasonable baseline requirement must be established in accordance with the Act.

TABLE III
EFFLUENT LIMITATIONS FOR THE FEEDLOT INDUSTRY

Technology level		*Limitations*
BPCTCA (existing sources)	(a)	No discharge except overflow from 10 year, 24 hr storm
	(b)	0·91 kg BOD_5 per 1000 ducks 400 MPN faecal coliform per 100 ml
BATEA (existing sources)	(c)	No discharge except overflow from 25 year, 24 hr storm
BADCT (new sources*)	(c)	No discharge except for overflow from 25 year, 24 hr storm

(a) All feedlot sub-categories except ducks.
(b) Duck sub-category.
(c) All feedlot categories including ducks.
* After 7 September 1973.

As a result of the intensive study on the feedlot industry and detailed analysis of possible pollution control technologies such as runoff control, biological treatment, solid waste management, utilization and land disposal, the requirements shown in Table III were implemented as the basis for the effluent limitations and standards for the feedlot industry. Details of the study, including an outline of the feedlot industry and categories, waste characterization, possible technologies, and logic in arriving at the requirement and limitations, have been published as a Development Document for the industry and other interested individuals.[3] The specific effluent limitations also have been published.[2]

The Act requires a serious consideration of achieving no discharge of point source pollutants if such technology is feasible for the industry. The obvious use of runoff control facilities and the general use of land disposal for the ultimate disposal of the collected runoff and solid matter in the feedlots indicated that a no-discharge requirement could be reasonable for this industry. It is, however, too impracticable to design a runoff control facility to contain all the runoff from the maximum potential storm. Thus, the term 'no discharge' is a relative requirement.

The limitations recognize the technical and economic practicalities by identifying the extent to which runoff control must be practised. As an example, for BPCTCA, the limitations note that:

> 'Process wastewater pollutants in the overflow may be discharged to navigable waters whenever rainfall events, either chronic or catastrophic, cause an overflow of process wastewater from a facility designed, constructed, and operated to contain all process generated wastewaters plus the runoff from a 10 year, 24 hr rainfall event for the location of the point source.'

A discharge is permissible provided that a process wastewater control facility has a certain capacity and that this capacity is kept available through proper 'operation' of the facilities such that runoff from smaller storm events than that noted can be contained. Definition and explanation of several key words and phrases in the limitation regulation help clarify the meaning of the exception to 'no discharge' as follows:

> 'The term "process wastewater" shall mean any process generated wastewater and any precipitation (rain or snow) which comes into contact with any manure, litter or bedding, or any other raw material or intermediate or final material or product used in or resulting from the production of animal or poultry or direct products such as milk and eggs.
>
> The term "process generated wastewater" shall mean water directly or indirectly used in the operation of a feedlot for any or all of the following: spillage or overflow from animal or poultry watering systems; washing, cleaning or flushing pens, barns, manure pits or other feedlot facilities; direct contact swimming, washing or spray cooling of animals; and dust control.
>
> The term "10 year, 24 hr rainfall event" shall mean a rainfall event with a probable recurrence interval of once in ten years.'[8]

As defined, these three terms prescribe the amount and composition of the 'no discharge' capacity which the EPA can reliably document in accordance with the Act. As a practical matter, the major thrust is towards minimizing water pollution from runoff due to chronic, low-level precipitation events (open lots for beef, dairy and swine) and hydraulic manure removal (pen flushing in swine and dairy operations and liquid manure systems for feedlot operations). The specified storm serves to establish the flexible end-point based upon existing technology.

Equally important to the success of the stipulated pollution control limit are certain terms not specifically defined in the regulation. These key terms 'designed' and 'operated', the interpretation and implementation of which are vital in generally achieving 'no discharge' of pollutants. An explanation of the intent of the EPA when using these terms may be informative.[6] 'Designed' has been used to convey to the feedlot operator an explicit concern for considering the soil, climatic, animal type and related circumstances pertinent to the specific site so as to devise plans for building, maintaining and managing control facilities. The control facilities, once constructed, must function in a manner which precludes subsequent discharges each and every time process wastewater enters them. The term 'operated' has therefore been used to convey the expectation that a deliberate attempt is to be made to ensure, within reason, (a) that the control facilities are available for recurring runoff, (b) that such actions as may be taken to ensure availability do not result in a direct discharge to navigable waters, and (c) that established principles of sound agricultural or waste disposal practice are not compromised to an extent that fosters ancillary problems of soil contamination, malodorous conditions or the like.

In this regard, facilities which never discharge regardless of conditions are not precluded because such facilities clearly satisfy the 'no discharge' restriction. Except for such performance in arid areas, however, there is little economic practicality for such a system. Fortunately, many facilities which have the required level of feedlot runoff control at reasonable cost are already in place and in use at feedlots throughout the United States.[5] The regulation reflects the reasonableness of the limitations in both having limitations that can be met by animal producers and in meeting the requirements of the Act.

Because the no-discharge technology is feasible for the feedlot industry, no discharge was established as the limitation for BATEA and BADCT as well as BPCTCA. After considering advanced technologies that may be utilized by new sources which are not constrained by existing locations and production practices or by existing feedlots in the future, no discharge continued to be the best technology to meet the three levels of limitations.

The Act requires that, where feasible, more stringent pollution control technology be considered and adapted for BATEA. Evaluation of the possibilities indicated that the overflow criteria should be based upon the overflow from a process wastewater control facility designed, constructed and operated to contain all process-generated wastewaters plus the runoff from a 25 year, 24 hr rainfall event. Such a standard continued to recognize the impossibility of containing the runoff from all storm events but decreased the probability of pollutant discharges from feedlots. This standard has caused no difficulties for the feedlot industry and, as noted in Table II, had been part of some feedlot pollution control systems that have been utilized.

The no-discharge requirements and runoff standards were not found to be applicable for duck production for BPCTCA. Extensive studies had demonstrated that while some 'no discharge' technology was in practice, biological treatment methods, such as aerated lagoons, should serve as the technology upon which limitations for BPCTCA should be based. Many duck producers had installed aerated lagoons and were achieving a high degree of pollutant reduction.[4] Other duck producers had reduced their water usage and had developed no-discharge approaches which generally involved some form of land disposal.

A no-discharge limitation was not immediately feasible for duck producers. Available technology was evaluated and the BPCTCA limitations established on the levels that could be achieved by well-designed and operated aerated lagoons. Actual data obtained from aerated lagoons treating duck production wastewaters were used to establish the limitations. As noted in Table III, the BPCTCA limitation for duck producers was in mass units for BOD_5 and in concentration units for faecal coliform. Insufficient data were available to establish limitations on other pollutant parameters.

The detailed review of the duck production industry and available technologies indicated that a no-discharge limitation was reasonable for BATEA where land disposal and hence no discharge of point source pollution is possible. By 1983, the date BATEA is required to be implemented by all producers, the duck producers should be able to modify their operations to achieve no discharge. Such modifications would include acquiring adequate land for disposal of their wastes in a non-pollutional manner and/or reducing their flow to the level of current 'dry' systems which are already achieving no discharge.

Thus, while specific effluent limitations are required of duck producers for BPCTCA, no discharge is the requirement for BATEA and BADCT.

SUMMARY

Federal water pollution control legislation in the United States has required point source effluent limitations of all industry including that producing agricultural products such as feedlots. The limitations are technology-based, and are to achieve three levels of control: best practicable control technology currently available (BPCTCA) and best available technology economically achievable (BATEA) for existing sources, and best available demonstrated control technology (BADCT) for new sources. The technology is to consider no discharge as a possible limitation wherever technically and economically feasible.

The feedlots limitations evolved from an intensive thirteen-month study of the feedlot industry including evaluation of possible pollution control technology and the costs associated with the application of the technology. The resulting effluent limitations are: (a) no discharge except for the overflow from a specified storm for BADCT and BATEA for all feedlots and for BPCTCA for all feedlots except duck producers; and (b) specific mass and concentration limits for BPCTCA for duck producers.

The limitations incorporated in the regulations for the feedlot industry are reasonable, are being met by current feedlots, are consistent with the technical and economic capabilities of the industry, and are within the intent of the Act.

REFERENCES

1. Federal Water Pollution Control Act, as amended, 1972 (PL92–500), 18 Oct. 1972. US Government Printing Office, Washington, DC., 1972.
2. Effluent Guidelines and Standards, Part 412, Feedlots Point Source Category. *Federal Register*, **39,** 5704–10, 14 Feb. 1974.
3. US Environmental Protection Agency, *Development Document for Effluent Limitations Guidelines and Standards of Performance for New Sources for the Feedlots Point Source Category*. US Government Printing Office Stock No. 5501–00842, Washington, DC, Feb. 1974.
4. Loehr, R. C. and Schulte, D. D., Aerated lagoon treatment of Long Island duck wastes. In: *2nd International Symposium for Waste Treatment Lagoons*, pp. 249–58. Federal Water Quality Administration, Kansas City, Mo., June 1970.
5. Loehr, R. C., *Agricultural Waste Management*. New York, Academic Press, 1974.
6. Denit, J. D., Effluent regulations for livestock and poultry feedlots. In: *Proceedings of 6th National Conference on Agricultural Waste Management*, pp. 51–58. Cornell University, Ithaca, NY, 1974.
7. Denit, J. D., Feedlot effluent limitations based upon exemplary operations. In: *International Symposium on Livestock Wastes*, University of Illinois, Urbana, Ill., 21–4 Apr. 1974.
8. National Weather Service, *Rainfall Frequency Atlas of the United States*, Technical Paper No. 40. US Government Printing Office, Washington, DC, May 1961.

8

Impact of Intensive Animal Production on Human Ecology

K. Bögel

Veterinary Public Health Officer, World Health Organization, Geneva, Switzerland

GENERAL CONSIDERATIONS

Animal populations provide reservoirs of many diseases transmissible to man. Moreover, animals may contribute to the inorganic pollution of the ecosystem, including natural resources and nutritional environment. The interactions between man through urbanization, management of domestic animals and wildlife on the one hand, and the dynamics of animal populations, their products and wastes on the other, form a most important complex of benefits and hazards in human ecology.

In planning comprehensive ecological approaches in (a) human health and wildlife; (b) human health and animals in rural areas; and (c) human health and animals in urban areas, major categories of animals have to be considered, namely, (a) animal industries; (b) animals used for work and transportation; (c) wildlife; and (d) pet animals.

Depending on the prevailing ecological conditions, these categories of animals play a different role in primary and secondary contamination (as sources of infection, injury, poisoning and chemical pollution) of our environment, including food. Hazards and benefits must therefore be seen for the whole cycle comprising animal feed, the live animal, foodstuff, dead animal carcasses, offal and manures.

LARGE FEEDLOTS

Changes in farm management and the development of animal production centres influence strongly the population structure and infrastructure of rural areas, and provide a basis for further urbanization and industrialization. These developments also result in an increasing transport, across national boundaries, of animal feed, live animals, animal products and wastes. Industries and services arise which, together with the mass production units, appear to be beneficial, if the ecosystems of areas of feed supply, feedlot sites and waste disposal can be kept in a balance independently, or by interaction. However, the complexity of factors involved is not yet fully understood and we are often alarmed by incidences of adverse effects on human and animal health. This urges us to improve surveillance at national and international levels.

The trend towards mass production of food animals (in some areas also of fur animals) involves separation of the animal production units from the source of feed and from areas of waste disposal. This complicates epidemiological conditions and sometimes makes measures of precaution and control very difficult. Zoonoses, such as salmonellosis, which are likely to be introduced into animal feedlots, may become ubiquitous and very difficult to control. On the other hand, the accumulation of large numbers of animals and certain types of farm management may facilitate surveillance and control of health hazards, *e.g.* brucellosis, tuberculosis, cysticercosis, hydatidosis.

In order to procure safe food and to protect the personnel of animal industries, veterinary public health services in collaboration with other disciplines have to watch carefully over the whole production chain of feed, animal, food, by-products and wastes.

The livestock raised for or kept in mass production units must be monitored intensively for infectious diseases. Hygienic requirements for animal feeds should be strictly observed, emphasis being placed on surveillance and treatment of feeds containing products of animal origin (*e.g.* bonemeal, meatmeal, fishmeal, dung) and other potential vehicles of infectious or toxic substances (water, plants). Feeds containing animal substances must be properly treated. One should stress the importance of investigating and controlling the application of antibiotics, antiparasitics, hormones, herbicides, insecticides and tranquillizers. The same applies to heavy metals such as copper, which may build up in soils on which pig manure is disposed of, or even accumulate in cattle fed partially on pig manure. It is essential to establish health criteria or tolerances of chemical residues in material used for animal feed. These requirements must also consider processing, storage and transport of such materials.

Appropriate health services should be consulted whenever sites for large animal feedlots are planned or modified. This applies also to other production units such as rendering plants, plants in which animal feed is mixed or pelleted, slaughterhouses and packing plants. In general, special supervision is required in any unit which utilizes animals or materials from other premises, or distributes products or wastes to various other units involved in animal production. These activities thus become closely linked with the food hygiene programme. One should draw attention to the fact that a number of infectious and non-infectious contaminations have been found to be particularly associated with feedlots, such as salmonella in cattle, brucellosis in pigs and packaging plant workers, ringworm in persons engaged in collective calf rearing, hormones in chickens, and antibiotics in calves. In some areas the odour nuisance restricts the development of new production units, particularly of pigs, because the public objects to the odour of the animal wastes.

Hygienic means of disposal of animal wastes and dead animals are of great importance in developing countries as well as in areas with highly developed and specialized animal industries. Hygienic disposal methods are prerequisites for the control of most epizootics, including zoonoses, and for meat hygiene.

In recent years, environmental health problems have been greatly aggravated in areas of intensive animal production due to negligence in appropriate disposal of large quantities of organic wastes. It is essential to establish environmental health guidelines based on a census of animal production, type of landscape, number and size (in surface units) of farms, number and species of animals per feedlot in relation to its surface area, percentage of dead animals, percentage of inedible offal per slaughtered animal of the various species, and effluents from slaughterhouses.

In some countries, detailed requirements for the land disposal of animal wastes have been established. It is hoped that these activities will lead to the formulation of guidelines for countrywide programmes on dead animal carcass and waste disposal. International collaboration is required in animal waste control programmes. National services are invited to harmonize and improve requirements for the international transfer of products prepared from animal carcasses and wastes.

It is suggested that a thorough investigation be carried out of all methods of recycling or other treatments of dead animal carcasses, offal and wastes in order to reduce the environmental pollution and to benefit from useful components in these materials. More attention should be paid to recent developments in this field. Health criteria should be established for the substances to be recycled in animal production or used for other purposes.

9

Management of Hygienic Problems in Large Animal Feedlots

DIETER STRAUCH

Professor of Veterinary Hygiene, University of Hohenheim, Stuttgart-Hohenheim, Federal Republic of Germany

Hygienic problems involved in large animal confinement feedlots must be considered under two aspects:

(a) Major problems in epidemiology of infectious diseases;
(b) Major problems in environmental health.

INFECTIOUS DISEASES

Epidemiological problems centre upon the identification, the utilization of factors inhibiting infections within the agricultural establishment and the prevention of contributing factors, the protection of rural animal husbandry against the danger of infection resulting from large animal confinement, and the protection of the general public against zoonoses or health hazards due to residues of drugs, active agents and additives as well as therapy-resistant microorganisms.

Epidemiological problems involved in large animal feedlots are closely associated with those of animal waste disposal.[1]

Incidence of latent infections increases when animals of homogeneous populations are concentrated in confinement. Most infected animals eliminate the pathogenic agent by way of urine, faeces or other methods so that germs ultimately come into contact with the floor of the buildings.

Conventional Waste Handling

Conventional livestock units where bedding is used do not cause special epidemiological problems because if proper management procedures are carried out, dungheaps develop such high temperatures as to destroy pathogens that may be present. After three weeks the dung and bedding mixture is considered as disinfected and can be used for agricultural purposes.[2]

Liquid Waste Handling

In modern animal confinement feedlots, bedding is no longer used, but instead liquid collection and handling methods are used. No matter what handling and storing methods are used, spontaneous generation of heat that could entail the destruction of pathogens will not occur in liquid manure systems.

Self-disinfection

Laboratory and field experiments have shown that *Salmonella* or stable forms of parasites remain alive in liquid manure both in summer and in winter over many months.[3] Table I shows that, under laboratory conditions, the viability of *Salmonella* can be 350 days in winter (8 °C) and approximately 180 days in summer (17 °C) in cattle liquid manure. Experiments carried out in farm liquid manure pits corroborate these findings. However, the marked differences between summer and winter storage periods were not as distinct as in the laboratory experiments.

Other pathogens with a viability of several months are viruses, especially

TABLE I

VIABILITY OF SALMONELLAE IN CATTLE WASTE SLURRY IN LABORATORY TESTS

Slurry mixture	*Organism*	*Days of survival*	
		8 °C	*17 °C*
Mixture I:	*S. enteritidis*	320	200
(11 % faeces, 9 % urine, 77·8 % water, 2·2 %	*S. gallinarum*	50	50
straw)	*S. typhimurium*	125	60
	S. paratyphi B	175	110
	S. cairo	350	150
Mixture II:	*S. enteritidis*	280	140
(11 % faeces, 9 % urine, 80 % water)	*S. gallinarum*	80	80
	S. typhimurium	50	50
	S. paratyphi B	50	20
	S. cairo	300	120
Mixture III:	*S. enteritidis*	300	100
(11 % faeces, 9 % urine, 70 % water, 10 %	*S. gallinarum*	120	80
sludge)	*S. typhimurium*	60	60
	S. paratyphi B	60	40
	S. cairo	280	100
Mixture IV:	*S. enteritidis*	300	160
(11 % faeces, 9 % urine, 79·8 % water, 0·2 %	*S. gallinarum*	120	120
subphosphate)	*S. typhimurium*	130	60
	S. paratyphi B	120	50
	S. cairo	300	150

if they are enclosed in tissue or faecal segments. Slurry storage of more than three months is needed to ensure virus disinfection.

Figure 1 shows that there are seven direct and indirect ways by which animal excreta pass not only to man but also to other animal production units. Successful prevention of infections is only possible if these ways can be blocked efficiently. In view of the many ways of disease transmission shown in Fig. 1, a great number of measures would be necessary to block all ways of transmission unless epidemiological measures are launched before germs are released into the environment. This should be done at the animal production facility.

Chemical Disinfection

Disinfection of slurry with chemicals is a sure way of controlling disease agents, but there are problems, both economic and technological. The less expensive chemical products such as caustic soda or calcium chloride render animal waste slurries inappropriate for subsequent agricultural application. Caustic soda or calcium chloride cause damage to plant

TABLE II

Disinfectant	*Dosage rate*	*Application time*
Fresh-slaked lime[a]	30 kg/m^3	7 days
Thick lime milk[b]	60 kg/m^3	7 days
Calcium cyanamide	20 kg/m^3	7 days

[a] Preparation: Fresh-burned lime is introduced uncomminuted into a voluminous vessel and sprinkled uniformly with water (about half the quantity of lime). Lime will disintegrate into powder while warming and swelling up.

[b] Preparation: To 1 litre of fresh-slaked lime are gradually added 3 litres of water, stirring all the time. If fresh-slaked lime is not available, lime milk can also be prepared by using slaked lime from a lime pit provided the topmost lime layer of the pit altered by the influence of air has previously been removed.

growth. Slurry treated with those products must logically be disposed of in municipal sewage or chemical treatment plants which are rarely located near feedlots.

Table II provides practical methods for the chemical disinfection of small quantities of slurry. Slurry can safely be disposed on land. However, this technique is not feasible for large feedlots.

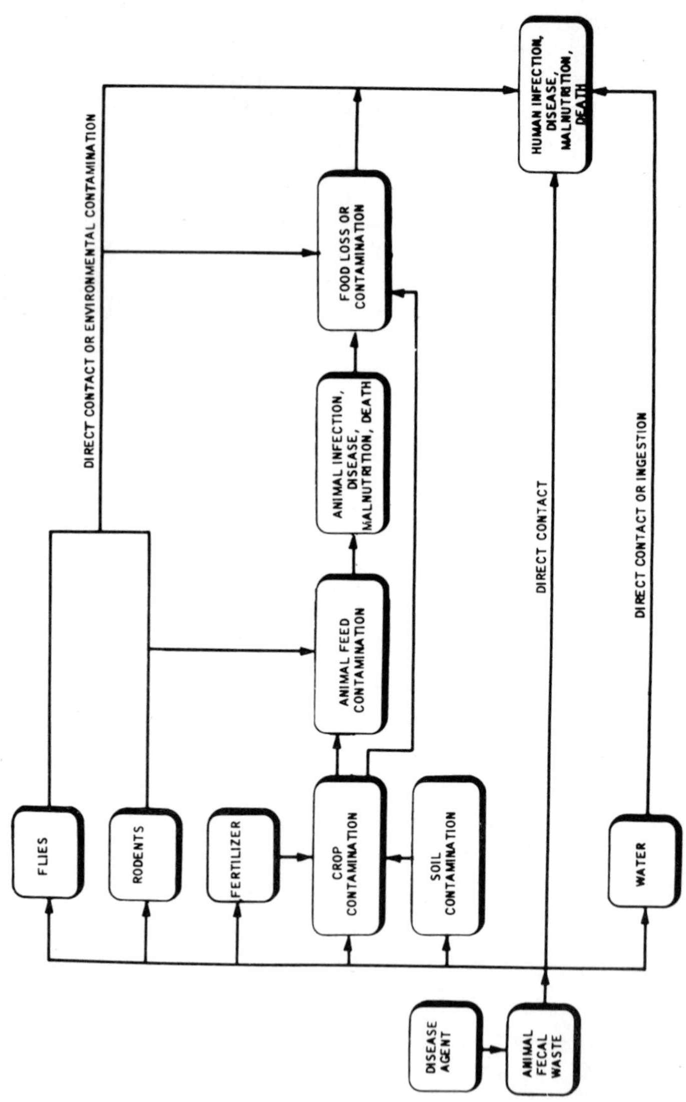

FIG. 1. Interactions between animal faecal wastes and infectious diseases in man and animals.

Physical Disinfection
Large amounts of liquid manure can be disinfected by using physical methods. One such method is aeration by submerged air outlets and high rate of mixing. In such systems, regulated aeration of the liquid manure and optimum distribution of atmospheric oxygen may cause thermophilic bacteria to activate. As a result, temperature rises into ranges of 45–65 °C with pH levels of up to 8·0 and higher can facilitate the destruction of pathogenic organisms.

Another method that has been used in practice is the oxidation ditch. Hygienic data are, however, not satisfactory because this system does not succeed in killing off pathogenic germs. Modifications are currently being tested[4] to make the oxidation ditch capable of purifying the wastewater. Electrolysis has been tried on an experimental laboratory basis but has not proved practical.[5]

Dehydration of manure is another method of disinfection which has been resorted to especially on large-scale poultry farms. These techniques operate with temperatures ranging from about 80 to 350 °C depending on whether the energy source is direct or indirect. Significant experimental data, however, on the pathogen survival are still lacking.[5]

COMMON HYGIENIC PROBLEMS

Other health problems with large animal feedlots are noxious gases, contamination of soil with heavy land applications of wastes, and water pollution.

Noxious Gases
Noxious gases may cause public nuisance through the emission of bad odours and health problems to humans and animals in totally confined feedlots.[6] Although odours by themselves do not cause disease, they do affect the health of humans and animals alike by creating discomfort and anorexia.[7] Inhaling of noxious gases emitted from animal wastes at high concentrations has resulted in death of humans and animals. The major fixed noxious gases around animal feedlots are ammonia, hydrogen sulphide, carbon dioxide and methane. Odours are produced by ammonia, hydrogen sulphide and by a large number of organic compounds which are intermediate products of biological decomposition of the organics in animal wastes.[6]

There are many ways of controlling or moderating odours. Such methods include aeration[8] to direct biological decomposition through the aerobic process. Other methods are dilution, ozonation, combustion of exhaust air, activated carbon, etc.[6]

Excessive amounts of liquid manure being applied over several years may overload the capacity of the soil to filter out and retain manure nutrients. Thus, some of these nutrients could reach ground and surface waters, thus causing pollution problems.[9]

Phosphoric acid contained in liquid manure diffuses more rapidly in the soil than that of commercial fertilizers because organic matter contained in liquid manure favours phosphate diffusion. When spreading liquid manure on highly water-saturated soils, phosphoric acid in relatively high quantity will be transported immediately after dressing to lower depths.

Although phosphate accumulation occurs in deeper layers, it has not been identified in groundwater.[9] Pollution of groundwater by phosphates can be avoided by intermittent application (several part-dressings) of large amounts of liquid manure into relatively dry soil.

Permanent application of highly water-diluted liquid manure or frequent rainfall after spreading accelerates the diffusion of phosphates and all the other nutrients into deeper layers and thus into groundwater.

Besides phosphate, another substance to take into consideration within the scope of environmental protection is nitrate.

Nitrate contents found in groundwater under fields treated with very high amounts of liquid manure over several years (160 m^3/ha) were ten times as high as under the non-treated fields. The ten- to twenty-fold higher contents of nitrate verified in drinking water are injurious to health.

To prevent soil contamination and protect public health from land application of animal wastes, maximum permissible livestock density per hectare in the Federal Republic of Germany is projected as follows:

Adult cattle	3
Young cattle, up to 2 years	6
Calves, up to 3 months	9
Adult horses	3
Horses, 1 year old	9
Breeding sows with 1-year progeny	6
Fattening pigs (places)	15
Pigs fattened per year	36
Sheep	18
Laying hens	300
Broilers (places)	900
Cockerels fattened per year	5400
Fattening turkey-hens (places)	300
Fattening ducks (places)	450

If the number of animals kept exceeds these standards, it is assumed that pollution of water might occur if all the manure generated is disposed on land every year. In Sweden, limitations on animal density are based on mean nitrogen disposal requirements.[10]

Water Contamination

Pathogenic organisms are excreted in such a way as to find their way into animal waste slurries through faeces and urine excrement. Even when these animal slurries are treated by settling out solids, aeration, or even sludge drying, pathogens are not eradicated. It is to be expected, therefore, that in cases of infection of animals in a large feedlot, large numbers of pathogens would be discharged into rivers. It must be added that in nearly all animal confinements some animals may discharge pathogens without showing signs of infection. This is especially true for salmonellosis.

Salmonella bacteria can multiply 100 000-fold in river water which contains 100 mg of organic substances per litre. Discharge of untreated animal wastewaters into surface or groundwaters creates health hazards to man and animals consuming these waters or coming in contact with them.

Salmonellosis is a disease ranging in severity from mild gastroenteritis to septicaemia, enteric fever, and meningitis leading to death.[11] Its severity depends on the infecting serotype (there are over 900 serotypes of *Salmonella* isolated from animals, man and their environment) and other factors. It is transmitted orally. Therefore, these organisms must be kept out of food or water consumed by humans or animals. Refeeding of animal wastes to animals would increase the incidence of animal *Salmonella* hosts, unless wastes are preprocessed to a degree that ensures the destruction of microorganisms.

Salmonella has been found in many water supplies in almost all parts of the world. If it gets into public water supplies, it can affect thousands of people within a short time. Chlorination of any discharges of wastewaters into public water supply resources should be mandatory.

Other hygienic problems which may be caused from discharge of feedlot wastewaters into streams are diseases from coliforms, leptospirosis, tularemia, foot-and-mouth disease, hepatitis, hog cholera, etc. Some of the coliform, *E. coli*, have manifested pathogenicity which could be hazardous to adult humans and animals and may be fatal to infants.[11] Leptospirosis is a disease of large proportions and is worldwide in distribution. Symptoms are similar to influenza, infectious gastroenteritis and enteric viral infections. Leptospirae originating from animal sources are frequently found in stream waters.

One of the most unusual causes of stream pollution is tularemia bacillus, which is normally found in rabbits and occasionally in sheep.[11] The organism is excreted in the wastes of these animals. Frequently, this organism will end up in streams where it may settle in the stream sediment, multiply, and become a threat to humans and animals.

It is expected that feedlot wastewaters discharged into river waters will have to be disinfected. Chlorine disinfection is the most common method. Chlorination dosages will vary depending on the type of chlorine compound used, organic matter content of the wastewater, pH, temperature, chlorine residual and contact time. A 15–30 min contact time is usually sufficient. Normally the chlorine residual required should not be less than 0·1–0·5 mg/litre. The major part of the chlorine added is not used up for the disinfection of microorganisms but to satisfy the oxidation requirements of sulphides, ammonia, reduced forms of iron and carbohydrates.

When properly applied and controlled, chlorination of wastewaters from animal feedlots for disinfection is an effective measure for improving the bacteriological quality of wastewater and protecting the human and animal population against transmission of enteric diseases via the water route.

ANTIBIOTICS IN ANIMAL WASTES

In the early 1950s the feeding of diets containing low levels of antibiotics was found to increase the growth of farm animals. This was a major breakthrough for commercial animal feedlots. It was not long, however, before scientists questioned the wisdom of this practice because the widespread use of antibiotics for non-therapeutic purposes resulted in the development of microorganisms which became resistant to antibiotics.

The situation was made worse by the fact that some bacteria became capable of transferring their antibiotic resistance to organisms which had not had contact with these antibiotics.

Antibiotic resistance, according to many scientists, represents development of genetic mutants in a bacterial population which are insensitive to the antibiotic being administered and consequently multiply and then predominate while susceptible bacteria are suppressed. Development of mutants is a normal phenomenon of life which contributes materially to the survival of a species and is unrelated to the presence of antibiotics. Development of antibiotic-resistant bacteria usually results from their repeated exposure to antibiotics in the form of therapeutic agents and as growth promotants, food preservatives, residues and contaminants.

In recent years, antibiotic resistance in bacteria acquired a new dimension when Japanese workers in 1959, and subsequently scientists all over the world, found that resistance could be transferred from one bacterial strain to another and even from one family to another simply by bacteria coming into contact with each other in the same environment. These findings immediately compounded the already recognized danger of antibiotic usage. Of special concern were reports of transfer of antibiotic resistance between *E. coli* and *Salmonella* organisms, because both have been incriminated as pathogens. Furthermore, *Salmonella* are considered zoonotic pathogens. Transfer of resistance has been shown to occur also in other bacteria such as *Pseudomonas*, *Klebsiella*, *Vibrio comma*, *Serratis marcescens* and *Shigella*. Use and distribution of antibiotics in animals is so widespread that the chance of resistance in human bacterial pathogens having an animal origin seems quite good.

The phenomenon of transfer of drug resistance from one bacterial cell to another has been explained as conjugation and as transduction. In conjugation, bacterial cells join together so that part of the nucleic acid of one passes to another, thereby passing a new genetic character such as specific antibiotic resistance. In transduction, the nucleic acid would be transmitted by a bacterial virus. Transferable drug resistance is also known as transmissible drug resistance, infective drug resistance and infectious resistance. The last two terms seem the least desirable, since resistance is not normally recognized as an infective or infectious quality.

Although scientists recognize that transmission of drug resistance poses a real threat to disease control in both animals and man, no one seriously recommends that use or production of antibiotics be discontinued. However, transferable resistance cannot be taken lightly. Everybody facing this problem should be aware of the potential hazards of transferable resistance because of the public health implications and because understanding the phenomenon of resistance helps explain why some patients do not respond to therapy. If transferable resistance proves to be more common than has been shown so far, and there is much evidence that this is so, then the discharges of feedlot wastes have to be given special consideration as a massive source of resistant bacterial strain.

Recent investigations in Czechoslovakia have shown that the surroundings of large animal confinements show a heavy contamination with all kinds of intestinal microorganisms and with up to 100% resistant *E. coli* strains. The self-cleaning process of the soil in the surroundings of these farms is considerably retarded so that health hazards to the population living in the vicinity have to be taken into consideration.

It has been feared that the transfer of resistance capability from coliform bacteria to *Salmonella* might be a setback for the control of certain infectious diseases in human medicine. Further tests have shown, however, that transfer *in vivo* does not occur as frequently as was to be expected according to tests *in vitro*. In addition, resistant *E. coli* bacteria of animal origin have proved not to resist for long to the intestinal flora in humans. Therefore it has been inferred that animals cannot be a major source of resistant *E. coli* bacteri for man since he is anyway the carrier of antibiotic-resistant *E. coli* bacteria of human origin. Only in the personnel of agriculture or slaughterhouses who are professionally in close contact with the animals may resistant strains of *E. coli* bacteria of animal origin be found sporadically.[12]

However, in order to minimize environmental hazards, official proposals have been made to the effect that antibiotics used in medicine for therapeutic purposes should be suppressed in animal feeding. The antibiotic to be prohibited first of all will be penicillin. Other antibiotics will follow in the prohibitory list. In the meantime, new products have been put on the market which are not to be used in human medicine. These substances are reported not to be resorbed, hence will not get into food of animal origin. Manufacturers also claim that these antibiotics do not produce transferable resistance.

REFERENCES

1. Mayr, A. and Rojahn, A., Infektionsfördernde und infektionshemmende Faktoren bei der Massentierhaltung. *Tierärztl. Umschau*, **23** (1968) 555–65.
2. Geissler, A., Rojahn, A. and Stein, H., *Sammlung tierseuchenrechtlicher Vorschriften.* Verlag R. S. Schulz, München, 1967.
3. Strauch, D., Feste und flüssige Abfälle aus dem landwirtschaftlichen Bereich. In: Roots-Haupt-Hartwigk, *Veterinärhygiene*, 2nd ed., pp. 241–59. Verlag Paul Parey, Berlin and Hamburg, 1972.
4. Müller, W., Hygiene landwirtschaftlicher und kommunaler Abfallbeseitigungssysteme. Hohenheimer Arbeiten, Bd. 69, *Tierische Produktion.* Verlag E. Ulmer, Stuttgart, 1973.
5. Strauch, D., Hygiene der Kot- und Harnbeseitigung. In: Comberg-Hinrichsen, *Tierhaltungslehre*, pp. 425–30. Verlag E. Ulmer, Stuttgart, 1974.
6. Norén, O., Noxious gases and odours (this volume, pp. 111–129).
7. Taiganides, E. P. and White, R. K., The menace of noxious gases in animal units. *Trans. ASAE,* **12**(3) (1969) 359–67.
8. Wolfermann, H.-F., Emissionen und Massnahmen zur Verhinderung von Immissionen. In: Comberg-Hinrichsen, *Tierhaltungslehre*, pp. 431–51. Verlag E. Ulmer, Stuttgart, 1974.
9. Vetter, H. and Klasink, A., Grenzen für die Anwendung hoher Flüssigmistgaben. In: Vetter, *Mist und Gülle*, pp. 19–49. DLG-Verlag, Frankfurt a.M., 1973.
10. World Health Organization, *Report on Consultations on the Hygienic Disposal and Recycling of Animal Carcasses and Waste.* WHO, Veterinary Public Health, Geneva, 1973.
11. Decker, W. M. and Steele, J. H., Health aspects and vector control associated with animal wastes. In: *Management of Farm Animal Wastes*, pp. 18–20. ASAE, St Joseph, Mich., 1966.
12. Bräuchle, B., Hygienische Probleme der Resistenzbildung durch Beifütterung von Antibiotika. Ldw. Dipl.-Arbeit, Universität Hohenheim, 1971.

10

Health Effects from Waste Utilization

J. Hojovec

Veterinary Medicine College, University of Brno, Brno, ČSSR

INTRODUCTION

Utilization of wastes from modern large animal feedlots may create public health problems. Transmission of infectious diseases to other animals and humans, environmental pollution, and health effects through the refeeding of the wastes are identifiable problems which veterinarians must manage. This paper will focus mainly on public health problems associated with the utilization of animal wastes in refeeding. Of course, refeeding includes both direct refeeding after drying or other processing, and indirect refeeding such as application on to grasslands which are grazed by animals.

UTILIZATION AS ANIMAL FEED

For waste to be eligible for use as supplemental feed to animals, three things must first be established: (a) its nutritive value, (b) its safety to animals, and (c) the safety of its residues in the animal products consumed by humans, assuming, of course, that the economic and technical feasibility of refeeding is assured.

For a waste to have a feed value, it must be high in concentration of proteins and/or vitamins, amino acids and/or energy ingredients. However, it must also be free of extraneous materials such as wire, glass, nails, toxic levels of heavy metals, a limited level of coliform organisms, and free of pathogenic bacteria and prescribed levels of drug residues.[1]

The nutritive value of liquid manure has been established.[2] Animal safety involves disease and interference with the performance of the animal in production and/or in breeding. Human safety is mainly tied to residues of drugs, heavy metals, and even pathogens in the animal products. Drug residues make their way into the waste via spilled feed and via the faeces which, on average, contain 50 % of the drug quantities contained in the feed.[1]

Research is still needed to document in detail the expected concentration ranges for every one of the components of the waste considered for animal refeeding. Generally, however, unless these various ingredients are concentrated by processing, they should be at levels no higher than those found in the feed itself. Since the feed is considered acceptable from the standpoint of animal and human health, the animal excrements should not present particularly severe health problems from the standpoint of metal and drug residue concentrations. Problems arise, however, when waste from animal species is fed to another type of animal. Abortions, for example, were observed in beef cows fed dried poultry litter which had oestrogenic activity greater than 10 μg DES equivalent per 100 g of litter.[3] Copper toxicity has been reported in ewes fed diets containing 25–50% poultry litter.[1] On the other hand, autoclaved, cooked dried waste from caged layer hens fed to steers, deposited no detectable residues of lindane, aldrin, dieldrin or heptachlor in the fat tissues of the beef animals.[4] However, the steers fed poultry waste had higher arsenic levels in their liver than the control animals.

A possible complication with refeeding is that diagnostic procedures may be affected.[1] For example, cattle fed poultry litter may be infected or sensitized to *Mycobacterium avium*, causing false positive reactions leading to diagnosis of tuberculosis.[4]

Processing of animal excreta can be effective in reducing toxic levels and/or eliminating pathogenic elements. Parasitic nematode larvae did not develop in beef cattle manure ensiled with hay, even though the raw manure did contain eggs of such parasites.[5] Seven days of ensiling poultry litter eliminated *Salmonella*.[1] The only method which completely sterilised poultry litter is dry heat at 150°C for 3 hr or longer.[6] To prevent nitrogen losses, however, it is best to acidify the litter beforehand. Pasteurization can be accomplished by dry heating the waste at a thickness of 1·25 cm at 125°C for 30 min or more.

Physiological observations, detailed necropsies and examinations of histological sections have revealed no detectable adverse effects for animals fed processed wastes up to a certain level, such as layers fed up to 40% dehydrated poultry waste,[7] cattle fed up to 40% ensiled feedlot waste,[8] or cattle fed up to 50% poultry litter.[6] Furthermore, feeding poultry litter to cattle would not necessarily markedly affect pesticide residue levels in their fat or liver, or the taste of the meat.[6] Quality of eggs as evaluated by taste panels, weight, shell thickness, etc., is not adversely affected by feeding hens rations containing up to 30% or 40% dehydrated excreta.[7] Milk production from cows receiving 20% of their protein from dehydrated hen manure is not affected. However, there are problems of acceptance of dehydrated manure by ruminant animals, particularly if the percentage of the manure in the feed is high.[9]

UTILIZATION AS FERTILIZER

Utilization of unprocessed feedlots waste presents problems of zoonosis, aesthetic problems and pathogen viability. Table I shows the large number of bacteria in swine liquid waste from existing swine feedlots. There is a great variability in bacterial count numbers between feedlots, times of the year, and even time of the day. Also, numbers in the range of 10^{18} have been reported in the literature. However, zoonosis hazards are based more on the viability of pathogenic organisms than on total counts of enteric bacteria.

TABLE I

TYPES AND TOTAL COUNTS OF BACTERIA IN LIQUID SWINE FEEDLOT WASTES

Bacteria	*No. of bacteria per litre of manure*	
	Minimum	*Maximum*
Mesophilic	$2{\cdot}0 \times 10^7$	$9{\cdot}6 \times 10^{12}$
Psychrophilic	$1{\cdot}3 \times 10^8$	$5{\cdot}8 \times 10^{12}$
Coliform	$4{\cdot}2 \times 10^5$	$1{\cdot}3 \times 10^8$
Coliform L+	$9{\cdot}0 \times 10^5$	$1{\cdot}3 \times 10^8$
Coliform L−	$3{\cdot}3 \times 10^6$	$1{\cdot}0 \times 10^3$

Salmonella viability is affected by temperature[10] plus other factors such as dilution of waste and solids concentration. *Salmonella* survives 30–50 % longer in dilute wastewaters than raw waste to which little water has been added. Therefore, from the standpoint of public health, storage of feedlot waste without the addition of dilution water is recommended.

Another disease agent of significance is the virus which causes foot-and-mouth disease in cattle and pigs. This virus is inactivated in 25–32 days when liquid manure is in the 12–18 °C temperature range. For storage under freezing temperatures, survival of the virus might be as long as 180–220 days, according to studies carried out in the Soviet Union.

To ensure control of infectious disease from animal feedlots it is recommended that liquid manure be disposed after a minimum of seven days of storage. This would mean that at least two storage facilities be available for each feedlot so that no fresh waste is added to the stored lagoon at least seven days before land disposal.

It is technically impossible to test liquid manure for pathogens before disposal. Negative tests for pathogens cannot necessarily be considered to mean complete absence of pathogens. Furthermore, the state of health of the animals is not a reliable criterion of the hygienic state of the liquid waste, because there is an incubation period from the time of infection to the appearance of clinical signs. During that time, pathogens can be excreted and contaminate the waste. This incubation period varies with the type of infection and the type of pathogen involved. The storage period of seven days is selected as the shortest incubation period for economically significant infections.

If infection of the animals with an infectious disease agent is verified, then liquid waste must be disinfected before disposal. Disinfection methods depend on the nature of the agent to be eliminated or inactivated, on the subsequent use of the waste and the liquid waste itself. For bacterial pathogenic agents, disinfection by chlorination, ozonation, formaldehyde, ionizing radiation, heating, pH alteration and the addition of other bactericidal chemical compounds may be used. All these methods are expensive, and their technical feasibilities in large feedlots have not been demonstrated.

Some very basic health and technical problems still exist with the utilization of feedlot wastes for animal feed and even for land disposal. Management of large animal feedlots must be fully aware of these problems, because uncontrolled disease can destroy the entire operation or cause panic among consumers of animal products.

REFERENCES

1. Taylor, J. C., Regulatory aspects of recycled livestock and poultry wastes. In: *Livestock Waste Management and Pollution Abatement*, pp. 291–2. ASAE, St Joseph, Mich., 1971.
2. US Dept of Agriculture, *Animal Waste Reuse: Nutritive Value and Potential Problems from Feed Additives*, ARS 44-224. USDA, Washington, DC, 1971.
3. Griel, L. C., Kradel, D. C. and Wickersham, E. W., Abortion in cattle associated with feeding of poultry litter. *Cornell Veterinarian*, **59**(2) (1969) 227–35.
4. Carriere, J., Alexander, D. C. and McKay, K. A., The possibility of producing tuberculin sensitivity by feeding poultry litter. *Canadian Veterinary J.*, **9**(8) (1968) 178–85.
5. Ciordia, H. and Anthony, W. B., Viability of nematode in wastelage. *J. Animal Science*, **28**(1) (1969) 133–4. (Abst.)
6. Fontenot, J. P. *et al.*, Studies of processing, nutritional value and potability of broiler litter for ruminants. In: *Livestock Waste Management and Pollution Abatement*, pp. 301–4. ASAE, St Joseph, Mich., 1971.
7. Flegal, C. J. and Zindel, H. C., Dehydrated poultry waste (DPW) as a feedstuff in poultry rations. In: *Livestock Waste Management and Pollution Abatement*, pp. 305–7. ASAE, St. Joseph, Mich., 1971.
8. Anthony, W. B., Cattle manure as feed for cattle. In: *Livestock Waste Management and Pollution Abatement*, pp. 293–6. ASAE, St Joseph, Mich., 1971.

9. Bueholtz, H. F. *et al.*, Dried animal waste as a protein supplement for ruminants. In: *Livestock Waste Management and Pollution Abatement*, pp. 308–10. ASAE, St Joseph, Mich., 1971.
10. Strauch, D., Management of hygienic problems in large animal feedlots (this volume, pp. 95–104).

11

Noxious Gases and Odours

O. Norén

Vice-Director, Institute of Agricultural Engineering, Uppsala, Sweden

INTRODUCTION

Degradation processes are continually working in manure and urine, resulting in the formation of gaseous compounds, some of which are immediately released and some of which are retained. The gases formed to some extent depend on whether degradation is aerobic or anaerobic, which in turn depends on the manure handling method used. In methods where a large portion of the urine is removed and with generous amounts of bedding (*e.g.* straw), aerobic conditions can be induced. However, even in these cases anaerobic degradation may occur locally. In methods using slurry the degradation is anaerobic, provided there are no special arrangements ensuring the supply of oxygen. In large-scale livestock production the slurry method is generally used because of handling and labour advantages and also because of difficulty in obtaining the large amounts of bedding needed for the solid manure handling methods.

The gaseous degradation products released under aerobic conditions are odourless and non-toxic (*e.g.* carbon dioxide). During anaerobic degradation, however, gases are formed which are toxic and/or malodorous. Such gases have frequently caused the poisoning of livestock and in some cases also of humans. It should, however, be mentioned here that with correct design and management of the systems no lethal concentrations of gases will occur. Inconvenience has been caused by malodorous gases in many cases. Problems with malodorous gases in exhausted ventilation air, from manure storage pits and during the spreading of manure increase with the size of the enterprise.

MANURE GASES

The air in a livestock building is known to contain a large number of different gases. However, it appears that a satisfactory total analysis of gas components in buildings for different animals has never been conducted.[7] Some report about 30 and others about 20 different gases. The presence of hydrogen sulphide, ammonia, carbon dioxide and methane has been fully documented. Measurements in pig units have revealed the presence of 10–12 carbonyl compounds.[6] The compounds identified were ethanol, propanal, butanal, hexanal, acetone, 2-butanone and 3-pentanone. In measurements of gas components from dairy animal waste the compounds tentatively identified were hydrogen sulphide, methanethiol, dimethyl sulphide, diethyl sulphide, propyl acetate, *n*-butyl acetate, trimethylamine and ethylamine.[23] Literature references on what is known about the presence of these gases are found in refs. 4, 15 and 19.

TOXICITY OF MANURE GASES

Of the gases mentioned above, hydrogen sulphide and ammonia are mainly those that can occur in toxic amounts.

Hydrogen Sulphide (H_2S)

There is considerable evidence that hydrogen sulphide is the main cause of poisoning cases. Hydrogen sulphide is slightly heavier than air, colourless

TABLE I

EFFECTS OF DIFFERENT CONCENTRATIONS OF HYDROGEN SULPHIDE WHEN INHALED BY HUMANS (from ref. 2)

Concentration of H_2S		*Effect*
mg/litre air	*ppm*	
1·2–2·4	800–1 600	Instantaneous death
0·6–0·84	400–550	Death within 0·5–1 hr
0·5	350	Hazardous when inhaled for more than 0·5–1 hr
0·24–0·36	160–250	Can be tolerated for 0·5–1 hr without negative consequences
0·12–0·18	90–130	Can be tolerated for 6 hr without essential symptoms

and has a very unpleasant odour at low concentrations (the smell of rotten eggs). The odour limit for H_2S is at 0·1 ppm. The odour intensity is then unchanged with moderate increases of the concentration. At high concentrations—above 30–50 ppm—the sense of smell is rapidly paralysed.[20] Thus at high concentrations the gas is sometimes unnoticed because the sense of smell can become rapidly fatigued. The influence on humans of inhaled concentrations of H_2S is illustrated in Table I.

Ammonia (NH_3)
Ammonia is lighter than air, colourless and has a pungent odour. An ammonia concentration of 50% by volume results in death within a few minutes.[8, 22] The odour limit is about 5 ppm.[22]

Carbon Dioxide (CO_2)
Carbon dioxide is heavier than air, colourless and odourless. The atmosphere contains CO_2 in a concentration of about 300 ppm. At concentrations of 7–8% by volume, suffocation occurs due to the reduction of the oxygen concentration.

Other Gases
These gases are perhaps toxic and may contribute to the poisoning process, but knowledge of whether they are of importance or not is very uncertain. Many of them, perhaps all, are malodorous and can cause odour problems.

There is also the possibility that additive or even multiplicative effects of different gases, *e.g.* hydrogen sulphide and ammonia, can occur in poisoning cases.[8]

PERMISSIBLE GAS CONCENTRATIONS

The authorities for occupational safety and health have set maximum permissible values for H_2S, NH_3 and CO_2 within industrial buildings where people work or are otherwise present for a maximum of 8 hr a day. This MAC (maximum allowed concentration) value used in many countries is given in Table II.

TABLE II

	Maximum allowed concentration (MAC)		
	ppm	*mg/m³*	*vol %*
Carbon dioxide	5 000	9 000	0·5
Ammonia	25	18	0·002 5
Hydrogen sulphide	10	15	0·001

THE TOXIFICATION PROCESS

Cases of livestock poisoning have occasionally been acute but in some the process has been more prolonged, leading to chronic poisoning.

The acute poisonings have occurred in connection with cleaning out or with agitating the manure, either in the livestock building or in a manure pit, from which the gases have entered the building. In acute cases the animals have frequently lost consciousness and died within a few minutes. Autopsies have revealed increased haemophilic trends and poorly coagulated blood. Dominant features are large amounts of liquid in the lungs and haemorrhages over large areas of the body. Microscopic investigations of tissue apparently normal to the naked eye reveal extensive haemorrhages. The most important course of treatment for acute cases of poisoning is to get the patient or animal into fresh air as quickly as possible.[2, 8]

In animals with chronic poisoning the illness is characterized by tender and damaged hooves, and some haemophilic tendency with swellings, etc., under the skin and around joints. Hoof alterations on affected animals leading to an altered stance, crossed forelegs and a curved back were sometimes typical symptoms.[2, 8]

CONCENTRATIONS OF NOXIOUS GASES

The gas concentrations differ widely between the different manure handling systems and during the various phases of manure handling. In buildings with solid manure systems, *i.e.* with urine drainage, generous supply of bedding and cleaning twice a day, hydrogen sulphide is hardly detectable. The problem of manure gases is thus primarily associated with slurry systems. The distribution of gases throughout a livestock building may be very uneven.

Occurrence of Gas Between Cleaning Operations

As long as the manure in cattle buildings remains static in gutters, channels or pits, concentrations of H_2S that are detectable in the ppm range[16, 20] do not develop. In pig units, on the other hand, H_2S is occasionally released from the surface of the manure under warm conditions. Figure 1 shows the concentration of H_2S measured about 2 cm above the slatted floor in a pig unit with a sluice gate cleaning system. The concentration of H_2S in the exhausted ventilation air was not detectable. Hydrogen sulphide can also occur in pig units with mechanical cleaning of slurry, despite the storage period being a maximum of 12–14 hr.

Carbon dioxide and methane occur in all livestock buildings. Both gases have a fairly large daily variation and the concentrations are also dependent on the capacity of the ventilation system. The highest concentrations occur when the activity is greatest, *i.e.* when the animals are moving about (see Fig. 2). Levels of CO_2 of between 800 and 4000 ppm are common in cattle buildings and levels between 800 and 2000 ppm are common in pig units. The concentration of methane in cattle buildings is about one-tenth of the

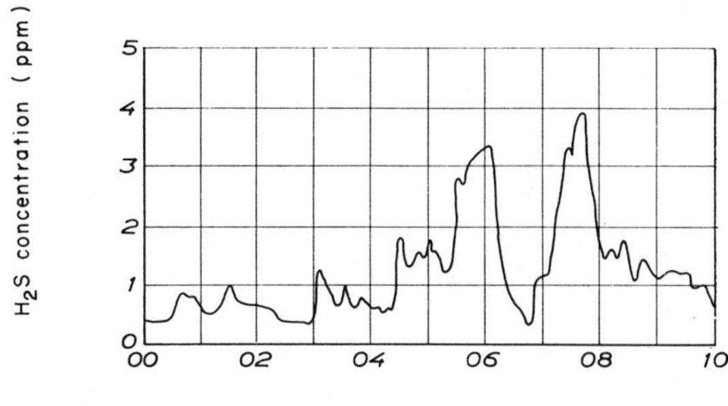

FIG. 1. Concentration of hydrogen sulphide 2 cm above slatted floor in a pig house with sluice gate system. The liquid has been stored for 14 days during the summertime.

carbon dioxide concentration. Methane occurs only occasionally and in very low concentration in pig units.

In cattle buildings, ammonia generally does not vary throughout the day. The concentrations are usually relatively low (seldom more than 20–30 ppm) and in buildings with slurry systems the ammonia level is sometimes almost zero. However, ammonia is always present in pig units. The concentrations vary widely, from a few ppm up to 30 ppm.

During the winter, when low rates of ventilation are used the concentration of carbon dioxide and methane is higher at roof level than at floor level, as is shown in Fig. 3. The density of the different gases has little influence on the distribution between floor and roof. Instead there are other factors, such as temperature differences, that determine the distribution.[16, 20]

The horizontal distribution of the gases depends on the air currents in the building. High concentrations of, for example, carbon dioxide can occur

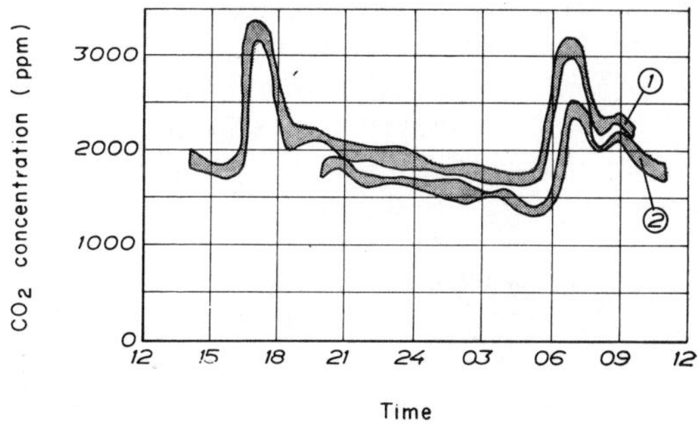

FIG. 2. Concentration of carbon dioxide in the exhausted air from a cow stable. (1) Stable with liquid manure handling; (2) stable with solid manure handling.

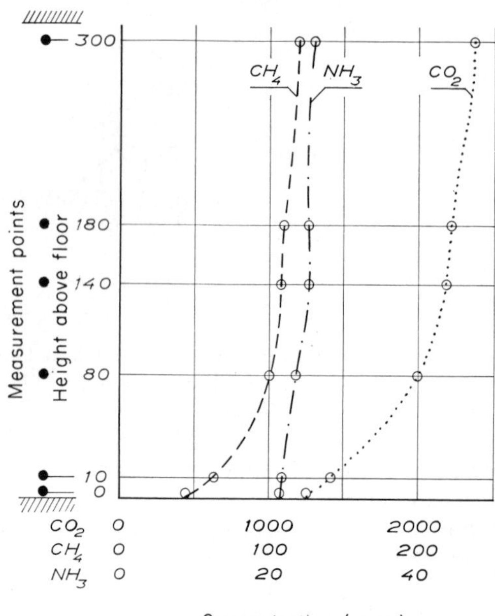

FIG. 3. Distribution of gases between floor and ceiling in a cow stable with liquid manure handling.

locally. The currents in the ventilation air are strongly influenced by temperature differences. Currents of cold air are of great importance for the distribution of gases. For example, if cold air flows in via cleaning drains and manure gutters, or if cold air from air inlets in the ventilation system flows directly down to the floor and then to manure gutters and channels, the gutters will be continually flushed, which will result in any released gases being moved up into the living zone.

In buildings with slatted floors, for both cattle and pigs, there are often upward currents of air from the manure pits at places where the animals are standing for the moment. These currents are due to the animal's generation of heat. At the same time, the areas where there are no animals have downward currents through the slats.

Occurrence of Gas During Cleaning and Agitation

Hydrogen sulphide is released immediately after slurry starts to move or is agitated. Carbon dioxide, methane, etc., are also released. The amount of gases released is to a certain extent dependent on the fluidity of the slurry. At high water contents the manure flows more easily whereby the movement is more turbulent and large volumes of gas are released. Figure 4 shows a typical pattern of gas release in connection with cleaning a pig unit. The curves show the concentrations in the ventilation air. Even a fairly negligible agitation of the manure can result in the release of hydrogen sulphide which moves into the living zone.

The distribution and concentration of gases in livestock buildings during cleaning and agitation of manure show very wide variations and local conditions have a very strong influence. There are considerable differences between the various methods of handling slurry.

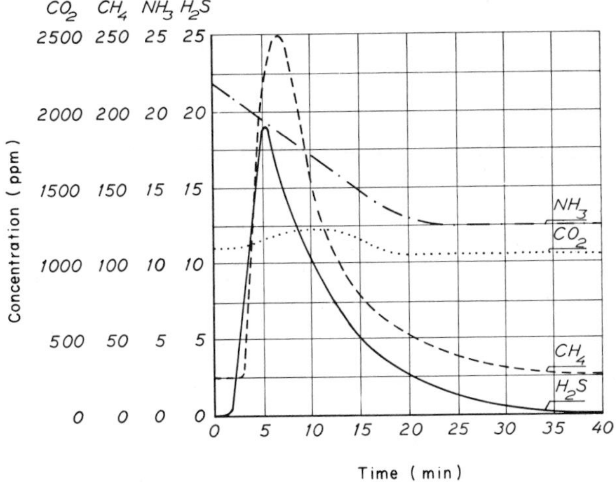

FIG. 4. Changes in gas concentration in a pig house during cleaning-out with a sluice gate system. The outlet channel has no water-seal and no evacuation.

The gases are primarily released at flushing outlets and sluice gates, and in drains and pits, *i.e.* where the manure is violently agitated. If the buildings are constructed so that the drains open directly into the collection pits outside the building, a strong current of air will flow into the building from the outside (provided the ventilation system does not work at high positive pressures, which is unlikely). The gases released are thus forced up from the manure channels into the living zone. Very high concentrations of hydrogen sulphide, for example, can occur locally. Figure 5 shows examples of hydrogen sulphide concentrations in a pig unit during cleaning where cold air forced hydrogen sulphide into the building. Hydrogen sulphide concentrations of 700 ppm, *i.e.* lethal concentrations, occurred in the proximity of the sluice gate. Installations of this kind must be fitted with gas traps and ventilation of drains on the lines illustrated in Fig. 6. The hydrogen sulphide concentration will then decrease to levels that are hardly detectable.

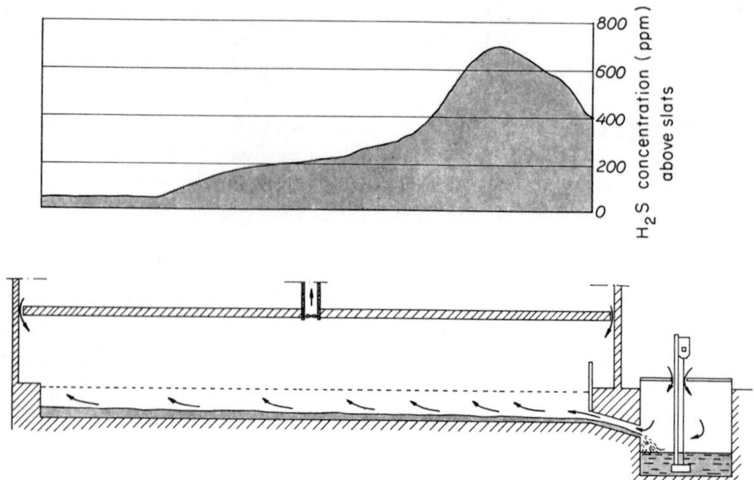

FIG. 5. When there is no water-seal and no evacuation of the cleaning-out culvert in a liquid manure system, dangerous concentrations of hydrogen sulphide can occur during cleaning-out, especially in pig houses.

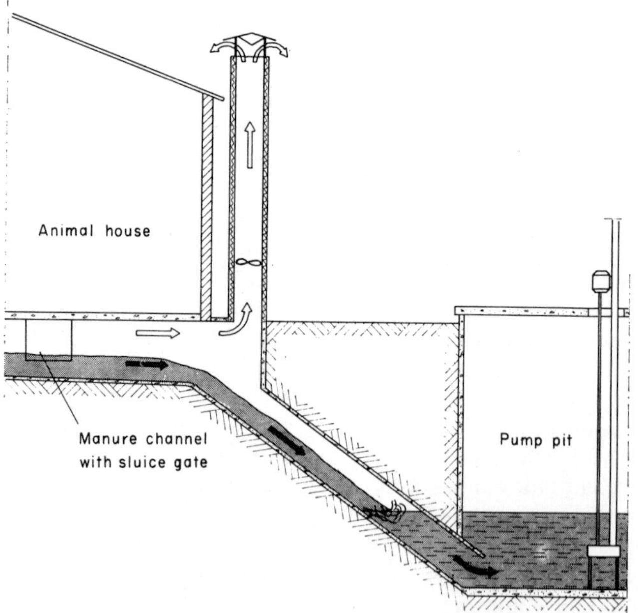

FIG. 6. Example of how it is advisable to arrange evacuation of the cleaning-out culvert when building a new installation with sluice gate system.

In cattle buildings the hydrogen sulphide concentrations are generally lower than in pig buildings, but in some situations may reach lethal levels.

In buildings using a floating manure system, hydrogen sulphide is released from the surface of the manure in the drains. The concentrations of hydrogen sulphide will, however, be proportionately low. Also, with this system it is necessary to ventilate the drains so that gas from pump pits and drains is prevented from entering the living zone of the animals.

Very large volumes of gases are released when manure is agitated in buildings using the slatted floor plus manure pit system. The concentrations of hydrogen sulphide may then reach lethal levels. This type of system should be avoided and only used if special ventilation measures can be arranged with adequate ventilation under the slatted floor.

In recent years mechanical scrapers have become more commonly used in livestock buildings to transport the slurry either directly to a pump pit or to a drain—a flush drain—from which the manure is flushed to the pit. Hydrogen sulphide also occurs in this system, primarily when the manure is scraped out. These systems should also be fitted with gas traps and separate ventilation so that the gases do not enter the building.

DESIGN OF SLURRY SYSTEMS TO PREVENT GAS POISONING

To prevent manure gases from entering the livestock building during cleaning, slurry systems should be fitted with gas traps and the drain ventilation suited to the system being used. Systems using sluice gates, or combined sluice gate and flushing systems, should be fitted as illustrated in Fig. 6. The floating system can use the method shown in Fig. 7. When flushing nozzles are used they should be placed close to the bottom of the gutter or channel (see Fig. 8). In buildings using combined scraping and flushing the system can be designed as shown in Fig. 9, where the flushing drain is ventilated by a fan.

Cleaning systems with sluice gates should have a drain that is large enough to avoid blockages when the gates are opened. The pump pit that collects the manure must be placed at a level so that it can hold all the manure from one channel. Fans for ventilating the drain should have a capacity of at least 5000 m^3/hr.

The ventilation should be designed to mix fresh air with the air in the building effectively at relatively high levels and to prevent cold air from flowing up through manure channels, gutters, etc. It is also of very great importance to have an emergency ventilation system if the fans should stop due to a power failure.

Large volumes of gas can, as mentioned above, generally be found near places where slurry is being agitated. Thus, there is a great risk of poisoning

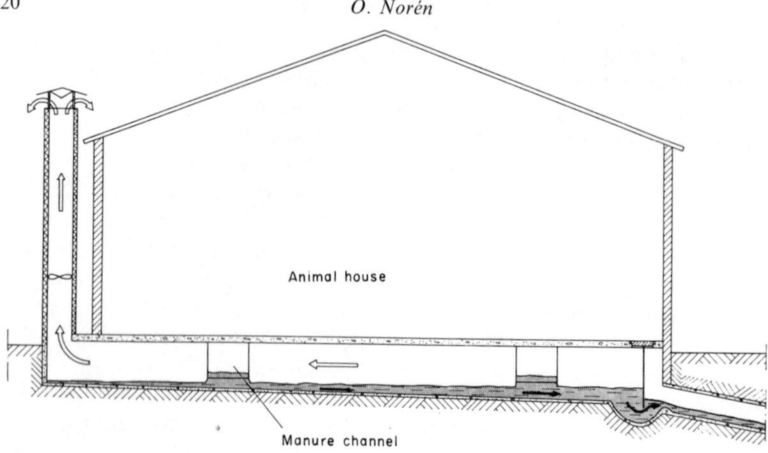

FIG. 7. Example of how to arrange a water-seal and evacuation in a floating system.

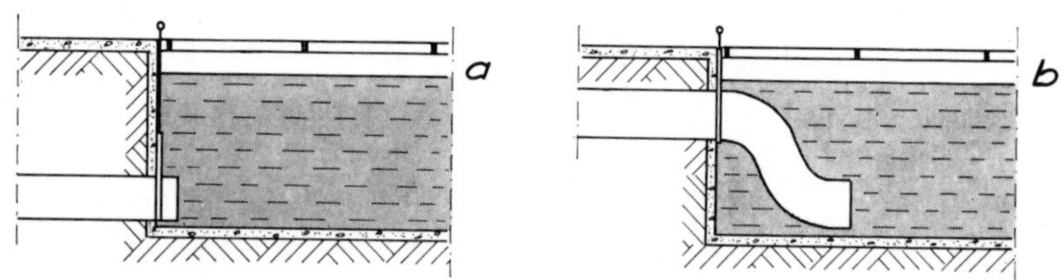

FIG. 8. In a flushing system the outlet nozzle should be positioned as low as possible. (a) Incorrect installation; (b) correct installation.

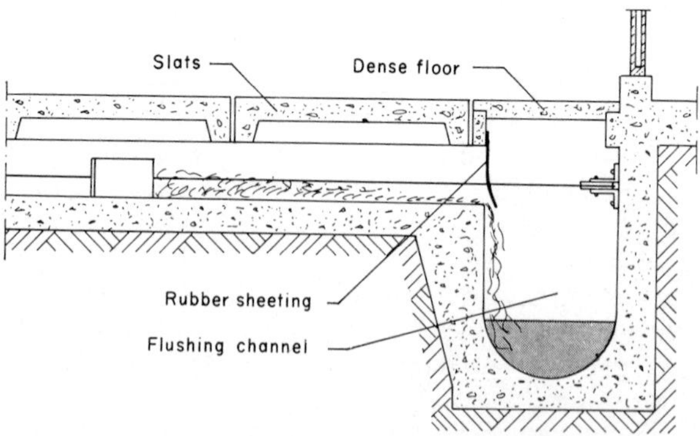

FIG. 9. Combined scraping and flushing system.

around pits and tankers when the slurry is being pumped or agitated. Naturally, there are very high concentrations of gases in drains, pits and covered containers.

When slurry is loaded into a tanker the gases released inside the tanker rise up through the filling hatch. Concentrations of 1000–2000 ppm have been measured here. The concentration may even be lethal in the neighbourhood of the filling hatch. Lethal accidents may occur if tankers, pump pits, manure containers, deep channels, etc., are entered before they have been thoroughly ventilated, or if an oxygen mask is not worn. Safety belts and ropes should be used and two men should be available to supervise the work in these places. Because of explosion risks due to the presence of methane, no open flames should be used in or near pits or tankers.

ODOUR NUISANCE

The odours are more a nuisance than a hazard to health. Complaints about odours usually come from people living near large livestock enterprises. The complaints frequently involve different psychological angles, resulting in wide differences of opinion between the parties concerned. Air that is as odour-free as possible is not only essential with regard to the neighbouring residents but is also just as important for the people who have to work in the livestock building. Noxious odours are more commonly found in connection with poultry and pig enterprises than with cattle enterprises. At present there is no quantitative base for evaluating the effect of noxious odours, but such a base would be useful.

ODOUR MEASUREMENT

The determination of odour emissions from livestock buildings can only be done by means of a panel. This is because no significant relationships have been found between the odorants in manure that can be measured objectively and the intensity of the odour. This is probably largely because the odour usually consists of a mixture of many odorants. In the following discussion the odour threshold value is used as a unit of measurement. The threshold value is given as the dilution factor, expressed as a logarithm, that is needed to dilute the polluted air with fresh air so that the threshold value is reached.[9]

ODOUR SOURCES AND STRENGTH

Noxious substances are formed both in livestock buildings and in storage places for manure and urine—either mixed, as slurry, or separated.

Odour is caused by a large number of different gases. Powerful organic odorants such as aliphatic amines, mercaptans, sulphides, organic acids and skatole are components of the odour of animal waste.[3, 5, 12, 13] Odour from feedstuffs such as silage is also included in the 'odour spectrum'.

Inside the buildings, odour comes primarily from degradation in manure and feed remnants, but also to some extent from the animals themselves. The concentration of the emission is considered[1] to be largely proportional to the phosphate and nitrogen contents of the faeces. Poultry slurry has on average a higher content of phosphate and nitrogen than pig slurry, which in turn has a higher content than cattle slurry. Thus, cattle enterprises are less potential sources of odour problems than pig and, in particular, poultry enterprises. This is in good agreement with practical experience.

Large amounts of gases are released from manure in connection with cleaning-out operations. Also, the longer the storage, the greater are the amounts of gases released.

Odorants from livestock buildings are introduced into the surroundings via exhausted ventilation air. Determinations of odour strength in ventilation air exhausted from pig units gave the odour threshold values shown in Table III.

TABLE III

STRENGTH OF ODOUR IN VENTILATION AIR EXHAUSTED FROM PIG UNITS (from ref. 11)

Manure handling and ventilation system	*No. of pigs*	*Summer conditions*		*Winter conditions*	
		Vol. of odour air (m^3/hr)	*Threshold as logarithm*	*Vol. of odour air (m^3/hr)*	*Threshold as logarithm*
Scraping system, high evacuation rate	500	35 000		15 000	
Mean value			3·78		3·42
Standard deviation			0·55		0·81
Sluice gate, combined high and low evacuation rate	550	41 000		15 000	
Mean value			3·95		4·18
Standard deviation			1·14		0·42

Calculation of the relative odour strength, with consideration given to both quality and quantity, can be done by multiplying an 'odour load unit' (OLU), given in Table IV, by an odour load factor given in Table V.

Large volumes of odorants are formed during the storage of manure,

TABLE IV

ODOUR LOAD UNITS (from ref. 21)

Type of animal	*OLU*	*Min. vol. of exhausted ventilation air (m^3/hr)*
Calves (for replacements)	0·2	13
Calves (fattening)	0·4	23
Breeding cattle, young	0·6	30
Fattening cattle	0·8	35
Dairy cattle	1·0	45
Growing pigs, 30–100 kg live weight	0·2	7
Breeding pigs	0·3	18
Poultry breeding and broilers	0·01	0·15
Laying hens	0·02	0·4

both solid and fluid. In general, stored manure does not cause odour problems other than in the immediate neighbourhood as long as the manure remains static. The odorants released are apparently rapidly dispersed in the fresh air and the odour threshold is not reached. However, during agitation, loading or pumping, large volumes of gases are released that may cause odour problems in the surrounding areas. In some cases, strong winds have caused such movement in large slurry pits (diameter > 25 m), giving rise to odour problems. Emission during transport and spreading of manure may also lead to troublesome odour problems.

TABLE V

ODOUR LOAD FACTORS FOR DIFFERENT CLEANING METHODS (from ref. 21)

Cleaning method	*Odour load factor*	
	Closed buildings	*Open buildings*
Manual and mechanical cleaning:		
not daily	1·5	1·0
one or more times a day	1·0	1·0
Deep litter	1·5	1·0
Slurry storage in the building	1·8	1·5
Sluice channels, slurry	1·5	1·0
Return flushing	1·8	1·5
Insufficient hygiene, unfavourable climate	2·5	1·8

FEEDLOT ODOUR CONTROL

Two main approaches to the problem can be considered. One is to reduce emission and the other, which is only applicable when new buildings are to be built, is to choose suitable sites.[16,18]

Reduction of Emission

It is important to reduce the formation of noxious odorants as much as possible. In this respect, the buildings should be kept clean, the manure should not be spread over a large surface and, when mechanical scraping is used, the buildings should be cleaned at least twice a day.

Another way to reduce the formation of odorants is to use a chemical or biochemical compound. Many such compounds have been tested and some are considered to give relatively good effects.[5, 14, 19] Ammonium persulphate and potassium permanganate are among the most promising, but the expense is considerable and there are technical and labour problems connected with their application. At present there do not seem to be any practically applicable methods of this type that can be recommended without reservations.

Another method of reducing emission is to treat the ventilation air before it is released. This can be done by adsorption, absorption, combustion (direct or catalytic) or by chemical destruction.[14, 19] In the adsorption method, use is made of beds or columns of active carbon, silica gel, pethium chloride, active aluminium, etc. A problem here is that the large amounts of dust often present in the ventilation air rapidly clog the filter.

In the absorption method the odorants in the ventilation air are removed by a cleaning fluid. The equipment may consist of a tower[17] through which all the ventilation air is drawn in the opposite direction to a stream of water. Relatively good results have been obtained with such towers, which are filled in the upper part with cellulose impregnated with phenol formaldehyde.

Rapid oxidation of odorants can be obtained by combustion, which occurs either directly with an open flame or by catalysis at a lower temperature. High temperatures are required if combustion by flame is to be complete: the smaller the contents of combustible components in the ventilation air, the higher the temperature necessary. Catalytic combustion does not require such high temperatures.

A method of chemically destroying odorants in air is to use ozone, which oxidises the odorants. The ozone can either be applied directly in the building or in a special mixing chamber placed under the ventilation outlet. The latter alternative allows ozone to be used at higher concentrations without involving health hazards.

All these methods of treating the ventilation air require fairly considerable investment and running costs. This applies particularly to the combustion method. Most promising among the methods mentioned here appears to be some kind of absorption, *i.e.* washing the odorants.

Localization

A completely different way of preventing the surroundings from being subjected to odour in such concentrations that they cause discomfort is to improve the atmospheric distribution of released odorants. If a new unit is to be built it can perhaps be sited so that there is a sufficiently large buffer zone between it and the nearest residential area.

The distribution of odours is highly dependent on the volume of odorants emitted, *i.e.* on the size of the enterprise. However, it is also highly influenced by wind conditions, topographic conditions and vegetation, such as trees. Calculations of the distribution can be made from data on wind conditions at the place in question given by a wind-rose. Mathematical meteorological diffusion models can then be used to calculate the necessary buffer zones with regard to the local conditions.[21] Use of a simplified calculation method enables the influence of the factors mentioned to be included in a nomogram, as shown in Fig. 10. An example has been drawn on this nomogram. To start with, calculations are made of the odour load units (Table IV) to be considered. The values for manure handling systems odour factors are found in Table V. The nomogram also include the height of the outlets for ventilated air, the vertical air speed of the ventilated air and the percentage of time the wind is blowing in the direction in question.

ODOUR REDUCTION DURING WASTE HANDLING

In connection with the loading, transport and spreading of slurry, particularly untreated slurry, large volumes of noxious odorants are released that may cause odour problems. Reduction of odour release can be achieved either by treating the manure, by incorporating the slurry directly into the soil, or by various management techniques.

Reduction of Odour in Manure

As mentioned earlier, different chemical and biochemical substances provide some possibility of preventing the formation of noxious odours or of masking them.[5, 19] Another method of reducing the release of odours from manure is by using aeration to create aerobic conditions so that aerobic degradation of the manure occurs. Placing the slurry under the soil surface during spreading is of course another step to prevent odour emission. Tables VI and VII show results of odour reduction when cattle and pig manure were treated and spread by different methods. Ammonium persulphate resulted in considerable reductions in odour release, while a chemical deodorizing product had no effect at all.[9]

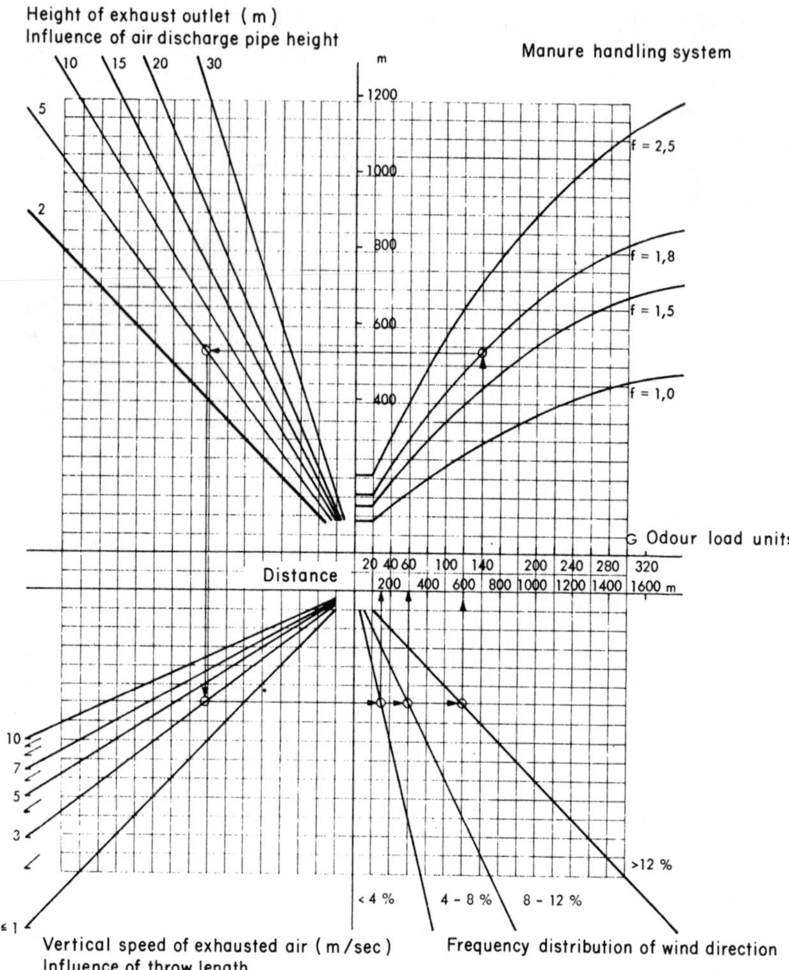

FIG. 10. Odour emission from farm building in relation to distance to nearest residential area (from ref. 21). Example: 140 odour load units (liquid manure), storage in closed barn ($f = 1{\cdot}8$), air shaft height 5 m, vertical air speed 3 m/sec.

Thus, incorporation of the slurry greatly reduces the emission of odour from the field. The means of incorporation, whether by a conventional tillage implement such as a plough or disc harrow, or by a slurry injector, appears to be of secondary importance. However, a slurry injector has the advantage that odour emission during incorporation is minimal, while with

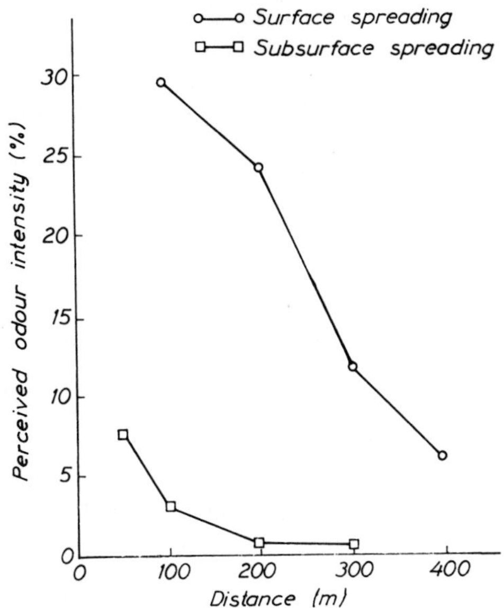

FIG. 11. Perceived odour intensity as a function of distance from the spreading area for two spreading alternatives (from ref. 10).

the other implements large areas must be spread before incorporation takes place.

Comparisons of odour emission at different distances from fields where manure was spread on the surface and where manure was injected directly gave the results shown in Fig. 11. After injection, no manure odour could be detected at distances between 150 and 200 m, while surface spreading gave the same odour level at 400 m as injection did at 50 m.

A most important method of preventing odour emission during the

TABLE VI

ODOUR THRESHOLD VALUES FOR LAND-SPREAD PIG MANURES[a]

	$(NH_4)_2S_2O_8$ *treated pig manure*	*Untreated pig manure*	*Pig urine*	*Unmanured soil surface*
Mean value	2·82	3·82	2·58	2·36
Standard deviation	0·23	0·41	0·43	0·53

[a] Average background contamination 1·84.

TABLE VII

ODOUR THRESHOLD VALUES FOR TREATED AND UNTREATED ANIMAL WASTES[a]

	Untreated pig manure				*Cattle manure*		
	Surface spread	*Buried with*			*Aerobically decomposed, surface spread*	*Untreated, surface spread*	*Unmanured soil surface*
		Injector	*Harrow*	*Plough*			
Mean value	3·45	1·50	2·12	2·30	1·67	2·26	1·71
Standard deviation	0·23	0·40	0·33	0·40	0·46	0·54	0·56

[a] Average background contamination 1·11.

transport of manure is to keep tankers and loading places clean so that manure is not spilled on the way to the field. Naturally, the prevention of leakages from the tanker is also important.

REFERENCES

1. Blanken, G., Helfen Trockenkotbatterien das Kotproblem lösen? Vortrag 27, Oct. 1970.
2. Blaser, E., Ein Beitrag zur Kenntnis der Schwefelwasserstoffvergiftung beim Tier durch Jauchegase. *Schweiz. Arch. Tierheilk.*, **88** (1946).
3. Burnett, W. E., Air pollution from animal wastes: the determination of malodors by gas chromatographic and organoleptic techniques. *Environmental Science and Technology*, **3**(8) (1969).
4. Burnett, W. E., Gases and odors from poultry manure: a selected bibliography. *Poultry Science*, **50** (1971) 61–3.
5. Burnett, W. E. and Dondero, N. C., Control of odors from animal wastes. *Trans. ASAE*, **13**(2) (1970).
6. Hartung, L. D., Hammond, E. G. and Miner, J. R., Identification of carbonyl compounds in a swine-building atmosphere. In: *Livestock Waste Management and Pollution Abatement*, pp. 105–6. ASAE, St Joseph, Mich., 1971.
7. Hillinger, H. G. and Matthes, S., Wirkung von Staub-, Keim- und Geruchsimmissionen auf Mensch und Tier. *Berichte Ldw.*, **50** (1972) 557–70.
8. Högsved, O., Gödselgaser—en litteraturgenomgång och erfarenheter från praktiken. *SFL Förhandsmeddelande*, No. 311, 1968.
9. Lindvall, T., Norén, O. and Thyselius, L., Luktreducerande åtgärder vid flytgödselhantering. Specialmeddelande 22, Swedish Institute of Agricultural Engineering, 1972.
10. Lindvall, T., Norén, O. and Thyselius, L., On the abatement of animal manure odours. In: *Proceedings of the 3rd International Clean Air Congress*, Düsseldorf, 1973.

11. Lindvall, T., Norén, O., Rosén, G., Thyselius, L. and Grennfelt, P., Källstyrkebestämningar i ventilationsluften från svinstallar. *JTI-Rapport*, No. 13, Swedish Institute of Agricultural Engineering, 1975.
12. Merkel, J. A., Hazen, T. E. and Miner, J. R., Identification of gases in a confinement swine building atmosphere. *Trans. ASAE*, **12**(3) (1969).
13. Miner, J. R. and Hazen, T. E., Ammonia and amines: components of the swine building odor. *Trans. ASAE*, **12**(6) (1969).
14. Missfeld, B., Geruchsminderung durch Haltungsverfahren. *KTBL-Schrift*, No. 183, 1974.
15. Muehling, A. J., Gases and odors from stored swine wastes. *J. Animal Science*, **30** (1970) 526–31.
16. Norén, O., Skarp, S.-U. and Aniansson, G., Nyare erfarenheter från JTI:s undersökningar över gödselgasproblemet. Cirkulär 20, Swedish Institute of Agricultural Engineering, 1967.
17. Phelps, A., Drown piggery odours. *Pig International*, **3**(4) (1973) 11.
18. Riemann, U. and Missfeld, B., Gerüche aus der Stallabluft—wie kann man sie vermindern? *Landtechnik*, **29**(1) (1975).
19. Schirz, S. A., Blanken, G., Kunze, D. M., Priewasser, J., Sebastian, D. and Wolfermann, H. F., Geruchbelästigungen durch Nutztierhaltung und die Möglichkeiten der Vermeidung und Abhilfe. *KTBL-Bauschriften*, No. 13, 1971.
20. Skarp, S.-U., Säkrare flytgödselhantering: gödselgas och luftströmning i djurstallar. Specialmeddelande 20, Swedish Institute of Agricultural Engineering, 1971.
21. Stuber, A. and Leimbacher, K., Geruchsemissionen aus landwirtschaftlichen Betrieben. *FAT-Mitteilungen*, No. 12, 1974.
22. Taiganides, E. P. and White, R. K., The menace of noxious gases in animal units. *Trans. ASAE*, **12** (1969) 359–62.
23. White, R. K., Taiganides, E. P. and Cole, G. D., Chromatographic identification of malodors from dairy animal waste. In: *Livestock Waste Management and Pollution Abatement*, pp. 110–13. ASAE, St Joseph, Mich., 1971.

12

Bio-Engineering Properties of Feedlot Wastes

E. PAUL TAIGANIDES

Professor of Environmental Engineering, Department of Agricultural Engineering, Ohio State University, Columbus, Ohio, USA

INTRODUCTION

For every kilogram of animal product consumed by humans, over 20 kg of wastes are generated in animal feedlots. The challenge, therefore, is to develop efficient animal feedlots to produce the necessary quantities of food to meet human requirements for animal protein, yet manage the large quantities of feedlot wastes in such a way as not to create environmental pollution.

Feedlot wastes are materials which are generated during the production of wool, hides or recreational/work animals, but mainly in the production of meat, milk or eggs, and which pose a threat to the production process if they are allowed to accumulate unprocessed on the feedlot premises. The term 'feedlot wastes' refers to semi-solid to solid mixtures of urine and faeces, with no bedding or other foreign materials added. 'Feedlot wastewater' refers to feedlot wastes to which enough water has been added for the wastes to become free-flowing liquids.

Feedlot wastes are affected by many factors, including animal species, size and age, animal feed and water intake, micro-environment on the feedlot, and climate.[1] These properties are further modified after excretion by the waste handling system and biochemical activities. It is therefore impossible to arrive at design values which would be suitable for all feedlot wastes. Numerical values which are given here are expected ranges and average values which may be used as guidelines in planning, not in detailed design.

Properties of feedlot wastes may be classified as physical, chemical and biological parameters.

PHYSICAL PARAMETERS

The most important physical parameters are quantity and solids content. Physical properties which are relevant are temperature, colour, taste, odour, specific gravity, bulk density, particle size, viscosity and hydrophilic properties. Some of the more important properties and the two physical parameters will be discussed in more detail.

The colour of wastes varies from light brown for fresh to black for old, septic wastes. Fresh wastes have a feed-like musty odour which turns into a 'rotten-egg'-like smell for old, septic wastes. The specific gravity of feedlot wastewaters can be assumed to be 1, the same as that of water.

Bulk Density

Bulk density is the ratio of the weight of the waste sample (wet or dry) to the volume of the sample. For water, bulk density is 1 g/cm^3; this value is approached by dilute feedlot wastewaters. Bulk density measurements are important in the design of storage facilities and pumping.

Compressing air-dried manure samples at a pressure of up to 1400 kg/cm^2 may increase their bulk densities from 0·20 g/cm^3 to as high as 2·1 g/cm^3.[2] Therefore, relatively dried manure can be pelletized into hard, stable, dust-free and essentially odour-free pellets without the use of binding agents.[3] Such pellets can be stored indefinitely or transported readily, as long as they do not come into contact with moisture. Costs for pelletizing include the economic and energy costs of first air drying the wastes and then compressing them into pellets.

As the solids content of feedlot wastes is increased by removing moisture, bulk density also decreases.[4,5] Bulk density is almost the same as that of water at solids content of as high as 20–40 %; bulk density then increases to over 1 g/cm^3 at 30–60 % solids content, and decreases to a range of 0·2–0·6 g/cm^3 at solids content of 90 %.[6] Generally, therefore, for feedlot wastewaters and raw wastes of up to 10–15 % solids content, the term mg/litre may be interchanged with ppm.[7]

Particle Size

Particle diameter and distribution are important in the design of solids separation facilities either through sedimentation or mechanical screening.[7]

Particles in dilute wastewaters can be in suspension, dissolved, or in colloidal state. Figure 1 shows that when the diameter of discrete particles in water exceeds 0·3 μm, the particles are in suspension. Particles with diameters between 0·003 and 0·3 μm remain in colloidal state, while particles smaller than 0·003 μm are dissolved and thus cannot be removed by sedimentation. One frequent way of removing dissolved particles in dilute wastewaters is to feed these particles to bacteria, which being about 1 μm in size and having a density of about 1·03 g/cm^3, can be settled out, as is done in biological treatment systems of feedlot wastewaters. Particles of 1 μm size and 2 g/cm^3 density settle at the rate of 0·004 cm/sec or 15 cm/hr in

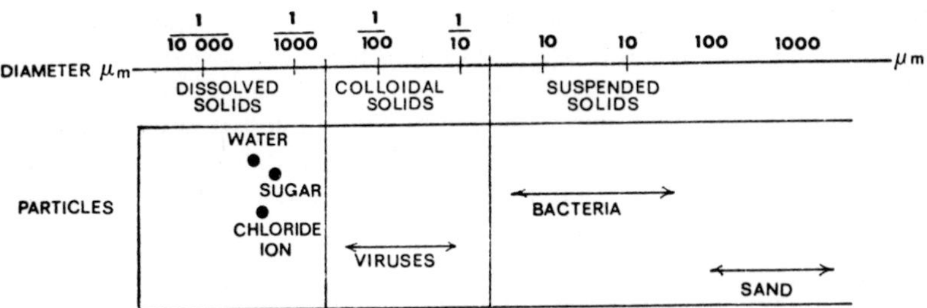

FIG. 1. Classification of particles in water according to their size.

still water at 25 °C. Bacteria settle out at smaller rates than 15 cm/hr because of their lower density.

Particle diameter and size distribution by weight in feedlot wastewaters vary with the feed and the animal. Feed ground finely such as in poultry rations will produce particles smaller in size than those in dairy animals where some of the feed ration is not ground at all. Figure 2 shows that approximately 50 % (by weight) of the particles in dairy wastes are finer than 0·2 mm and therefore would go through a mechanical sieve of openings not exceeding 0·2 mm. For poultry manure, over 60 % of the particles will go through a 0·2 mm screen.

Viscosity

Besides being important in the overall design of waste treatment facilities, moisture content of feedlot wastes is important in planning of the waste

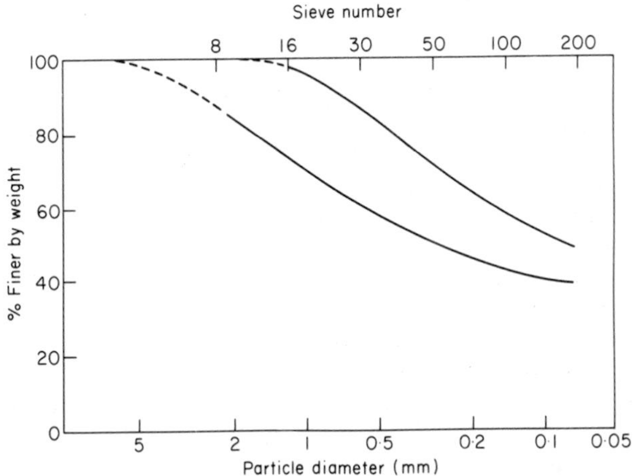

FIG. 2. Particle size distribution in fresh feedlot wastes (from ref. 4).

management system because moisture content determines handling methods. As is shown in Fig. 3, liquid handling is suggested for moisture contents exceeding 88 %, while semi-solid to solid handling systems are suggested for moisture contents below 88 %. Feedlot wastes may be considered liquids when the solids content is below 10 % (*i.e.* water content exceeds 90 %). Actually, feedlot wastewaters frequently contain less than 1 % solids (99 % moisture).

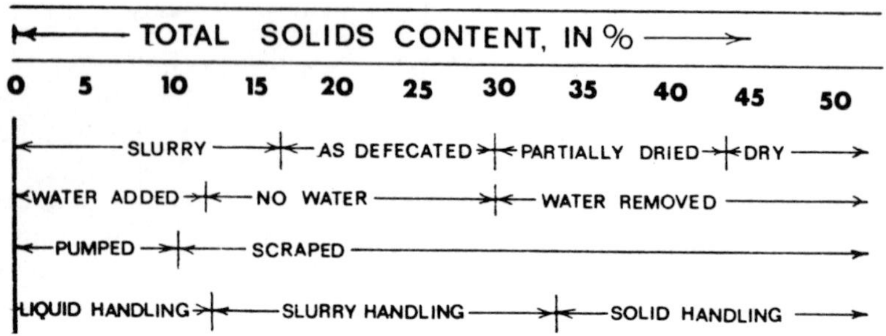

FIG. 3. Classification of feedlot wastes into liquid and solid according to dry matter content (from ref. 8).

Pumping rate decreases as solids content of feedlot wastes increases.[9-12] Viscosity of animal slurries increases in direct proportion to the second power of the dry matter content,[13] once dry matter content exceeds 10 %. It is difficult, however, to measure accurately the viscosity of non-Newtonian fluids such as semi-liquid feedlot wastewaters. Special devices may be used to estimate the pumpability or flowability of feedlot wastes.[11]

Waste Quantities

For planning purposes, waste production may be assumed to be directly proportional to the live weight of the animal.[1] As is shown in Table I, daily excretion of wastes ranges from 3·6 % of total live weight (TLW) for sheep to 9·4 % of TLW for dairy cows. In other words, a 50 kg sheep will excrete 1·8 kg of manure per day, while a 500 kg milking cow will produce 47 kg of waste per day.

Expressing properties of wastes as percentages of live weight is better than as 'kg/animal'.[14] Animals vary in size and live weight according to their age, as is shown in Fig. 4. Feedlot wastes quantities can be calculated by multiplying the waste generation values of Table I with the TLW estimated using Fig. 4 or by weighing the animals. The latter is too cumbersome to be practical.

There is a great variability in the published data on the quantities excreted by various animals.[14] The values given in Table I are presented as

TABLE I
BIO-ENGINEERING PARAMETERS OF FEEDLOT WASTES
(from ref. 19)

Parameter	*Symbol*	*Units*	*Pork pigs*	*Laying hens*	*Beef cattle*	*Feedlot sheep*	*Dairy cattle*
Wet excreta waste	TWW	% TLW/day	5·1	6·6	4·6	3·6	9·4
Total solids	TTS	% TWW	13·5	25·3	17·2	29·7	9·3
		% TLW/day	0·69	1·68	0·70	1·07	0·89
Volatile solids	TVS	% TTS	82·4	72·8	82·8	84·7	80·3
		% TLW/day	0·57	1·22	0·65	0·91	0·72
BOD	BOD	% TTS	31·8	21·4	16·2	8·8	20·4
		% TVS	38·6	29·4	19·6	10·4	25·4
BOD/COD		%	30·7	23·2	17·4	7·8	13·8
N		% TTS	5·6	5·9	7·8	4·0	4·0
P_2O_5		% TTS	2·5	4·6	1·2	1·4	1·1
K_2O		% TTS	1·4	2·1	1·8	2·9	1·7

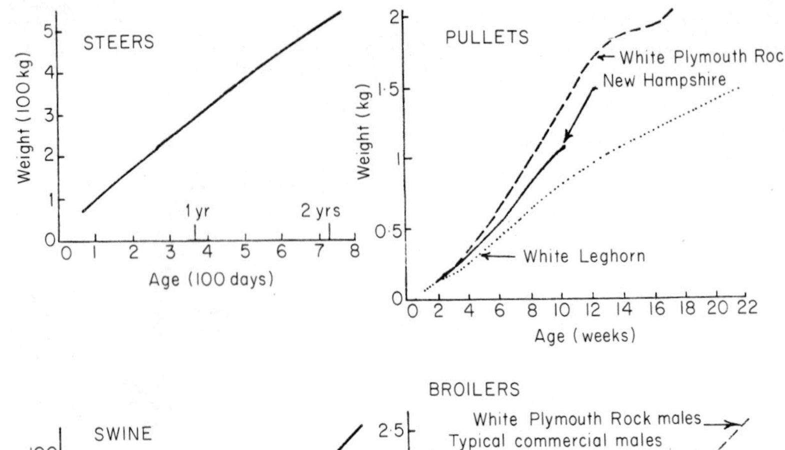

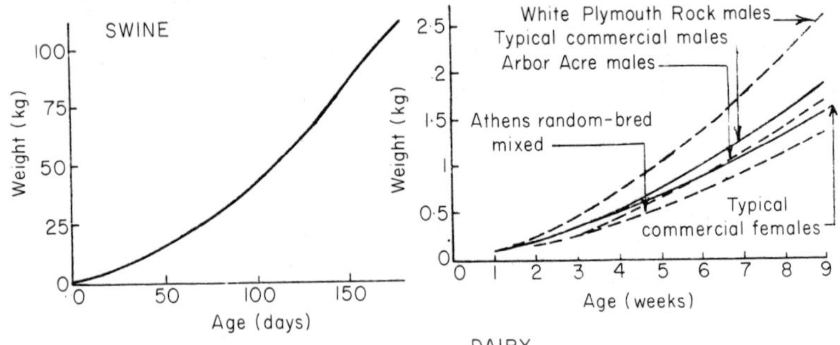

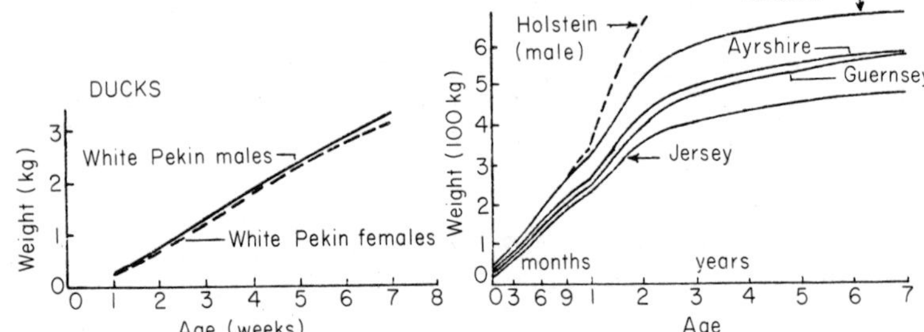

FIG. 4. Age versus weight of animals (from ref. 35).

guideline figures. Actual excretion rates may vary within ±20% of the values of Table I.[15]

The quantities to be handled will depend on how the excreta are collected. Liquid manure handling systems are amenable to volume calculations based on average waste production values. If the urine is collected separately, the quantities of urine will be roughly 30% of cattle excretions and 37 % of swine and sheep.[3] In open feedlots, water may evaporate or the organic matter be oxidized, thus reducing the quantities to be handled. Water may be added due to rainfall. Also, in scraping waste solids from open feedlots, soil is sometimes scraped along with waste matter, thus increasing the quantity for handling.[16,17]

Obviously, the design engineer or planner must consider not only waste excretion rates but also post-excretion changes in arriving at an accurate estimate of feedlot waste and wastewater quantities.

Solids Content

Solids content is the most important physical parameter of feedlot wastes. Solids content can be described by the following matrix:

$$\begin{array}{ccccc} (\mathrm{TTS}) & = & (\mathrm{TVS}) & + & (\mathrm{TFS}) \\ \| & & \| & & \| \\ (\mathrm{TSS}) & = & (\mathrm{VSS}) & + & (\mathrm{FSS}) \\ + & & + & & + \\ (\mathrm{TDS}) & = & (\mathrm{VDS}) & + & (\mathrm{FDS}) \end{array}$$

where TTS = total solids
TVS = total volatile solids
TFS = total fixed solids
TSS = total suspended solids
VSS = volatile suspended solids
FSS = fixed suspended solids
TDS = total dissolved solids
VDS = volatile dissolved solids
FDS = fixed dissolved solids.

There are standard procedures for determining each of these parameters.[18] TTS is the weight of dry matter remaining after evaporation of the moisture at 103°C until constant weight is reached. TFS is determined by weighing the ash remaining after ignition of the TTS at 600°C until complete combustion; TVS is calculated by subtracting TFS from TTS.

TSS and FSS are determined by first filtering out on a filter paper particles larger than 0·3 μm (see Fig. 1). The solids are dried and weighed as above for TSS followed by ignition to determine FSS. VSS is computed by subtracting FSS from TSS. All values for dissolved solids are computed by subtraction; they are not measured directly.

Typical values for TTS and TVS for feedlot wastes may be calculated from Table I. These parameters are measured routinely. Measurement of TSS and FSS is cumbersome and thus is usually made when treatment of feedlot wastewaters is involved.

Total Solids (TTS)
As was stated previously, TTS determines the flowability, pumpability and viscosity and characterizes the general physical state of animal wastes. It is usually expressed as a percentage of the total wet waste (TWW) or as a percentage of the TLW of the animal, as is shown in Table I.

TTS of feedlot wastes varies from a low of 9·3 % of fresh wet excrements for dairy cattle wastes to a high of 29·7 % for sheep wastes, as is given in Table I. Because of their low solids content, dairy and swine feedlot wastes are more amenable to liquid manure handling than poultry, sheep or beef cattle wastes.

Pigs produce the smallest quantity of solids per unit of live weight (0·69 % of TLW) per day, while chickens excrete wastes at the rate of 1·68 % of their TLW per day. Daily production of feedlot wastes from 1000 kg TLW animals is calculated from Table I as 6·9 kg for pigs, 16·8 kg for hens, 7·9 kg for beef cattle, 10·7 kg for sheep and 8·9 kg for dairy cattle.

Moisture content is computed by subtracting the percentage of solids (TTS) from 100.

Total Volatile Solids (TVS)
As is shown in Table I, the dry matter of feedlot wastes is over 80 % organic in nature, except for chicken waste which is 72·8 % volatile. Interestingly enough, however, in terms of rates of volatile matter excretions per unit of live weight, chickens excrete 12·2 kg/1000 kg TLW/day, while pigs excrete 5·7, beef cattle 6·5, sheep 9·1 and dairy cattle 7·2 kg of TVS per day for each 1000 kg of live weight animals in the feedlot.

TVS is the organic matter of the solids content of the waste. It is this component which must be stabilized during waste treatment. When TVS is reduced to the point where it constitutes 50 % of TTS instead of 80 %, then it may be assumed that stabilization of the waste has begun.

In feedlot wastewaters, volatile solids may be divided into suspended (SVS) and dissolved (DVS) solids. The percentage of TVS being in suspension depends on the degree of dilution, the type of waste and the degree of biochemical treatment.

Fixed Solids (TFS)
Fixed solids constitute the inorganics consisting mainly of the fertilizer elements such as N, P, K, plus Ca, Cu, Zn, Fe, etc. TFS represents less than 20 % of the total solid matter for most waste except for chicken waste where TFS might be 27·2 % of TTS, as can be computed from TVS data given in Table I. TFS solids can be suspended or dissolved. In dilute feedlot wastewaters, minerals are mainly dissolved and thus their removal constitutes a difficult problem.[21,23]

Suspended Solids (TSS)
TSS in feedlot wastewaters may range from 30 % to 80 % of TTS, while their

concentration is in the range of 1500–12 000 mg/litre from open beef cattle feedlots,[24] the average being around 2500 mg/litre TSS for both beef and dairy cattle wastewaters.[17]

Suspended solids may be classified as volatile suspended solids (VSS) and as inorganic suspended solids (FSS). The majority of TSS are VSS.

TSS are separated into settleable and non-settleable. Settleable solids are measured by using an Imhoff cone where a litre of wastewater is placed and the ml of solids settled at the bottom of the cone after 45 min are considered the settleable solids. Settleable solids measurements indicate the quantities of TSS which would be expected to be removed by sedimentation in a practical treatment system. Clarifiers and sedimentation tanks are designed on the basis of TSS and settleable solids measurements.[22] A scum of floating materials develops when feedlot wastewaters are stored under quiescent conditions. The scum can solidify, but it is usually unstable.

Dissolved Solids (TDS)

All dissolved solids are chemicals and thus are discussed under chemical properties.

BIOCHEMICAL PARAMETERS

Biochemical parameters of significance in feedlot wastewaters may include: biochemical oxygen demand (BOD), chemical oxygen demand (COD), total organic carbon (TOC) and soil oxygen demand (SOD). Of these, TOC, BOD and COD have been standardized.

TOC includes organic carbon such as that contained in lignin, cellulose and other chemical compounds, but in such a form that it is biodegradable in a biological treatment plant. TOC is not related to degradation rates but may be related to BOD and COD of wastes which have uniform composition. As such, TOC serves as a meaningful quantitative test for evaluating biochemical treatment process efficiencies and for measuring pollution levels in streams and effluents. Recently developed carbonaceous analysers use catalytic combustion at 900–1000 °C to oxidize all carbonaceous materials to carbon dioxide which is measured by infrared analysers. This makes TOC an extremely rapid test. TOC is therefore useful in monitoring efficiencies of treatment plants. It is not used as a design parameter.

Biochemical Oxygen Demand (BOD)

Theory

BOD is defined as the amount of oxygen required to stabilize decomposable organic carbonaceous material under aerobic conditions. The total oxygen demand is exerted in three stages as follows:

1. The immediate demand is exerted by the presence of certain reducing chemicals.
2. The carbonaceous demand is exerted by the decomposable carbonaceous material.
3. The nitrogenous demand is exerted by protein, urea and various other reduced forms of nitrogen.

In water pollution analyses we are largely concerned with the carbonaceous ultimate BOD.

For most organic wastes it has been shown that 80–90 % of the ultimate carbonaceous oxygen demand has been satisfied by the fifth day and, furthermore, that in 7–8 days a second-stage demand (nitrogenous demand) begins to appear.

The test was therefore standardized to ensure reproducibility and comparability between tests conducted by different people at different times and at different places. The standard test is performed at 20 °C for 5 days with settled sewage serving as seed and the dilution wastes containing an excess of mineral nutrients other than carbon to ensure that the only limiting nutrient to halt the reaction is the organic carbon being measured.

The standard equation describing the BOD curve for dilute wastewaters is

$$y = L(1 - 10^{-kt})$$

where y = BOD at time t (mg/litre) (normally BOD_5), L = ultimate BOD (mg/litre), k = rate of the reaction (day^{-1}) and t = time of reaction (days). The significance of the reaction rate is paramount. It differs for different wastes. Several wastes may have the same 5-day BOD, but because of different reaction rates the ultimate BOD would also be different for each waste. It is important, therefore, to determine all the components of the above equation for full BOD characterization of a waste or river water. In engineering terms, the water pollution potential of feedlot wastewaters is measured by the standard BOD test.[18]

BOD Values

Table I shows that pig wastes exert the largest oxygen demand per unit of total solids, 31·8 g of BOD/100 g of TTS. A 100 kg pig would produce waste with a BOD of 0·22 kg/day. If it is assumed that the BOD of human waste is 91 g/day, then the waste from a 100 kg pig will have a population equivalent (PE) of 2·4.

Comparing animal wastes with city sewage on the basis of BOD is irrelevant, however. The 5-day BOD for animals if not directly comparable to that of human sewage because it represents only 16–60 % of the ultimate BOD, while sewage is supposed to represent 80 % of carbonaceous BOD.[25]

Furthermore, the value of the BOD test is questionable for the characterization of animal wastes for two fundamental reasons: animal manures are solid wastes, not wastewaters, and ultimate disposal of animal manures is on land, not in streams. With the advent of liquid manure systems and other recent developments, particularly the application of sewage treatment techniques for the biodegradation of wastes from confinement units, data on manure BOD can be useful in the same way that they are for other concentrated industrial organic wastes.

The basic chemical technique of measuring the dissolved oxygen in water is the Winkler method.[18] The azide modification of the Winkler method should be used because it removes the interference caused by the presence of nitrite ions in the sample.

A serious disadvantage of the standard bottle test is that high dilutions of the sample are required. There are inherent hazards in diluting wastes which are neither chemically nor physically uniform. The dilutions necessary for the 5-day BOD test for animal wastes are in the range of 20–30 mg of total solids per litre of dilution water, which amounts to a dilution of 1·0 in 12 000.[25]

In the Warburg method,[26] higher concentrations, and therefore more representative samples of the waste, may be used. On the other hand, data resulting from the Warburg method for BOD determination have not had as wide an application in actual design as the standard dilution method.

Polarographic techniques are recommended as an alternative to iodometric titration techniques for industrial waste when titrimetric methods could be subject to serious error in the presence of high concentrations of waste in the samples being tested. There are several polarographic instruments available for measuring dissolved oxygen.

As is shown in Figs. 5, 6 and 7, poultry, swine and beef wastes have similar BOD curve characteristics, while dairy and sheep wastes show similar relationships at the lower end of the range.[25] The rate of oxygen demand is lower at 20°C than at 24°C. The BOD reaction rate ranges between 0·007 and 0·085. This is much lower than municipal sewage in which the reaction rate coefficient is in the range of 0·1–0·3. The 5-day BOD of feedlot wastes ranges between 16% and 26% of total BOD measured at 56 days.[25]

Chemical Oxygen Demand (COD)

COD is not normally considered as a design parameter for biological treatment processes. It is frequently used as a monitoring parameter because it can be completed in 1–3 h, whereas BOD takes 5 days. The COD test measures the oxygen consumed in the oxidation of organic carbon under a high-temperature, strongly acidic chemical digestion process. It bears little or no relationship to the biological oxidation process. Under COD test conditions, organic carbon which is quite resistant to biological degradation is oxidized. COD values are therefore larger than BOD, as

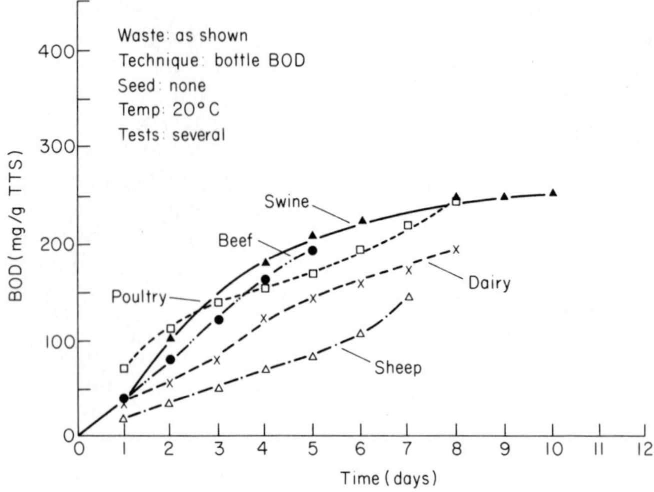

FIG. 5. Typical standard bottle BOD curves for unseeded feedlot wastes (from ref. 25).

shown in Table I. For example, the BOD of sheep feedlot wastes is 7·8% of its COD; this means that if a sheep waste has a BOD of 10 000 mg/litre, its COD is computed at 128 200 mg/litre. The ratio of BOD to COD is 1·0 to 1·8 (*i.e.* 31·8% of COD) for swine feedlot wastes, according to computations based on data given in Table I.

BOD/COD

Typical values of BOD and COD expressed as milligrammes of BOD or COD per gramme of total solids (mg/g TTS) are given in Table II.

TABLE II
TYPICAL BOD AND COD AVERAGE VALUES AND RANGES
(from ref. 25)

Animal faeces	*BOD_5* Average (mg/g TTS)	*BOD_5* Range (mg/g TTS)	*COD* Average (mg/g TTS)	*COD* Range (mg/g TTS)
Swine	181	114–296	1 409	870–2 280
Poultry	123	82–165	887	775–960
Beef	110	50–190	1 438	1 054–2 500
Sheep	103	80–128	1 109	1 000–1 210
Dairy	94	24–208	1 387	910–2 520

Expressing BOD and COD in terms of mg/g TTS enhances the usefulness of such data.[14,25] In other words, if one determines the TTS of a swine wastewater sample (a more routine and much easier test than the BOD test) and finds that TTS = 4% of TWW, that would mean that in a litre of this wastewater there are 40 g of TTS. Since, according to Table II, swine wastes exert 181 mg BOD/g TTS, the BOD of this wastewater is 7240 mg/litre, *i.e.* $40 \times 181 = 7240$ mg BOD/litre of wastewater.

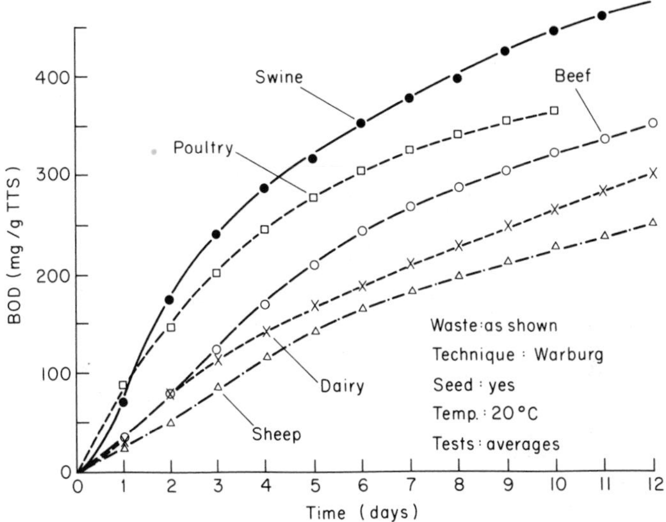

FIG. 6. Typical standard Warburg BOD curves for seeded feedlot wastes (from ref. 25).

The ratio BOD/COD varies from 0 to 1·0. It is a useful measure of the degradability of the organic carbon in the sample. High BOD/COD ratios indicate that the sample will degrade rapidly in a surface water body or holding pond and thus create nuisance conditions more rapidly.

Soil Oxygen Demand (SOD)

Excessive application of manures into soil would cause deterioration of the quality of the soil environment by depleting soil oxygen supply. If this rate of oxygen depletion can be assumed to be proportional to the pollutional potential of waste, then the quantity of animal wastes which can be incorporated in soil may be determined by the SOD test.[25] SOD is an experimental procedure. It is not a standard test and thus is recommended only for research programmes.

In the SOD test, waste is placed in an air-tight Warburg flask containing air-dried soil. Water is added to bring soil moisture to field moisture level. As microbes decompose the waste, oxygen is consumed and carbon dioxide is released. The carbon dioxide is absorbed by a solution of 40% potassium hydroxide which is placed in a vial in the flask. The quantity of oxygen consumed is determined by measuring changes in the pressure in the flask and calculating the change in quantity of gas, using the ideal gas law.

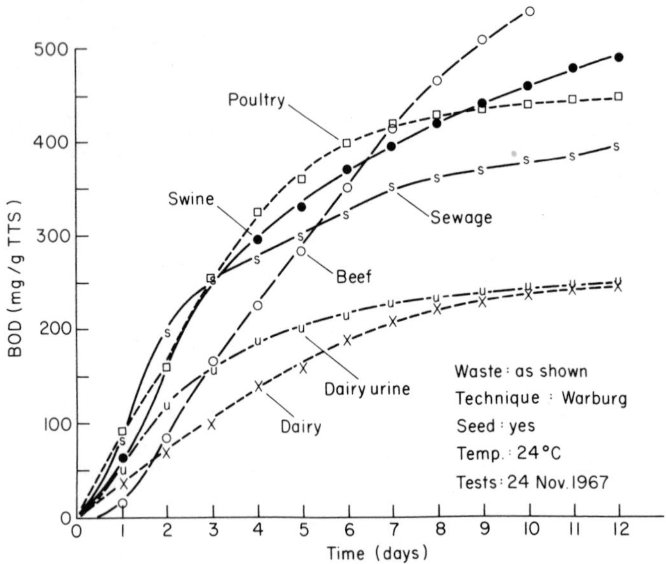

FIG. 7. Typical Warburg BOD curves for seeded feedlot wastes and human sewage at 24 °C (from ref. 25).

Beef, poultry and dairy exhibit 5-day soil oxygen uptakes of approximately 110 mg of oxygen per gramme of total solids, while 5-day oxygen demand of swine waste is approximately 140 mg of oxygen per gramme of total solids. The oxygen demand curve in soil has a longer lag phase than in BOD curves. This is because when samples are air dried and are rewetted at the time of the test, some adjustment time is needed for the development of bacterial cultures. The SOD curves tend to level off much sooner than BOD curves. In a water environment, waste materials may readily dissolve in water and therefore are available sooner to bacteria. In a soil environment, large portions of organic matter may be inaccessible to bacteria or to soil water to dissolve all soluble portions of the waste immediately.[27]

CHEMICAL PARAMETERS

Important chemical parameters of feedlot wastes are: pH, conductivity, fertilizer plant nutrients, toxic metals and other chemicals with biocidal properties.

Metals

Unlike municipal and other industrial wastes, feedlot wastes do not contain metals or biocidal chemicals in concentrations high enough to be toxic. Obviously, if wastes contain toxic concentrations, the symptoms would have appeared in the animals which would be affected first. However, metals tend to accumulate and can reach toxic levels under certain conditions. Metals which need to be watched are copper, zinc, boron, iron and manganese. Copper is used as an antibiotic in some feed formulations for both preventive and therapeutic purposes. Under certain conditions such as in anaerobic digestion, copper may accumulate to concentrations affecting significantly methane gas formation.[28] In feedlot wastes, quantities of metals may range between 2 and 100 mg Cu/kg of TTS. Ranges for other metals are 10–400 mg Zn/kg TTS, 60–33 000 for Fe, 15–270 for B.[29] These concentrations are within the range of values required for minor fertilizer nutrients for good plant growth.

Fertilizer Elements

The major fertilizer nutrients are N, P_2O_5 and K_2O. Secondary nutrients are S, Mg and Ca, while micronutrients are Fe, Mn, B, Cl, Zn, Cu, Mo, etc. The availability of all these elements as plant nutrients is significantly affected by the pH of the soil and whether the soil is mineral or organic.[30] Figures 8 and 9 indicate that while nitrogen can be available in both organic and inorganic soils in a wide range of pH values, phosphorus ceases to be available at pH below 6 in mineral soils, but its availability begins to increase as the pH in organic soils decreases from 6 to 5. Potassium is relatively easily available at pH above 6.

Approximately 72–79 % of the N, 61–87 % of P and 82–92 % of K in the feed is recovered in the animal wastes.[3] As is shown in Fig. 10, even though urine is less than 40 % of TWW, it is more concentrated with respect to N and K than faeces. Phosphorus is excreted mainly in the faeces, except for swine which have considerable phosphorus in their urine.[3]

Nitrogen

Nitrogen is found in feedlot wastes and wastewaters in many forms, as NH_3, NO_3, NO_2 and other organic and inorganic forms. The N content of animal wastes is extremely variable. Nitrogen in NH_3 form is volatile and can easily be lost to the atmosphere. However, predictions of N loss rates are difficult because NH_3 volatilization is affected by many environmental

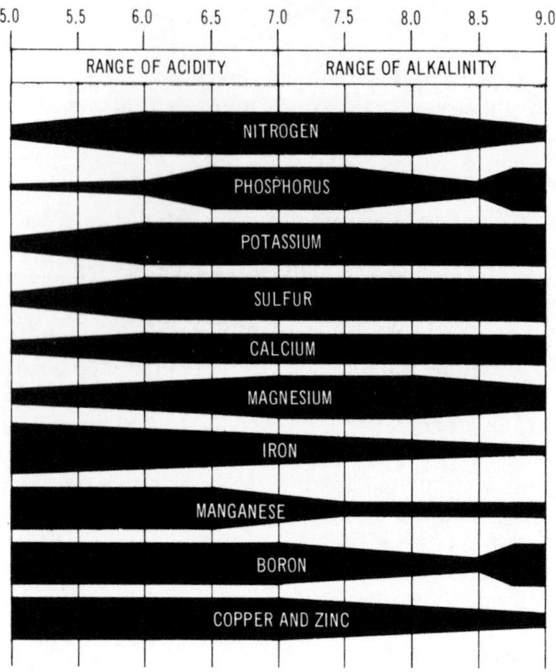

FIG. 8. Relative availability of elements essential to plant growth at different pH levels for mineral soils (from ref. 30).

factors and waste collection and handling methods. For solid handling systems, N losses will range from 20 % in deep pit systems to 55 % for open feedlots.[15] For liquid handling systems, N losses range from a minimum of 25 % in anaerobic systems to 80 % for aerated systems. High temperatures, wind and exposure to the environment accelerate nitrogen losses.

Total nitrogen in feedlot wastes ranges from 4 % of TTS in dairy wastes to 7·8 % of TTS in beef cattle wastes, as shown in Table I. However, the range for beef wastes, for example, may be from less than 1 % to 11 % of TTS. For dairy wastes the range may be from 1·5 % to 5·1 % of TTS.

Phosphorus

As is shown in Table I, phosphates range from 1·1 % of TTS for dairy to 4·6 % of TTS for laying hens. Generally, swine and poultry are higher in phosphorus concentration than cattle. Most phosphorus in feedlot wastes is bound to or is part of the solids and thus not much phosphorus is lost either through the waste handling or treatment processes.

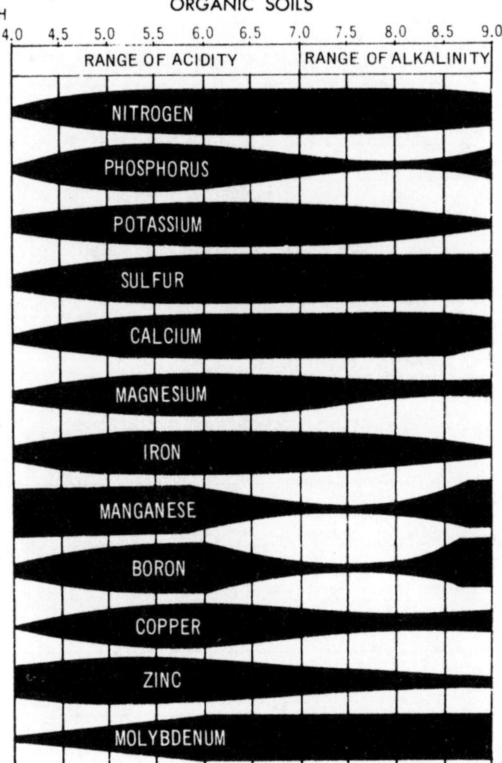

FIG. 9. Relative availability of elements essential to plant growth at different pH levels for organic soils (from ref. 30).

Potassium
Potash in feedlot wastes ranges from 1·4 % of TTS in swine to 2·9 % of TTS in beef cattle wastes. Vegetative plant parts have a higher concentration of potassium than the grain itself. Therefore, animals fed silage and roughage rations will consequently excrete more potassium than if the same animals were fed high-concentrate rations.

Potassium salts tend to accumulate in soils when applied in excess quantities and thus might interfere with magnesium uptake by plants and cause other salinity problems.[30]

Electrical Conductivity
Electrical conductivity (EC) is an estimation of the inorganic soluble salt concentration in feedlot wastewaters. The most important salts

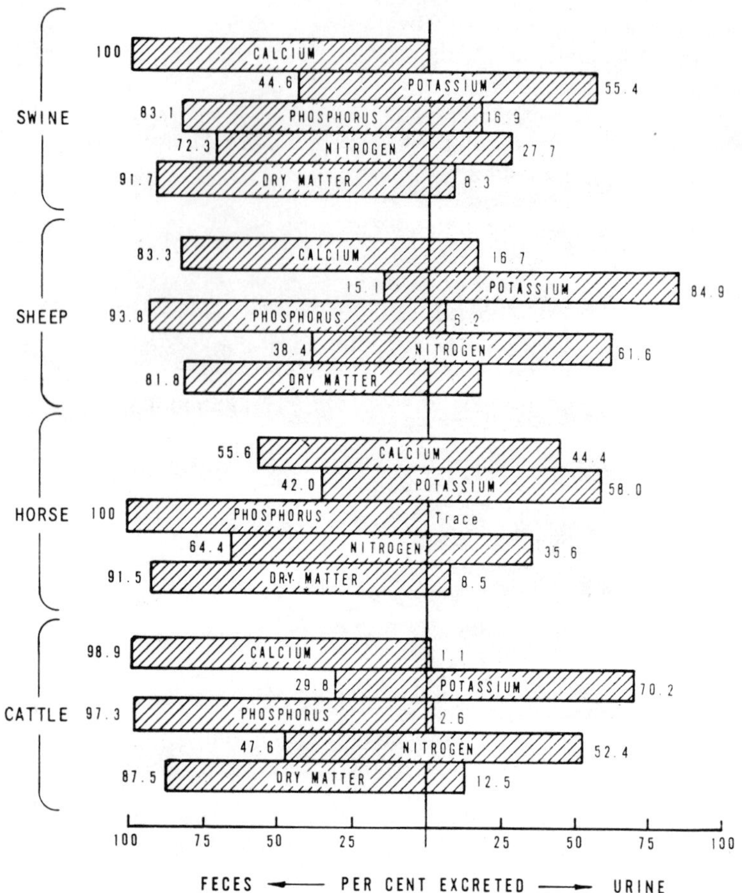

FIG. 10. Distribution of dry matter, nitrogen, phosphorus, potassium and calcium in faeces and urine of cattle, horses, sheep and swine (from ref. 3).

contributing to high EC values are salts of sodium, potassium, magnesium, etc. The EC of feedlot wastewaters may range from about 1 to 7 mmho/cm. In areas of deficit rainfall over evaporation, EC of feedlot wastewaters controls the quantity of wastes to be applied on the land, so as to avoid salt accumulation and soil salinity problems.[29]

BACTERIOLOGICAL HEALTH PARAMETERS

The major bacteriological parameters are total faecal coliforms, total count of bacteria, and identification of pathogens.

Identification of pathogenic organisms is quite cumbersome and requires specially trained personnel. Types of pathogens and disease transmission have already been discussed.

Total counts from direct microscopic examination show 250 to 2000 million bacteria per gramme of wet beef faeces.[31] However, only up to 9% of the total direct count can grow on glucose–yeast–tryptose agar. Most of the bacteria are facultative anaerobes. Coliform counts on EMB agar from beef wastes range from 340 000 to 560 000 bacteria per gramme of TWW. The majority of the coliform bacteria are typical *E. coli*.

In cattle feedlot wastewaters, high concentrations of total coliforms, faecal coliforms and faecal streptococci can be found.[32] MPN counts range from 3·3 to 790 million organisms per 100 ml. The highest counts are found during warm weather, and under conditions which induce high solubility of solids. Faecal coliforms are usually present in larger numbers than faecal streptococci.

A median MPN value of 4·6 million coliforms per 100 ml can be expected from raw duck feedlot wastewaters,[33] while the median MPN for cattle feedlot wastewaters is 79 to 130 million coliforms per 100 ml.[32]

More than 50 species of flies can be attracted to and breed in wastes accumulated in poultry confinement feedlots, while only two species are attracted to poultry manure droppings in the field[34] because chickens in cages cannot control fly breeding by eating the larvae. In cattle feedlots, seven species may be attracted as opposed to 40 species in 'cowpats' in pastures. However, of the 40 species in the field, only two are considered serious pests, while in confinement units all seven species of flies are obnoxious.

The breeding and development of flies are highly dependent on moisture control.[3] A good fly control programme could be designed into manure moisture management systems. There is no oviposition from any fly at moisture contents below 33%. Furthermore, there is no fly development below 35%. It is best to design waste handling systems which would store manure at moisture contents either above 85% or below 35%.

MISCELLANEOUS PARAMETERS

There are several other feedlot waste parameters which may not be classified by the chemical nature of the property they quantify but are useful in feedlot waste management planning. Some important parameters are population equivalent (PE), animal equivalent (AE) and animal unit (AU).

Population Equivalent (PE)

In planning for municipal sewage treatment plants, engineers use the term 'population equivalent' which is taken to be equal to 91 g BOD/day/capita.

Such parameters are useful in overall planning and assigning priorities. Unless feedlot wastes are discharged into the sewer system, PE numbers are meaningless. For example, in a city the BOD from the residential area might be 1000 kg/day, 2500 kg/day from various manufacturing firms and may be 3000 kg/day of BOD from an animal feedlot. Therefore the total PE for the city would be 71 429, of which PE 32 967 will be due to the feedlot.

Also, PE based on BOD may be used in developing water quality policies only when feedlot wastewaters are discharged directly into waterways. BOD is a measure of stream water pollution potential. Therefore, PE based on BOD would estimate the potential contribution of feedlot wastes to national water pollution for direct discharges of such wastes.[19]

PE may also be based on other properties such as solids (TTS), nitrogen, volatile solids (TVS), etc. For example, if wastes from human and animal facilities are to be discharged on to land, PE may be more meaningful if it is computed on the basis of TTS, N, TVS or SOD.

Animal Equivalent (AE)

Because feedlot wastes are not managed in the same manner as human wastewaters, it is better to compare animal wastes among animal species rather than with human wastes. The term 'animal equivalent' may be arbitrarily designated as a beef cattle animal weighing 500 kg TLW.[19] Using Table I, the quantities of various waste parameters for a 500 kg beef animal can be calculated as is done in Table III. Table III shows that 10 000 kg live weight of chickens, which would be about 5000 layers, would have an AE of 29 on the basis of TWW but an AE of 42 on the basis of TTS. When compared on the basis of BOD, these chickens would be equivalent 56 AE while dairy cattle with the same live weight on the basis of BOD would be equivalent to only 28 AE or 50 % less.

If nitrogen application per hectare is limited, for example, to 500 kg

TABLE III

ANIMALS AND BIO-ENGINEERING PARAMETERS PER ANIMAL EQUIVALENT

Parameter	*kg/day/AE*	*kg/year/AE*	*AE/1000 kg TLW of various animals*				
			Beef	*Pigs*	*Hens*	*Sheep*	*Dairy*
TWW	23	8 400	2	2·2	2·9	1·6	4·1
TTS	4·0	1 460	2	1·7	4·2	2·6	2·2
TVS	3·2	1 170	2	1·8	3·8	2·8	2·2
BOD	0·64	234	2	3·4	5·6	1·5	2·8
COD	3·6	1 310	2	2·0	4·3	3·4	3·6
N	0·32	117	2	1·2	3·1	1·3	1·1
P_2O_5	0·05	18	2	3·4	15·5	3·0	2·0
K_2O	0·07	26	2	1·4	5·0	4·4	2·2

N/ha/year for environmental protection purposes, then this could be translated into 4·3 AE/ha/year. From Table III it is computed, then, that for chicken feedlots the land area required would be $(4{\cdot}3/3{\cdot}1) \times 1000 = 1387$ kg TLW or approximately 700 chickens each weighing 2 kg. This means that the land limitations would be set at 700 layers/ha for poultry feedlots. In terms of dairy cattle, the land constraints on the basis of nitrogen applications per year would be 3900 TLW of dairy cattle/ha or about 8 Ayrshire or Guernsey milking cows/ha (see Fig. 4). Another way of expressing these results is to say that on the basis of same live weight, poultry feedlots would require 64% more disposal land area for the nitrogen in their wastes than dairy feedlots.

Animal Unit (AU)

Animal unit is defined as a comparison between different animals on the basis of their live weight. An animal weighing 500 kg TLW is frequently defined to be 1 AU. Using Fig. 4, it can be calculated that it would take 10 pigs 110 days in age to make 1 AU, and that two one-year-old Jersey cows would make 1 AU.

REFERENCES

1. Taiganides, E. P. and Hazen, T. E., Properties of farm animal excreta. *Trans. ASAE*, **9** (1966) 374–6.
2. Hafer, A. A. *et al.*, *Physical Properties of Farm Animal Manures*. University of California Agricultural Experiment Station Bulletin, Davis, Calif., 1974.
3. Azevedo, J. and Stout, P. R., *Farm Animal Manures*. Manual 44, University of California Agricultural Experiment Station, Davis, Calif., 1974.
4. Sobel, T., Physical properties of animal manures associated with handling. In: *Management of Farm Animal Wastes*, pp. 27–32. ASAE, St Joseph, Mich., 1966.
5. Surbrook, T. C. *et al.*, Drying poultry manure. In: *Livestock Waste Management and Pollution Abatement*, pp. 192–4. ASAE, St Joseph, Mich., 1971.
6. Hart, S., Thin spreading of slurried manures. *Trans. ASAE*, **7** (1964) 22–5, 28.
7. Overcash, M. R., Reddell, D. L. and Day, D. L., Chemical analysis (of animal wastes). ASAE Paper 74-4546, St Joseph, Mich., 1974.
8. Loehr, R., *Agricultural Waste Management*. Academic Press, New York, 1974.
9. Hart, S. A., Moore, J. A. and Hale, W. F., Pumping manure slurries. In: *Management of Farm Animal Wastes*, pp. 34–8. ASAE, St Joseph, Mich., 1966.
10. Taiganides, E. P. *et al.*, Properties and pumping characteristics of hog wastes. *Trans. ASAE*, **7**(2) (1964) 123–4, 127, 129.
11. Grimm, K. and Langenegger, G., Measuring method for evaluating the ability to pump semi-liquid and manure. In: *Livestock Waste Management and Pollution Abatement*, pp. 138–41, 145. ASAE, St Joseph, Mich., 1971.
12. Staley, L. M., Bulley, N. R. and Windt, T. A., Pumping characteristics,

biological and chemical properties of dairy manure slurries. In: *Livestock Waste Management and Pollution Abatement*, pp. 141–5. ASAE, St Joseph, Mich., 1971.

13. Sewell, J. I., Manure slurry irrigation system receiving lot runoff. *Trans. ASAE*, **16** (1973) 350–3.
14. Taiganides, E. P. and White, R. K., Typical variations encountered in the measurement of oxygen demand of animal wastes. In: *Proceedings of Waste Management Conference*, pp. 327–35. Cornell University, Ithaca, NY, 1969.
15. Midwest Plan Service, *Livestock Waste Facilities Handbook* (MWPS-18). Ames, Iowa, 1975.
16. Gilbertson, C. B., Characteristics of manure accumulations removed from outdoor, unpaved, beef cattle feedlots. In: *Livestock Waste Management and Pollution Abatement*, pp. 56–9. ASAE, St Joseph, Mich., 1971.
17. US Environmental Protection Agency, *Effluent Limitations Guidelines for Feedlots*, EPA 440/1-73/004. Washington, DC, 1973.
18. American Public Health Association, *Standard Methods for the Examination of Water and Wastewater*, 12th ed. APHA, New York, 1965.
19. Taiganides, E. P. and Stroshine, R. L., Impact of farm animal production and processing on the total environment. In: *Livestock Waste Management and Pollution Abatement*, pp. 95–8. ASAE, St Joseph, Mich., 1971.
20. Taiganides, E. P. and White, R. K., Performance of an automated waste treatment and recycle system. In: *Proceedings ISLW '75*. ASAE, St Joseph, Mich., 1975.
21. Mehta, B. S. and Taiganides, E. P., Tertiary treatment of wastewaters by reverse osmosis membranes. In: *Proceedings ISLW '75*. ASAE, St Joseph, Mich., 1975.
22. Shutt, J., White, R. K., Taiganides, E. P. and Mote, R., Evaluation of solids separation devices. In: *Proceedings ISLW '75*. ASAE, St Joseph, Mich., 1975.
23. Loehr, R. C., Nutrient control applicable to animal wastes (this volume, pp. 253–69).
24. Miner, R., Water pollution potential of cattle feedlot runoff. Ph.D. thesis, Kansas State University Library, Manhattan, Kans., 1967.
25. Taiganides, E. P., White, R. K. and Stroshine, R. L., Water and soil oxygen demand of livestock wastes. In: *Livestock Waste Management and Pollution Abatement*, pp. 176–9. ASAE, St Joseph, Mich., 1971.
26. Ludwig, H. F., Oswald, W. J. and Gotaas, H. B., *Manometric Technique for Measurement of BOD*, Institute of Engineering Research, Berkeley, Calif., 1951.
27. Stroshine, R. L., The development of a test to measure soil oxygen demand. MS thesis, Agricultural Engineering Dept, Ohio State University, Columbus, Ohio, 1971.
28. Taiganides, E. P. *et al.*, Anaerobic digestion of hog wastes. *J. Agricultural Engineering Research*, **8**(4) (1963) 327–33.
29. Environmental Protection Agency, *Research Status on Effects of Land Application of Animal Manures*, EPA-660/2-75-010. Washington, DC, June 1975.
30. Ohio Cooperative Extension Service, *Ohio Agronomy Guide*. OCES, Columbus, Ohio, 1975.

31. Witzel, S. A., McCoy, E. *et al.*, Physical, chemical and bacteriological properties of farm wastes. In: *Management of Farm Animal Wastes*, pp. 10–14. ASAE, St Joseph, Mich., 1966.
32. Miner, R. *et al.*, Stormwater runoff from cattle feedlots. In: *Management of Farm Animal Wastes*, pp. 23–7. ASAE, St Joseph, Mich., 1966.
33. Davis, R. V., Cooley, C. E. and Hadder, A. W., Treatment of duck wastes and their effects on water quality. In: *Management of Farm Animal Wastes*, pp. 98–105. ASAE, St Joseph, Mich., 1966.
34. Anderson, J. R., Biological interrelationships between feces and flies. In: *Management of Farm Animal Wastes*, pp. 20–3. ASAE, St Joseph, Mich., 1966.
35. American Society of Agricultural Engineers, 1975 *Agricultural Engineers Yearbook*. ASAE, St Joseph, Mich., 1975.

PART II

Technologies for Processing and Treatment of Animal Wastes

13

Collection, Storage and Transport of Cattle Wastes

M. VELEBIL

Director, Agricultural Engineering Research Institute, Prague, ČSSR

INTRODUCTION

Waste disposal from cattle barns is among the most tiresome of drudgeries in small-size dairy operations. It is unthinkable even to consider using human labour for the handling of wastes from modern large cattle feedlots. Manure handling is mechanized and in some instances automated in order to minimize human labour and drudgery.

The selection of systems for the collection, storage and transport of cattle wastes is affected by the type of housing and by the concentration of animals.

The housing system determines if solid (mixture of excrements and bedding) or slurry handling equipment will be used. The concentration of animals determines the quantities of wastes which in turn determine the size and type of equipment to be used.

Present methods of dairy cattle housing include the conventional stanchion barns, loose housing systems and free-stall systems with or without slatted floors.

The basic waste handling methods can be divided into stationary and mobile systems. Stationary systems have the following advantages: (a) they are always ready for operation, thus requiring little time to start; (b) they require less maintenance and offer better flexibility in terms of designing systems to fit in with housing layout; (c) they offer a good possibility of automation and programming.

Disadvantages of the stationary equipment are: (a) the necessity for special construction such as construction of the dunging alley, strengthening the building foundations to accommodate the power unit, construction of the discharge apertures, etc.; (b) the limited use of the machine in only one barn; and (c) the fact that it is confined to a limited number of stall rows.

Mobile systems have the advantage of servicing several barns and also of being used for other tasks. Disadvantages include high labour requirements and the limitation of their use to certain types of housing layout. Another disadvantage is that mobile systems require frequent opening of doors during cold winter months. However, the most serious disadvantage is the impossibility of automation with mobile systems.

STATIONARY SOLID HANDLING SYSTEMS

The most widespread stationary equipment for straw manure is the circulating endless chain cleaner equipped with paddles. Principally, it is a chain conveyor placed in the centre of shallow concrete gutters, running the full length of the gutter, behind the cattle stalls. In one or two opposite corners there are gears with electric motors driving the chain. In the remaining corners or deflection points the chains are led by pulleys. At least one of them is sliding and facilitates the tightening of the chains. The fall-through aperture is usually situated in the corner near the drive unit. Manure falls through the aperture on to a conveyor or stacker which conveys the manure out of the building. The power requirements of such gutter cleaners are about 3 h.p. with a capacity of 3 tons/h.

The chain and paddle cleaner is supplemented with a conveyor for loading manure on trailers or for stacking manure in layers on the farmstead manure heap. Depositing manure in layers is facilitated by the jib. The jib is mounted so that it can swivel and is able to add layers to the heap where the ground plan is a circle or section of a circle. The jib includes a circulating or reciprocating conveyor. Circulating conveyors have a capacity of 4 tons/hr while reciprocating systems can handle 15 tons/hr.

Other stationary systems for disposing of straw manure (shuttle stroke barn cleaners, pushing bars with flaps) are used in small-capacity barns. In large barns with minimal bedding, bars with flaps may be used to move manure out.

MOBILE SOLID MANURE HANDLING SYSTEMS

In some barns, mobile units for daily disposal of straw manure are used. These are either tractor-mounted front loaders or tractor-mounted scrapers with a special blade. This system has the advantage of high reliability. Disadvantages include noise and air pollution in the barn, high space and labour requirements, and pollution of the dungway.

STATIONARY SLURRY HANDLING EQUIPMENT

Use of reciprocating barn cleaners to remove slurry from beneath slotted floors results in excessive corrosion and frequent malfunctioning. The so-called towed cleaners are preferred. They consist of pushing bars or cables with hinged flaps on one or both sides. When in operation the flaps of the cleaner lean against the walls of the gutter, pushing manure out. During the idle run the flaps close towards the bar at a sharp angle as a result of the friction on the gutter bottom. The design of the pushing bar with flaps is simple and offers high operational reliability. Pushing bars with flaps have been found satisfactory for gutters up to 1 m in width. Improvements are being made which would permit the use of such cleaners in 3 m wide gutters.

Up to now, pushing bars with flaps have been used in gutters under the slatted floors; economical advantages are proved when using them as surface units on dunging areas. Such a system can be utilized in free-stall barns and also in stanchion barns. Both initial investment and operating costs are low.

MOBILE LIQUID HANDLING SYSTEMS

In some barns, especially those with stalls using no bedding, tractors mounted with blades are used to evacuate slurry and push it into a collecting storage. The width between stalls need not be increased because the width of the dunging alley is sufficient for the passage of the tractor. Also, the animals are not disturbed because the barn is cleaned during their absence (while in the milking parlour or the yard). This system has higher labour requirements and is not as suitable for automation as the stationary systems.

HYDROMECHANICAL WASTE REMOVAL

The first practical hydromechanical systems of slurry removal have been the flushing systems. The excrements are held up in the gutter for 3 days and are then flushed either by clean water or by recycling the supernatant from a sedimentation tank following the flushing operation.

All flushing systems require initial discharge at the head of the gutter of sufficient quantities of water to ensure an average velocity of at least 1 m/sec in the channel. This is the minimum bottom velocity necessary to scour manure solids on the floor.[1] However, the flow in the channel should also be of sufficient depth, preferably 5–8 cm, so that the flow has the capacity to carry with it the waste load. A bottom velocity of 1 m/sec is capable of moving gravel 30 mm in diameter.[2] In determining the necessary hydraulic parameters, equations for travelling or translatory waves, plus Manning's equation for open channel flow and equations for sediment-carrying capacity of the flow, should be consulted.

Manning's formula for velocity in the gutter is as follows:[3]

$$V = \frac{R^{2/3} S^{1/2}}{n}$$

where V = velocity (m/sec), R = hydraulic radius (m), S = slope (m/m) and n = roughness coefficient. The value for n for concrete varies from 0·012 for smooth concrete to 0·018 for rougher concrete. Hydraulic radius is the ratio of the wetted area of flow over the wetted perimeter. Wetted perimeter is, of course, a function of the depth of flow and the dimensions of the channel.

In practical systems, a slope of 1·5% in the gutter, which could vary along the length of the gutter, and 30–40 litres of flushing water per animal per day have been found adequate. In all cases, the depth of flow of the flushing stream should be no less than half the height of the manure piles in the gutter.

Flushing frequency is determined by the type of operation (more frequent when animals come and go as they please) and by what is necessary to prevent high odour levels in the building. For cattle wastes a frequency of anywhere from once in 4 hr to once in 24 hr is the recommended range. If the waste is stored in channels underneath slotted floors, the frequency may be extended to once in several days.

Flushing devices used in cattle operations are dosing siphons, tipping buckets or trap-door tanks.[4] The latter are the most common for dairy operations, while the first two are mainly used in swine operations.

The trap-door tank is a stationary tank installed on the floor at the top of the gutter channel. The trap-door is released by a float when the level of the water in the tank reaches a predesigned height. In dairy operations these tanks can be emptied manually because the flushing is not frequent. The tipping bucket is a slow-filling but vapid discharge device. It rotates around a shaft and tips when the water level reaches such a height as to change the centre of gravity of the bucket. Once emptied, the bucket returns to its original position due to the distribution of the weight of the bucket. Dosing siphons are similar to those used in sewerage lines.

In barns where no bedding is used, slurry may be collected in channels underneath slotted alleys or in storage pits. As long as the total solids content of the slurry does not exceed 11%, the slurry can be made to flow out without the addition of extra water. A disadvantage of this method is that the channel length cannot exceed 25–30 m. Slurry can usually remain in the channels for 3–4 weeks. These channels are emptied into pumping pits or into storage tanks by rapidly removing the dam at the end of the channel.

STORAGE

The volume of storage required depends on the quantities of waste generated, on the amount of bedding or water added, and on the number of days of storage required due to weather constraints or to suit the disposal system practised. In all cases, storage facilities should be located so that there is easy access to them not only from the animal buildings but also for the disposal operations. Special attention should be given to the traffic patterns for the various other operations on the feedlot so that waste handling operations do not interfere or are not interfered with by the feeding, milking or other operations around the feedlot. Surface runoff from rain should not be allowed to enter the storage area. Also, the storage facility should be designed to accommodate more than one method of handling in case of malfunctioning in the conventional system. Bedding volume is reduced by 50% during use in the barn.[4] It is best to fence out an area of 1 m^2 per cow for solid straw manure storage and to clean out the area weekly during the fly breeding season.[5]

Storage facilities for liquid manure handling may be below or above ground. Most commonly, storage tanks are underground. Recently, however, above-ground storage tanks have been gaining popularity due mainly to the lower costs of construction. Above-ground storage tanks are circular and constructed from concrete-coated steel. Costs savings are realized from the fact that above-ground tanks do not need strong covers and problems of water table can be minimized.

The dimensions of storage tanks are constrained not only by structural design considerations, but also by the effective lift of the pump, the agitating capacity of the agitator device, and the water table and soil properties. Depths of 3–4 m and diameter not exceeding 12 m are recommended for proper agitation and pumping. Specially designed and constructed silos for above-ground storage may be used with diameters of 10–30 m and over 6 m in height.[4] If at all possible, storage silos and underground tanks should be well ventilated to avoid excessive concentrations of noxious gases or explosive levels of methane gas. A gas trap should be installed between the sewer line and the storage tanks.

Waste storage tanks should be located at least 15 m from the milking parlour and no less than 30 m from any well which provides potable water to the feedlot. In determining such distances, regulatory agencies should be consulted as well as local soil experts.[4]

TRANSPORT OF WASTES

The method of transporting feedlot wastes from their point of generation or storage to disposal sites or to treatment units is determined by the solids content of the wastes. Generally, solid methods of transport are used for wastes with solids exceeding 20%.

Solid Waste Transport

The most common method of handling solid dairy manure containing straw is with the conventional-type manure spreader. The usual capacity of large conventional spreaders is 10–15 m^3 or about 20 tons. They are either tractor-drawn or mounted to lorries. Spreader mechanisms include paddles, flails and augers.[4]

Liquid Manure Transport

Dairy manure usually contains sufficient quantities of fibrous materials which must be chopped before pumping in order to avoid pump clogging. Manure pumps are usually equipped with chopper attachments yet can maintain a pumping rate of 4–10 m^3/min when total dynamic head does not exceed 5–6 m and manure slurries contain less than 20 % total solids.[1] These pumps are also equipped with a movable agitation pipe which can be swivelled in any desired direction so as to agitate the tank contents. This aperture is valved so that when the tank contents are uniformly mixed, the valve is closed and the tank contents are pumped to the tank wagon for transport to the field or to treatment units.

The types of pumps used are mainly PTO (power take-off) driven submerged centrifugal pumps. However, piston, helical rotor, diaphragm, etc., pumps can be used, provided solids content is low, preferably below 10 %, and a high rate of pumping is not essential for short lifts. If solids content is below 2·5 %, pipe head losses can be assumed to be equal to those of water.[6, 7] Therefore, standard formulae for pipe friction losses may be used.

Power requirements for pumping manure slurries with less than 2·5 % solids can be estimated by the following formula:[3]

$$\text{h.p.} = \frac{0{\cdot}786QH}{10^{10}E_p}$$

where h.p. = horsepower delivered to pump, Q = pump discharge (m^3/min), H = total head (m) and E_p = pump efficiency as a decimal fraction.

Efficiency for centrifugal pumps, E_p, may be 0·75 to 0·20. Usually a value of 0·60 may be assumed for well-operated pumps around feedlots. Internal combustion engines and electric-driven motors are the two most common ways of delivering power to pumps. The efficiency of gasoline engines is about 60 % for air-cooled and 70 % for water-cooled engines, and 80 % for diesel engines. The efficiency of electric motors varies from 85 % to 100 % depending on whether belt or direct drives are used.

Tank wagons for the transport of cattle slurries may vary in size from 2 m^3 to 40 m^3.[1] The latter are used in beef cattle feedlots of 10 000 cattle and more capacity. For dairy cattle operations, the size is in the range of 4–10 m^3. Although vacuum-type wagons have been used for cattle wastes, the most commonly used tank wagons are pump loaded and gravity unloading. Some tank wagons operate agitating during transport for uniform field transport.

The cost of tank tractor-drawn wagons may be estimated at \$500/m^3, and at \$1000–\$10 000/m^3 for 5–10 m^3 tank wagons and attached power units. Pump costs range from \$500 to \$1500 for low heads, but from \$6000 to \$15 000 with high-rate centrifugal pumps equipped with power units.[1]

NEW DEVELOPMENTS

At our Agricultural Engineering Research Institute we have developed a new system which combines the advantages of straw-less housing buildings with the advantages of solid disposal of the waste. Slurries from the building are removed by any of the conventional methods of liquid manure collection and removal. Once out of the building, the liquid waste is put through a bywash where cut straw is added. At the bywash outlets minerals and disinfectants are added, before the manure is put into heaps for storage as straw manure. In this way flies, odour nuisance and water pollution problems are minimized.

SYSTEM COSTS

It is impossible to provide specific costs for the various systems because of local design differences and variations in design and prices. Table I gives general relative costs for the most common methods of waste handling. These relative costs figures should be helpful in comparing the economic advantages of each method with those of ease of operation, maintenance and pollution prevention.

TABLE I
RELATIVE COST INDEX FOR DAIRY CATTLE MANURE SYSTEMS
(from ref. 5)

System	*Relative cost index*	
	Initial investment	*Operating cost*
Daily hauling (stanchion barns)	1·0	1·0
Stacking (stanchion barns)	1·7	0·7
Liquid manure (stanchion barns)	4·5	1·1
Stacking (free stall)	1·4	0·7
Liquid manure (free stall)	4·2	1·2
Liquid manure (free stall, slotted floor)	4·8	1·2

REFERENCES

1. Taiganides, E. P., *Livestock Waste Management*. Agricultural Engineering Dept., Ohio State University, Columbus, Ohio, 1973, 42 pp.
2. Jones, E. E. *et al.*, Improving water utilization efficiency in automatic hydraulic waste removal. In: *Livestock Waste Management and Pollution Abatement*, pp. 154–8. ASAE, St Joseph, Mich., 1971.
3. Schwab, G. O. *et al.*, *Soil and Water Conservation Engineering*, 2nd ed. John Wiley, London, 1966.
4. Midwest Plan Service, *Livestock Waste Facilities Handbook*. MWPS, Ames, Iowa, 1975.
5. Midwest Plan Service, *Dairy Housing and Equipment Handbook*. MWPS, Ames, Iowa, 1971.
6. Hart, S. A., Moore, J. A. and Hale, W. F., Pumping manure slurries. In: *Management of Farm Animal Wastes*, pp. 34–8. ASAE, St Joseph, Mich., 1966.
7. Staley, L. M., Bulley, N. R. and Windt, T. A., Pumping characteristics, biological and chemical properties of dairy manure slurries. In: *Livestock Waste Management and Pollution Abatement*. pp. 142–5. ASAE, St Joseph, Mich., 1971.

14

Collection, Storage and Transport of Swine Wastes

T. JELÍNEK

Swine Breeding Research Institute, Kostelec nad Orlicí, ČSSR

Waste handling is one of the major problems resulting from the industrialization of swine production. The main factors affecting systems for the removal, transport, storage and utilization of the wastes are public health considerations and technical feasibility requirements, including economic factors.

When determining the term 'production of manure', it is necessary to differentiate between the production of waste which is faeces and urine, and wastewater which contains additional water.

Production of wastes (faeces and urine) from young animals and pigs being fattened is given in Table I. These data indicate that young pigs of

TABLE I

AVERAGE VALUES OF WASTE PRODUCTION AND PROPERTIES FROM SWINE ANIMALS OF DIFFERENT LIVE WEIGHT

Parameter	*Unit*	*Animal weight (kg/animal)*			
		5–15	16–30	31–65	66–100
Total excrement	% TLW	7·2	8·5	6·3	4·9
Liquid/solids ratio	%	51·5	54·6	51·1	48
Dry matter in faeces	%	32·5	28·6	28·2	26
Dry matter in urine	%	4·0	4·0	4·4	5·0
TS in total excrement	%	12·8	12·2	11·9	11·5

5–15 kg live weight generate almost 50 % more excrement per unit of live weight than fattened pigs weighing 66–100 kg. However, there are small differences in the properties of the waste generated by young or fattened pigs.

The production of total wastewaters from seven modern large swine feedlots is given in Table II. Waste produced per mean weight of the pigs in the feedlots studied ranges between 5·5 and 10·5 kg of waste per day per pig

TABLE II

WASTEWATER QUANTITIES FROM LARGE CONFINEMENT SWINE FEEDLOTS

Feedlot No.	*Pig-days*	*Wastewater (kg/100 kg LW)*	*Dilution water*[a]
1	381 294	16·3	61
2	1 406 422	10·2	31
3	657 730	12·3	49
4	331 405	10·8	37
5	505 817	11·8	38
6	599 188	15·2	60
7	1 289 180	10·9	39

[a] Added water is calculated as the difference between the measured wastewater quantity and the quantity of excrements computed on the basis of Table I.

fattened. However, the mean wastewater production is 7·4 litres/day/pig, or 10·7 % of pig live weight per day of the pigs. Average values given in the literature for swine feedlots range from 5–10 % of live weight[1] to 20 %.[2] Added water amounts to approximately 50 % of the total wastewater generated on the feedlot. The range is from 31 % to 61 %.

To reduce excessive water content in the excrements, water feeders and methods of cleaning the pens with a minimum use of water are recommended. A water feeder which operates by having the pigs suck water from a pressure-free piping, by means of a specially adapted nozzle, reduced water quantities in wastewaters by 30 % per pig. The water feeder is shown in Fig. 1.

FIG. 1. Water feeder (Czechoslovak patent).

Pen cleaning is mostly carried out using high-pressure hoses discharging hot water to which disinfectants have been added. The use of high-pressure hoses when compared with flushing with pressure-free water combined with mechanical cleaning reduced water use by 20% and yet resulted in better cleaning from the standpoint of health and aesthetics.

The average total solids content in wastewaters from 265 samples taken in the course of one year from the waste canals in 12 pig fattening farms was

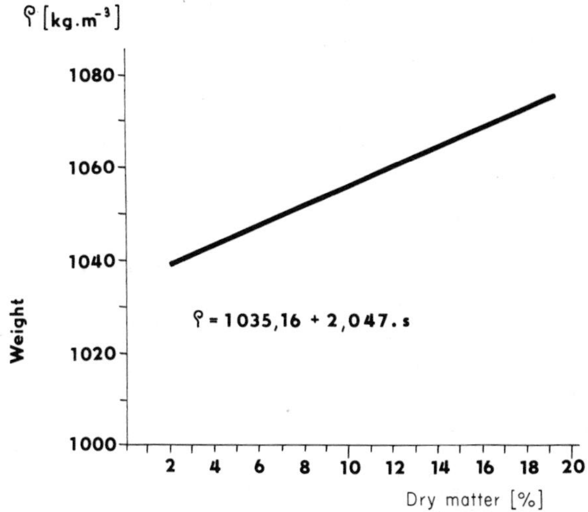

FIG. 2. Correlation between solids content and density.

8·4%, with the range being ±4·5%. The relatively high fluctuation from the mean value indicates variability in the amount of water added during cleaning.

Correlations between total solids content and density, solids content and viscosity, and viscosity and limiting viscosity stress, developed from samples from 12 modern swine feedlots, indicate a direct proportional correlation between solids content and these parameters, as is shown in Figs. 2, 3 and 4.

In housing units where no bedding or straw is used, mechanical floor scrapers of the type shown in Figs. 5 and 6 may be used. These scrapers operate in alleys of 0·5–3 m in width and 200–300 m in length. They are made of steel, wood, cast iron or rubber. Speeds may vary from 3 m/min to 8 m/min. Power requirements are 1–3 kW. The double-arm scrapers have a larger capacity than the single-arm scraper, but also cost 15–20% more and have higher construction-associated costs. Recently, attachments have been added to these mechanical scrapers (see Fig. 6) to speed up the time and distance the scraper needs to open up to the 'working' position.

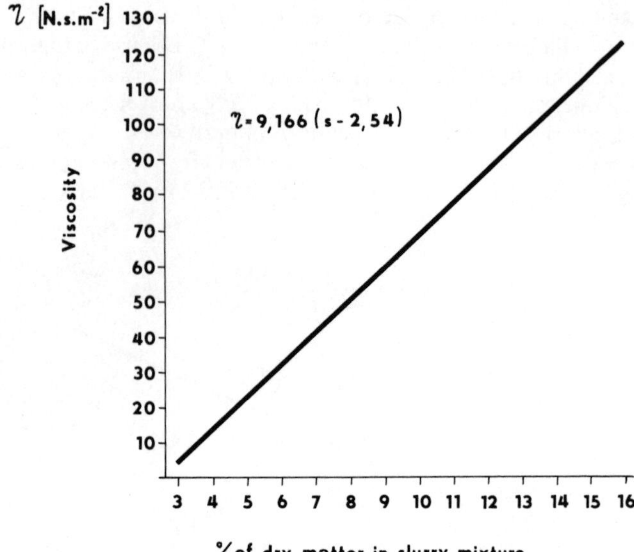

FIG. 3. Correlation between solids content and viscosity.

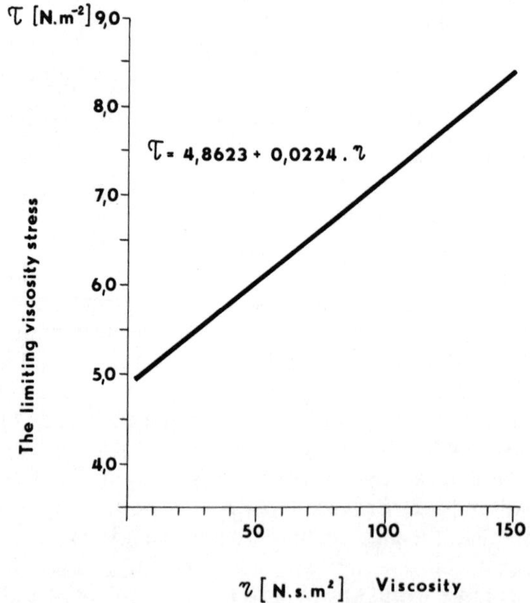

FIG. 4. Correlation between viscosity and limiting viscosity stress.

FIG. 5. Single-arm scraper.

Removal of liquid manure by gravity flow can be accomplished either by collecting waste in pits for 3–5 weeks and then releasing it or by continuous overflow canals in which a 15 cm gate is installed at the end of the channel and water is added to a depth of 15 cm, prior to operation. Continuous overflow canals should be narrow. From collection channels, gravity flows can reach $1{\cdot}2\ m^3/sec$, as is shown in Figs. 7 and 8, during the first 6 sec of opening the gate.

With storage of 5–25 days, the pH of the liquid decreases at the rate given in the following equation:

$$\mathrm{pH} = 7{\cdot}47 - 0{\cdot}03t$$

where t = time in days.

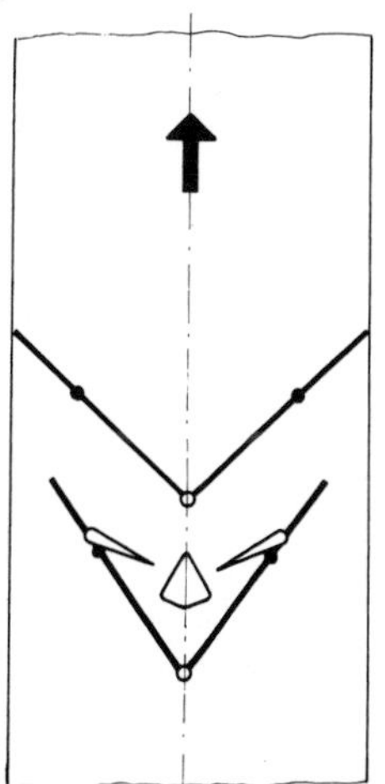

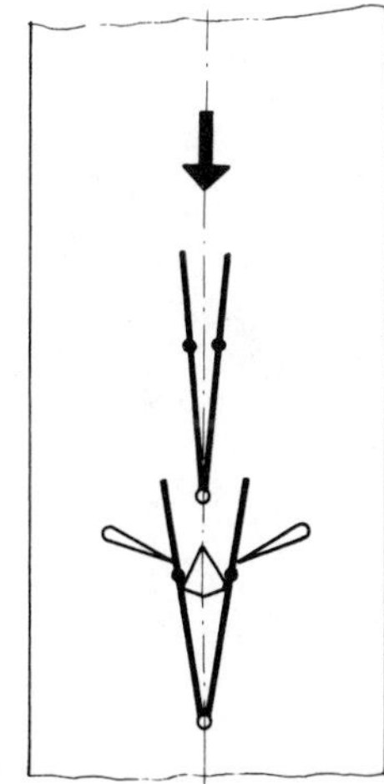

FIG. 6. Forced opening of double-arm scraper.

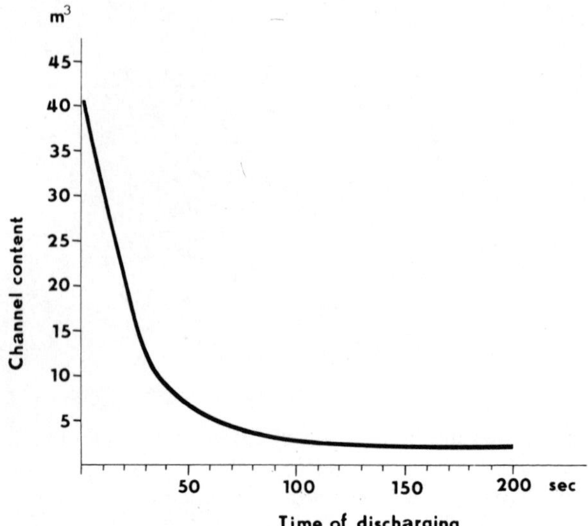

Fig. 7. Emptying the collection channel.

Flushing is an effective way of removing waste collected in gutters or in canals under slotted floors. A flushing system is shown in Fig. 9. Design principles and formulae for flushing systems, including water tanks, gutter slopes and pump design, are similar to those for cattle waste flushing, and have already been discussed.[6] In most countries, flushing water must be clean potable water. If liquids from lagoons are used, prior approval by

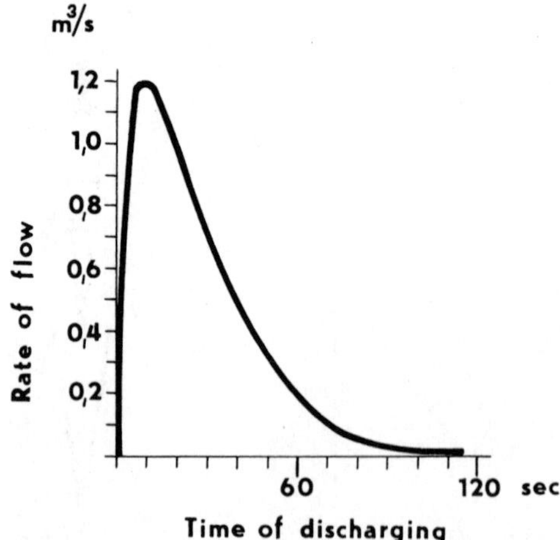

Fig. 8. Characteristics of passage on emptying the collection channel.

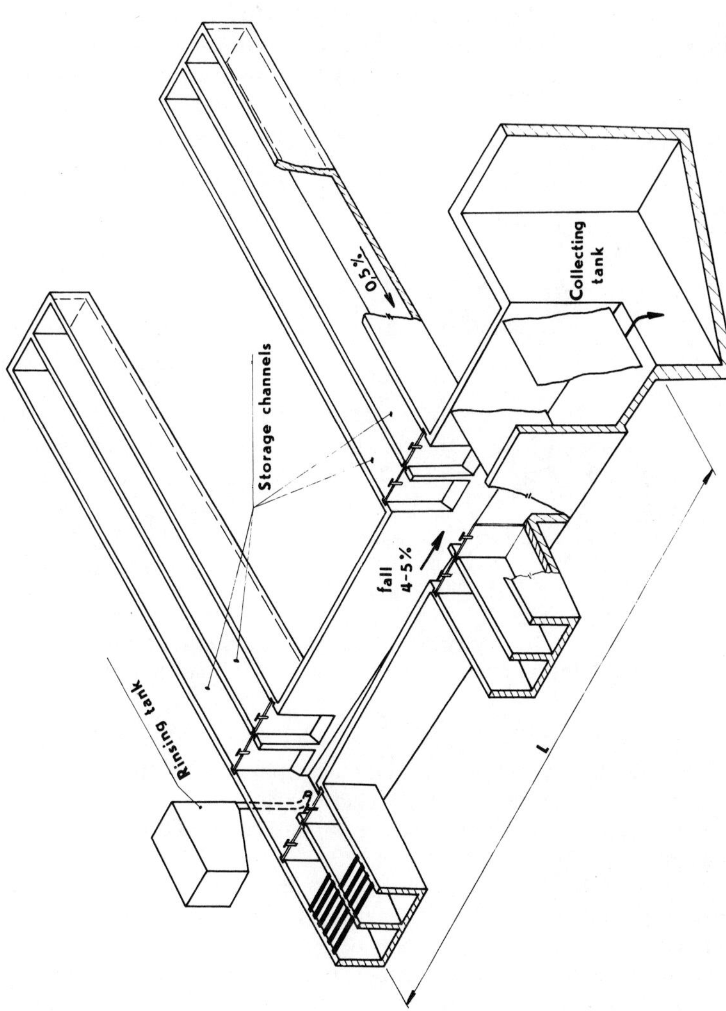

FIG. 9. Continuous gravity flow flushing system.

veterinary health service offices is essential. However, in several feedlots ranging in capacity from 500 to 1000 pigs, lagoon or treated wastewater has been used for flushing without any disease outbreak problems.[7, 8]

To keep solids from settling in pipes or canals, minimum velocities recommended are 0·8–1·2 m/sec. Care should be taken to keep pipes from freezing during winter operations; they should have adequate vents and, to avoid excessive friction losses, pipe diameter should be at least 8 cm.

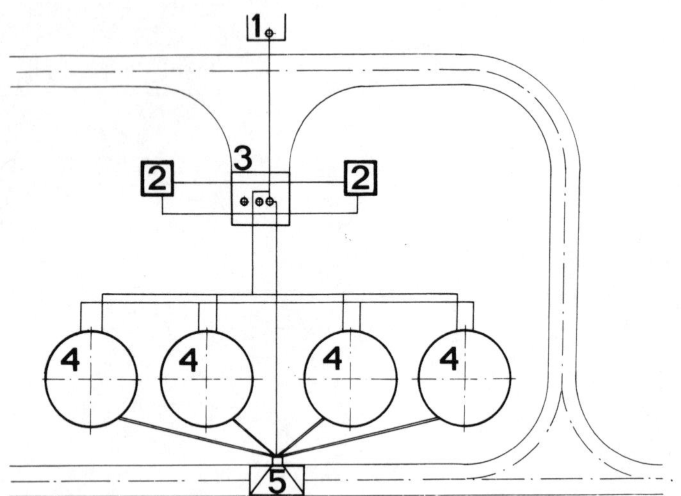

FIG. 10. Waste storage facility with surface tanks. (1) Collection tank. (2) Primary ditches. (3) Pumping plant. (4) Storage tanks. (5) Loading platform.

Recommended storage capacity for swine wastewaters is 3 months, including a mandatory storage for 7 days in a separate pit without new waste entering the pit during that week to meet the required time for pathogen control. The sump from which manure is pumped should have a 1–2 day storage capacity.

Table III and Fig. 10 show dimensions and layout of waste storage facilities for swine feedlots of 5400, 7000 and 10 200 pigs capacity. Collector pits of 50–100 m^3 capacity should be made of reinforced concrete, rectangular in shape, and should be covered with prefabricated panels. In feedlots of 5000–10 000 pigs capacity, the sedimentation pits should be of 400–650 m^3 capacity and should have two compartments, with one of them being capable of holding 7 days of wastewater, as is required by veterinary regulations.

Mixing of the liquid waste before pumping and/or transport to the field can be accomplished mechanically, hydraulically or pneumatically.

TABLE III

CAPACITIES OF STORAGE FACILITIES FOR VARIOUS SIZES OF SWINE FEEDLOTS

Number of pigs	*Production of waste (m^3)*			*Capacity of tanks (m^3)*		
	1 day	*7 days*	*90 days*	*Collector pit*	*Sedimentation pit*	*Storage tank*
5 400	48·6	340·5	4 378	50	400	4 500
7 000	64·6	·452·1	5 813	70	500	6 000
10 200	92·1	644·4	8 285	100	650	9 000

Stationary or mobile propeller agitators are common methods of agitation. Chopper pumps and compressed air are also used. Agitation may take from a few minutes to 3 hr, depending on the size of the tank and the capacity of the agitating equipment. Compressed air should be released at 1·5–2 kP/cm^2 for 15–30 min daily with blowers pumping at the rate of 150–300 m^3/hr. Large blowers are too noisy, sometimes exceeding 100 dB noise levels.

Both underground and above-ground tanks may be used. The trend is towards above-ground tanks because they are more economical, can be made leak-proof more readily, and are easier to construct than underground tanks, as is shown in Fig. 11. Construction costs per m^3 of storage space decrease as the size of the tank increases. Costs also vary with the construction materials used.

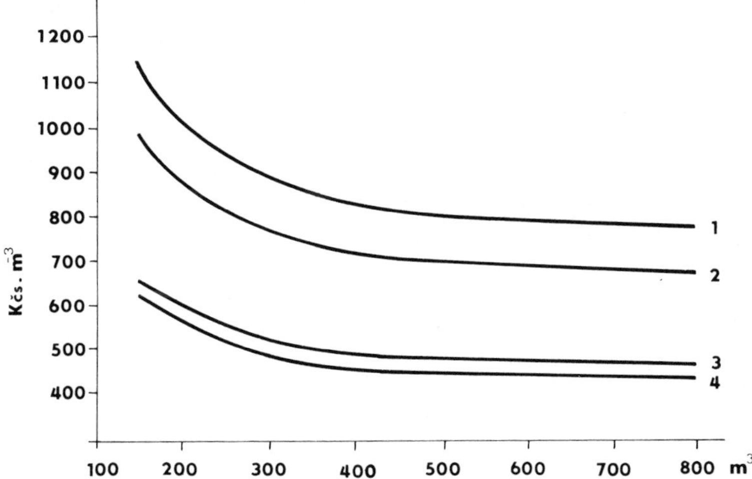

FIG. 11. Investment costs for different collection tanks. (1) Concrete underground tank. (2) Steel surface tank. (3) Wooden surface tank. (4) Concrete surface tank.

Transport systems include tank wagons varying in size from 3 to 10 m^3, tractor-drawn or mounted on lorries. Both vacuum and mechanically loaded wagons are suitable for swine wastewaters.

REFERENCES

1. Hrstka, J. *et al.*, *Liquid Manure Tank for Underground and Above-ground Storage*. Agroprojekt Study, České Budějovice, ČSSR, 1974.
2. Hörnig, G., *Druckverlusttabelle für das Fördern von Rinder und Schweinegülle in Druckrohrleitungen*. Institut für Mechanisierung, Potsdam-Bornim, German Democratic Republic, 1971.
3. Jelínek, T., Evaluation of the gravity removal of excrements from swine fattening feedlots. Research Report No. C11-329-003-01-03, Pig Breeding Research Institute, Kostelec nad Orlicí, ČSSR, 1972.
4. Jelínek, T. and Vítek, M., Daily production of excrements and dry matter content from pigs of different weight. *Živočišná Výroba*, **17** (1972) 781–5.
5. Velebil, M. *et al.*, Methods of pumping and storing excrements. Research Report No. ZT-0.1/3, Research Institute of Agricultural Machinery, Prague, ČSSR, 1972.
6. Velebil, M., Collection, storage and transport of cattle wastes (this volume, pp. 157–64).
7. Smith, R. J., Hazen, T. E. and Miner, J. R., Manure management in a 700-head swine finishing building. In: *Livestock Waste Management and Pollution Abatement*, pp. 149–53. ASAE, St Joseph, Mich., 1971.
8. Taiganides, E. P. and White, R. K., Automated recycle system for livestock waste treatment. In: *Symposium on Processing Agricultural and Municipal Wastes*, ed. G. Inglett, pp. 75–83. AVI Publishing Co., Westport, Conn., 1973.

15

Collection, Storage and Transport of Poultry Wastes

V. PETER

Poultry Research Institute, Ivánka pri Dunaji, ČSSR

and

F. ZACHARDA

Agricultural Technology Research Institute, Rovinka, ČSSR

INTRODUCTION

Inadequate solutions to the problems of waste handling from large poultry farms are restricting the expansion of poultry production in some countries. Poultry production as practised today is specialized, is concentrated, uses high-performance birds which are fed rations of high nutritive value, and requires special, skilled management. The management of the wastes from these large poultry units requires the use of modern technology in order to meet high standards of sanitation, disease and parasite control, conservation and proper utilization of waste products, and, furthermore, to protect the quality of the environment as is stipulated by environmental laws and regulations enacted since 1970 in almost every country in the world.

All commercial poultry production is carried out in confinement. Almost 100% of hens and broilers are raised in housed feedlots, while commercial ducks and turkeys may be raised in open feedlots. The main types of housing and major operations involved with each type of housing have already been delineated previously.[1] This paper examines waste handling techniques for commercial egg-laying poultry operations. In this paper, the term 'poultry wastes' refers hereafter to manures and wastewaters from large, modern commercial egg-laying operations only.

COLLECTION METHODS

Waste handling methods which have been practised widely may be classified as dry and liquid systems. Dry systems have the advantages of low quantities to be handled, control of water pollution potential, abatement of odours and less frequent waste removal. Major disadvantages include the need for good management, difficulties in automating the removal process, and also the fact that dust problems may arise. Of course, dry poultry manure can easily be dehydrated and/or composted for utilization. It is difficult to maintain dry systems unless excellent control of overflow from bird waterers is exercised. Thus, dry systems of waste handling limit the type of waterers which may be used. Overflow from waterers may add as much as 200 times more water than the layers excrete in their manure.[21] Also, dry systems have special ventilation requirements which would have to be engineered to provide good air circulation over the manure.[2]

Liquid systems have advantages which include less management attention than dry systems, ease of automation of transport and treatment, and thus they may require less labour and can be economical if adequate land for disposal is accessible and particularly if lagoons can be tolerated at the site. Major disadvantages include odour problems, water pollution potential, and larger quantities to be handled than in dry systems. During the last ten years new liquid handling equipment has been developed which is now readily available to the large number of farms which are using liquid systems. This equipment includes chopper pumps for feathers,[3] self-loading tank wagons, soil-injecting tank wagons, scraper mechanisms, and many commercially marketed pumps specially designed for liquid manures.

DRY SYSTEMS

Layers excrete 100–150 g of manure per day, of which approximately 25 % is solid matter. Such excreta require slurry handling systems rather than dry handling, which is suitable for wastes with over 30 % total solids content.[4] The handling system must therefore be designed to reduce the moisture content of the fresh excreted manure from 75 % to at least 65 % and preferably to 50 %. Manure above 30 % moisture cannot be stored without further aeration, if malodours are to be prevented.[5] Neither can manure be dried inside the poultry house much below 30 % moisture, without creating dust problems in mechanically ventilated houses.

Open Housing

In warm climates, where open houses of caged layers are used, manure can be dried successfully with good natural air circulation. In such systems, coning of the manure enhances drying by exposing a large surface area to the air[6] and by absorbing some of the moisture of the fresh droppings by the dry base of the cone.

Slat System

In mechanically ventilated, totally enclosed houses, the use of framed timber slats in two levels below the cages, the lower row being offset, improves the rate and degree of drying considerably.[7] Manure falling from the layers is retained on wooden slats. In such a system, drying efficiency results from the formation of tall cones or columns, from the fact that air is exhausted below the slats through windproof outlets in the pit walls, thus permitting ventilation air to pass over the columns, and from the heat emitted from the metabolic activities of the birds. The slat frames are pivoted on central posts and supported from a catwalk by cables. At the end of the laying period the slat frames are swung down, depositing the dried manure on the pit floor. Manure removal can be accomplished by tractor with front loader and/or by scrapers. Obviously, this slat system is suitable to high-rise (deep-pit) poultry buildings.

Narrow slats accelerate drying at the early stages, but in such cases the gaps between slats must also be narrower, thus resulting in 'bridging' of manure between slats, which retards the drying process. The best results are achieved with slats 10–15 cm in width.[7] With 10 cm slats, manure removal should be done every 6 months. However, if manure is to accumulate for a year, slats 15 cm wide should be used. Manure moisture content can be reduced to 12–15 % within 6 months with the slat system while maintaining ammonia and odour levels at concentrations half of those found in non-slatted deep-pit poultry houses.[7]

In-House Drying

Drying underneath cages may be accelerated by the use of fans placed every 10 m along the side of the building and by agitating the droppings with a stirring device similar to a spike-tooth harrow.[5] With such systems, moisture may be reduced to less than 45 % during winter when ventilation rates are low (less than 1 m^3/hr/bird). From April to September, manure moisture is reduced to less than 30 %. Thus, the amount of waste to be removed from the building is reduced by a factor of 3. In other words, instead of handling 3 tonnes, only 1 tonne of manure is removed. Air velocities of 150–300 m/min over the droppings are effective.[5]

Similar in-house drying systems utilizing ventilation fans to circulate air over cones of poultry droppings have been tested.[8, 9] Special systems tested included mechanical devices to turn over droppings without forced air ventilation, forced air systems, etc.[9] All such systems reduced moisture levels to below 60 %.

Dehydration

Once moisture is reduced to below 60 %, then dehydration with heated driers becomes economically attractive.[5, 10, 11] It costs twice as much to dehydrate manure when the initial moisture is 75 % than when it is 60 %.[12] Cost is five times greater if the initial moisture is 90 %. Therefore, it is essential that no water be added if the manure is to be dried. Manure can be dried to 10% moisture. At such a moisture level, the volume of the original droppings has been reduced by a factor of 7. However, in view of the fact that energy costs may account for from 36% to 73% of the total cost of drying manure,[11, 13] in-house drying is more attractive than heated-air dehydration. On the other hand, dried poultry manure has monetary value because it may be used in animal feed rations[11, 14, 15] or marketed in retail stores as garden or pot plant fertilizer.[16, 17]

Removal Equipment

Equipment for the removal of droppings varies with the method of drying and amount of storage. It may consist of a small garden-type tractor with a side arm for removing wastes accumulating below cages. For deep pits, farm tractors equipped with front-end loaders are often used. For shallow pits, many types of mechanical scrapers and conveyor cleaners are used.

For field disposal, conventional or flail-type manure spreaders may be used. For transportation to long distances lorry trucks should be used.

LIQUID SYSTEMS

In liquid handling systems, bird droppings and overflow from waterers and from other operations such as egg washing, etc., may be collected in shallow or deep pits underneath the cages. If the pits are shallow, a covered tank of sufficient capacity to store wastes and wastewaters for at least 100 days, in cold climates, must be provided. Furthermore, since stratification of settleable, floatable and suspended solids occurs in such storage tanks, equipment to agitate the tank contents before pumping is necessary. The typical PTO-driven liquid manure pumps equipped with a chopper may be used to agitate the tank contents and to lift the liquid to a tank wagon for transport to the field. Unless large volumes of water are added and feathers are either removed, chopped up, or prevented from entering into the storage tank, direct pumping for irrigation could result in severe clogging problems in the pipes and nozzles.

Dams

Several techniques can be used to move the waste/wastewaters from the shallow channel underneath the cages to the storage tank. One such technique is the floating-dam flooding system. The dam may be constructed with lumber and covered with rubber sheets. It is constructed to be as wide as the channel but shorter in height, so that overflows go into the channel itself rather than on to the sides of the pit channel. The dam is placed at the upper end of the channel, and water is pumped behind it. As the liquid depth behind the dam increases, it pushes the dam along the channel, scraping solids as it moves. If the amount of the waste in front of the dam is high, then the dam stops which adds more height to the water level behind the dam, since water is pumped at a constant rate. This additional water adds more force to the dam, and spillage over the top of the dam dilutes the wastes and makes them easier to move as well as providing lubrication for the dam. Thus the wastes are pushed into a cross-conveyor channel or into a pit, and the dam is returned to the upper end of the channel.

Flushing Gutter

Another technique for liquid manure removal is the flushing-gutter system which has been successfully operated in swine buildings,[18,19] using tipping buckets, dosing siphons, trap-door tanks, etc.[20] In poultry units, flushing is used to transport the wastes hydraulically in the shallow pit/channel underneath the bird cages.[21] The design of flushing gutters have been outlined previously.[22] The slope of the channel and the quantity of flushing water should be large enough to give a velocity of 1 m/sec in the channel and a flow depth of 5–8 cm.

Treated wastewater may be re-used for flushing the channels beneath the pits. This operation can be carried out automatically by pumping the treated effluent into flushing siphons[18–20] located at the upstream end of the channels. Wastewater flushed out of the buildings may be treated minimally before recycling, either in an anaerobic lagoon[21] or in aerobic systems as are described elsewhere in this book.[23] More extensive treatment may include surface aeration of faculative lagoons, oxidation ditches, or other extended aeration systems.

Transport Equipment

After a slow and reluctant start, many farm equipment manufacturers invested in the development of liquid manure equipment and are now marketing special chopper pumps, scrapers, conveyor belts and tank wagons. The latest development in tank wagons are those equipped with soil injection implements which deposit liquid manure below the soil surface and thus control both odours and fly breeding. Tank wagons range in capacity from 2 m^3 to as much as 40 m^3. The usual range is 2 m^3 to 10 m^3.

Storage

Storage pits may be built above or below ground. Above-ground tanks are becoming somewhat more popular in recent years and are constructed of poured or pre-stressed concrete. However, most poultry units use underground pits to which wastes may be scraped mechanically or hydraulically or with conveyors. Storage time must be designed to suit local climatic conditions, cropping patterns and land availability. Nutrient losses, particularly nitrogen volatilization, take place during storage. Such losses are desirable if the waste is to be aerobically treated subsequent to storage,[24] but constitute an important loss if the wastes are to be utilized as crop fertilizers. The extent of these losses cannot be estimated. Nitrogen losses may range from 30 % to 65 % of original nitrogen content in deep-pit storage, to as much as 70 % to 90 % in aerated systems.[25]

Treatment of poultry wastes is expensive. Therefore, the development of durable, inexpensive equipment for the handling, transport and uniform field application of liquid and solid wastes will go a long way towards alleviating present problems with livestock wastes.

REFERENCES

1. Runov, B. A., Animal feedlots: development, trends, problems (this volume, pp. 11–22).
2. Esmay, M. L., Layout and design of animal feedlot structures and equipment (this volume, pp. 49–72).
3. VanEe, G. R., Chopper pump. US Patent 3,807,644, 30 Apr. 1974.
4. Taiganides, E. P., Bio-engineering properties of feedlot wastes (this volume, pp. 131–53).
5. Bressler, G. O. and Bergman, E. L., Solving the poultry manure problems economically through dehydration. In: *Livestock Waste Management and Pollution Abatement*, pp. 81–4. ASAE, St Joseph, Mich., 1971.
6. Ostrander, C. E., Techniques that are solving pollution problems for poultrymen. In: *Managing Livestock Wastes*, pp. 71–3. ASAE, St Joseph, Mich., 1975.
7. Elson, H. A. and King, A. W. M., In-house manure drying: the slat system. In: *Managing Livestock Wastes*, pp. 82–3, 92. ASAE, St Joseph, Mich., 1975.
8. Esmay, M. L. *et al.*, In-house handling and dehydration of poultry manure from a caged layer operation: a project review. In: *Managing Livestock Wastes*, pp. 468–72, 477. ASAE, St Joseph, Mich., 1975.
9. Sobel, A. T., Undercage drying of laying hen manure. In: *Proceedings of Waste Management Conference*, pp. 187–200. Cornell University, Ithaca, NY, 1972.
10. Esmay, M. L., Dehydration systems for feedlot wastes (this volume, pp. 197–211).
11. Koskuba, K., Design of poultry manure drying system for 155 000-layers egg factory. In: *Managing Livestock Wastes*, pp. 78–82. ASAE, St Joseph, Mich., 1975.
12. Brian, D., Moisture: a key to much-drying profits. *Poultry Industry*, **3** (1975) 12–13.
13. Akers, J. B., Harrison, B. T. and Miller, J. M., Drying of poultry manure: an economic and technical feasibility study. In: *Managing Livestock Wastes*, pp. 473–7. ASAE, St Joseph, Mich., 1975.
14. Peter, V., Technológia a technologické zariadenia na niektorých hydinárskych farmách v USA. In: *Pokroky hydinárskej vedy a praxe*, pp. 85–96. ČVTS-Dum Techniky, Brno, ČSSR, 1975.
15. Hodgetts, B., The effects of including dried poultry waste in the feed of laying hens. In: *Livestock Waste Management and Pollution Abatement*, pp. 311–13. ASAE, St Joseph, Mich., 1971.

16. Scholz, H. G., Systems for the dehydration of livestock wastes: a technical and economic review. In: *Livestock Waste Management and Pollution Abatement*, pp. 27–9. ASAE, St Joseph, Mich., 1971.
17. Jordan, H. C., Marketing converted poultry manure. In: *Livestock Waste Management and Pollution Abatement*, pp. 197–8. ASAE, St Joseph, Mich., 1971.
18. Taiganides, E. P. and White, R. K., Performance of an automated waste treatment and recycle system. In: *Managing Livestock Wastes*, pp. 564–7, 575. ASAE, St Joseph, Mich., 1975.
19. Smith, R. J., Hazen, T. E. and Miner, J. R., Manure management in a 700-head swine-finishing building: two approaches using renovated waste water. In: *Livestock Waste Management and Pollution Abatement*, pp. 149–53. ASAE, St Joseph, Mich., 1971.
20. Midwest Plan Service, *Livestock Waste Facilities*. Iowa State University, Ames, Iowa, 1975.
21. Fehr, R. Z. and Smith, R. J., Management of a flushing-gutter manure-removal system to improve atmospheric quality in housing for laying hens. In: *Managing Livestock Wastes*, pp. 437–40. ASAE, St Joseph, Mich., 1975.
22. Velebil, M., Collection, storage and transport of cattle wastes (this volume, pp. 157–64).
23. White, R. K., Lagoon systems for animal wastes (this volume, pp. 213–32).
24. Loehr, R. C., Nutrient control applicable to animal wastes (this volume, pp. 253–69).
25. Vanderholm, D. H., Nutrient losses from livestock waste during storage, treatment, and handling. In: *Managing Livestock Wastes*, pp. 282–5. ASAE, St Joseph, Mich., 1975.

16

Solids Separation and Dewatering

K. BLAHA

Agricultural Engineering Research Institute, Prague, ČSSR

INTRODUCTION

Separation of solids from the insoluble fraction of animal wastewaters is becoming more important as new methods of processing wastes are developed. Solids separation first became important when biological treatment of animal wastes became necessary in modern large feedlots. Separation is also necessary as a pretreatment of wastes before thermal dehydration, or when application is by conventional irrigation systems, or when wastes are transported hydraulically and the solids may cause clogging or other operational problems.

Currently, a broad range of mechanical separators are being built and studied, including different types of decanting and desludging centrifuges, automatic filter presses, vibrating separators and gravitational straining screens. Gradually, this list of suitable equipment for feedlot wastes will be narrowed down as more experience is gained, particularly from long-term operations at large-capacity feedlots. However, due to the general character of agricultural production, it is not likely that any one separator will be used in all cases. The time-tested method of solids separation by settling in sedimentation tanks is and will be used in animal feedlot wastewaters, particularly when highly dilute wastewaters are treated.

Advantages of solid–liquid separation schemes include: volume reduction, concentration of solids for separate treatment or re-use, reduction or complete removal of settleable solids, reduction of the pollution potential of wastewater, prevention of clogging, ease of hydraulic handling, moderation of odours, etc. Major disadvantages are high energy and investment costs for mechanical devices, as well as labour requirements for maintenance and operation.

Solids separators may be broadly classified as mechanical, thermal, physical and bio-physical. However, there are two basic modes of solids separation. In one mode, differences in the density of the solids and the liquid are used, such as in sedimentation and centrifuging. Separation may also be effected by the physical dimensions of the solids, such as diameter, shape and size, as is done in screening and filtering.[1] In thermal separation it is the liquid that is separated from the solids, the separation being effected by evaporation of the liquid fraction.

Another classification of separators is on the basis of whether the separator device dewaters, clarifies the liquid, or classifies the solids. In feedlot wastewater applications, separation of solids, dewatering and clarification are the usual goals. Classification would be important only if extensive treatment is to follow in which the shape and size of the particles are important. Production, for example, of single-cell protein from wastewaters would require classifying-type separators.

Efficiency of separators is measured as amount of solids retained in the solid separated fraction compared with the quantity of solids in the influent. The liquid effluent from the screening operation is called the fugate liquid. In the food industry, where solid–liquid separation is more common and needs to be more precise, special efficiency formulae have been developed.[2]

CENTRIFUGAL SEPARATORS

Centrifuges produced for use in the chemical and food industries are the most commonly used separators. Horizontal-type centrifuges have a rotating cylindrical, conical drum with built-in screws rotating at different speeds than the drum itself. The solids are entrained along the axis of the drum and the screw transports them out of the machine. These centrifuges are designed for specific concentrations of incoming suspension, typically a concentrated suspension with a medium or coarse dispersion of the solid phase.

Centrifuges designed for separation of suspensions with a low concentration of highly dispersed solid phase require a long rotor and the highest possible rotational velocities. Also, research in the Soviet Union has shown that better performance can be achieved by increasing the length to diameter (l/d) ratio of the drum.[3] A high-speed centrifuge with an l/d ratio >2 is effective in separating highly dispersed suspensions of low concentration, while a medium-speed centrifuge with an l/d ratio <2 is suitable for separating concentrated suspension with medium to coarse dispersion of the solid phase.

There are also vertical types of desludging centrifuges, based on the so-called 'plate' system. At preset time intervals, operational pressure drops rapidly so that the rotating plate of the drum sags and opens a restricted space. This causes evacuation of the separated solid particles without interrupting the rotation of the machine. This type of centrifuge is unsuitable for feedlot wastewaters because of the high time requirements for maintenance and cleaning.

A variation of this design are centrifuges with two cones. They consist of two plates which are closed in the course of filling and centrifuging. The separated material, *i.e.* the solid particles, are hurled against the sides of the cones where, in view of the incoming new material, they form a layer from the tip to the centre of the internal space of the cones. The liquid phase concentrates along the longitudinal axis and leaves the space by way of lateral outlets. As soon as the solid material reaches the outlet, supply of the raw material is stopped, both cones are shifted by means of a hydraulic system, and the solid phase is automatically ejected from the machine.

Two-stage centrifuging seems to be best suited for solids separation of wastewaters and for thickening solids for refeeding or other uses.

Solids removal efficiencies vary with the type of machine, centrifugal forces applied, time of centrifugation, solids concentration of the entering slurry, type of solids, temperature, and other factors.[4, 5] Centrifuges with capacities ranging from less than 1 to more than 11 m^3/h have been operated around feedlots. Power consumption varies, but is in the range of 4 kWh/m^3 of fugate. Using waste with total solids content of 2–7 %, the centrifuged solids can concentrate up to 27 % total solids content, while the fugate solids could reduce to less than 1·8 %.[3] As the solids content of the influent slurry is increased, the amount of solids and solids concentration in the centrifuged solid fraction increases.[4] Concentrations of as high as 40 % solids in the centrifuged solids can be reached with low-capacity centrifuges.

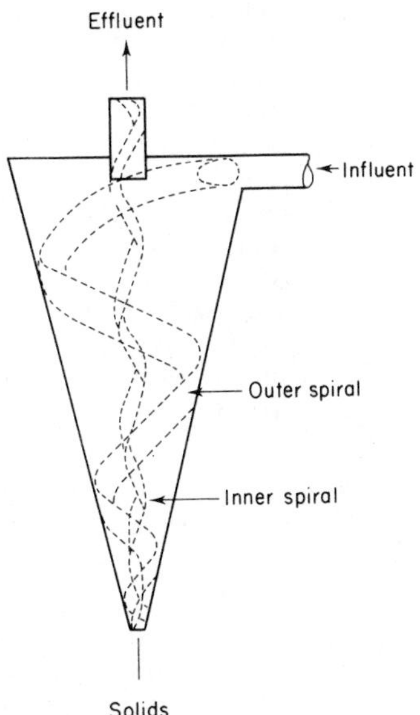

FIG. 1. Solids separation mechanisms in a liquid cyclone (from ref. 6).

TABLE I

SOLIDS REMOVAL AND CONCENTRATION PERFORMANCE OF LIQUID CYCLONE (PERCENTAGE OF INDICATED PARAMETER RETAINED IN UNDERFLOW DISCHARGE WITH INFLUENT CONCENTRATION OF 0·5% TSS)
(from ref. 7)

Parameter	*Units*	*Underflow nozzle diameter (cm)*		
		0·32	0·47	0·64
68 litres/min flow rate and 1·4 kg/cm² pressure drop:				
Flow volume	% Inflow	2·2	10·4	10·0
TS volume	% Inflow	20·3	14·7	26·1
TS concentration	% Wet basis	7·5	3·0	3·1
TSS volume	% Inflow	11·3	4·5	30·2
TSS concentration	% Wet basis	7·1	0·3	2·6
TVS volume	% Inflow	5·5	2·3	2·3
88 litres/min flow rate and 2·8 kg/cm² pressure drop:				
Flow volume	% Inflow	2·1	4·8	11·2
TS volume	% Inflow	26·5	14·1	24·1
TS concentration	% Wet basis	8·4	2·4	1·6
TSS volume	% Inflow	38·8	20·7	34·4
TSS concentration	% Wet basis	8·2	2·2	1·3
TVS volume	% Inflow	5·8	1·7	0·7
111 litres/min flow rate and 4·2 kg/cm² pressure drop:				
Flow volume	% Inflow	0·9	5·1	9·6
TS volume	% Inflow	10·5	23·5	27·0
TS concentration	% Wet basis	9·0	3·4	2·0
TSS volume	% Inflow	15·7	34·0	40·7
TSS concentration	% Wet basis	9·1	2·9	1·7
TVS volume	% Inflow	6·6	2·6	1·4
127 litres/min flow rate and 5·6 kg/cm² pressure drop:				
Flow volume	% Inflow	1·8	5·0	10·4
TS volume	% Inflow	15·8	17·1	24·2
TS concentration	% Wet basis	5·8	3·7	1·6
TSS volume	% Inflow	23·5	21·1	34·5
TSS concentration	% Wet basis	5·5	3·3	1·3
TVS volume	% Inflow	4·1	2·7	1·1

BOD reductions of 50–95% may be achieved in the fugate.[3] Solids separation can be significantly increased by adding flocculating agents such as aluminium sulphate.

Centrifuges cost money ($12 000–$20 000 per m^3/hr capacity), energy (20–30 h.p. per m^3/hr capacity),[4] and have high labour requirements for cleaning and maintenance. A less expensive device than centrifuges, but one which makes use of centrifugal forces, is the liquid cyclone shown in Fig. 1. Typical results obtained with a liquid cyclone are given in Table I.

SCREENS

There are two major categories of screens used in feedlot wastewater treatment: stationary and vibrating. Typical screens of the two types are shown in Figs. 2 and 3. Typical efficiencies of solids separation for stationary and vibrating screens for various loading rates and screen sizes are given in Tables II and III. Solids removal efficiencies are affected by the properties of the solids in the wastewaters, flow rate, temperature, screen size and configuration, screening device, etc. Therefore, it is best to test screens under the conditions they will be used. As can be noted in Tables II and III, screen efficiency can be measured in many ways and with different

TABLE II

RESULTS OF EFFICIENCY OF SOLIDS REMOVAL BY STATIONARY SCREEN (EFFICIENCY REPRESENTS PERCENTAGE RETAINED ON SCREEN AND THUS DIVERTED TO SOLIDS DIGESTER)

(from refs. 7 and 8)

Parameter	Loading rate		Units	Size of openings (cm)	
	litres/min	litres/min/m²		0·10	0·15
Flow volume	123	352	% Inflow	2·1	0·4
TS volume			% Inflow	35·2	3·0
TS conc.			% Wet basis	9·1	10·9
TVS volume			% Inflow	21·5	17·7
BOD volume			% Inflow	62·2	13·9
COD volume			% Inflow	69·1	11·4
Flow volume	183	526	% Inflow	2·9	0·3
TS volume			% Inflow	25·8	4·2
TS conc.			% Wet basis	7·2	6·4
TVS volume			% Inflow	30·5	0·9
BOD volume			% Inflow	38·4	4·1
COD volume			% Inflow	63·8	11·5
Flow volume	235	675	% Inflow	2·5	0·3
TS volume			% Inflow	27·5	9·8
TS conc.			% Wet basis	7·6	6·0
TVS volume			% Inflow	33·7	5·3
BOD volume			% Inflow	51·7	34·0
COD volume			% Inflow	71·1	24·2
Flow volume	313	897	% Inflow	0·6	0·4
TS volume			% Inflow	11·3	4·2
TS conc.			% Wet basis	6·9	6·0
TVS volume			% Inflow	13·7	5·6
BOD volume			% Inflow	13·7	5·6
COD volume			% Inflow	51·6	24·1

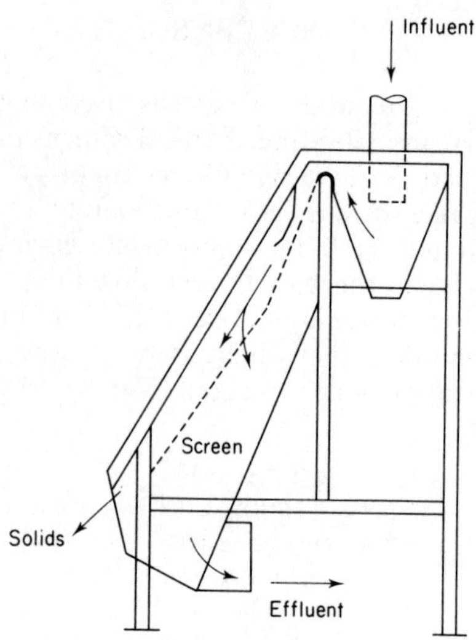

FIG. 2. Stationary rundown screen.

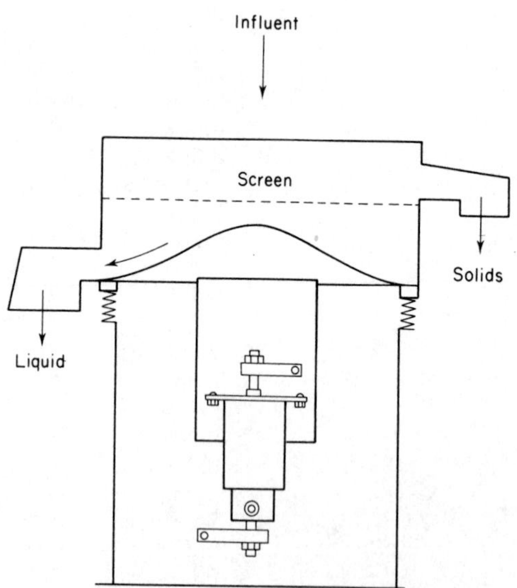

FIG. 3. Vibrating screen.

TABLE III

RESULTS OF SOLIDS REMOVAL EFFICIENCIES BY VIBRATING SCREEN (EFFICIENCY REPRESENTS PERCENTAGE RETAINED ON SCREEN AND THUS DIVERTED FROM MAIN FLOW PATH)
(from refs. 7 and 8)

Parameter	*Loading rate*		*Units*	*Size of openings (cm)*			
	litres/min	*litres/min/m²*		0·012	0·017	0·021	0·039
Flow volume	41	248	% Inflow	0·2	0·5	2·8	0·4
TS volume			% Inflow	2·5	5·8	14·3	12·6
TS conc.			% Wet basis	5·7	3·9	8·7	12·2
TVS volume			% Inflow	3·3	8·4	12·4	4·3
BOD volume			% Inflow	—	—	4·5	—
COD volume			% Inflow	—	3·3	16·3	10·0
Flow volume	67	411	% Inflow	1·2	0·7	0·7	0·6
TS volume			% Inflow	13·8	14·0	9·8	22·2
TS conc.			% Wet basis	8·5	10·9	10·8	16·4
TVS volume			% Inflow	18·5	17·0	12·9	28·1
BOD volume			% Inflow	—	—	2·4	—
COD volume			% Inflow	—	15·3	8·9	16·1
Flow volume	110	670	% Inflow	2·1	0·8	0·9	1·6
TS volume			% Inflow	18·7	1·8	7·0	12·3
TS conc.			% Wet basis	4·8	1·9	4·8	4·9
TVS volume			% Inflow	42·8	2·4	9·0	17·2
BOD volume			% Inflow	—	—	3·9	—
COD volume			% Inflow	—	12·5	10·7	12·2

parameters. The most common parameter used in screen efficiency measurements is total solids reduction. Total solids reductions of up to 60 % can be achieved.[3, 7–9]

FILTERS

Solids may be separated from feedlot wastewaters by filtering devices. The most common filtering methods used in municipal sewage sludge thickening operations are the vacuum filter and filter press. Such devices are expensive to install and to operate in terms of initial investment, in energy consumption and in the use of materials to aid the filtering process.[3, 4, 10]

Vacuum filters used in diluted poultry manure, aided by the addition of anionic polyelectrolytes followed by cationic polyelectrolytes, yielded a clear fugate and a filter cake of over 30 % total solids concentration.[10] Without chemically conditioning them, chicken wastes are more difficult to dewater than municipal sludges. The use of ferric chloride as a chemical conditioner for chicken manure is not practical because it results in excessive foaming without significant improvements in the dewatering rate of the waste. Polyelectrolyte dosages of 1·9–3·6 % of total solids are required to produce good filtering results.[10]

Using the filter press to dewater feedlot waste, aluminium sulphate and/or fly ash may be used to aid in the formation of the filter cake. Aluminium sulphate is effective as a dewatering aid when added to the waste at the rate of 1 % on volume basis. Fly ash should be added at the ratio of 1:0·5–1 fly ash to waste slurry. Under such conditions, the specific filtering resistance of the waste is reduced, filtering rate is increased by a factor of up to 5, and a filter cake with a solids concentration of 64 % may be produced, while the solids content in the filtrate is reduced to less than 1 %.

EVAPORATORS

One way to concentrate solids from wastewaters is to remove the liquid by evaporation. Using an electrically heated evaporator equipped with a stirring mechanism, wastewater of 6–10 % solids contents can be thickened to 88 % solids concentration.[3] To achieve this, fly ash and quicklime had to be added at the rate of 5–20 % of the waste volume, temperatures were maintained at 69–91 °C, while operating pressure was 0·4–0·6 kP/cm^2. The distillate output was only 0·4–3·7 kg/min. The energy consumption is 1117 kcal/kg of water removed, which was reduced by 17 % and 43 %, respectively, when quicklime and fly ash were added. There are problems with this type of solids separation which need to be researched and resolved before the method can become practical.

SEDIMENTATION

Suspended solids in wastewaters settle from a liquid medium in four different ways, as is shown in Fig. 4. Particles in the Class 1 region of Fig. 4 are completely dispersed and have no tendency to cohere on collision. Because these particles are discrete, they settle at their own characteristic rate until they strike the bottom of the basin. Particles on the right-hand side of Fig. 4, in the Class 2 clarification region, would tend to coagulate and have a negligible tendency to settle alone.

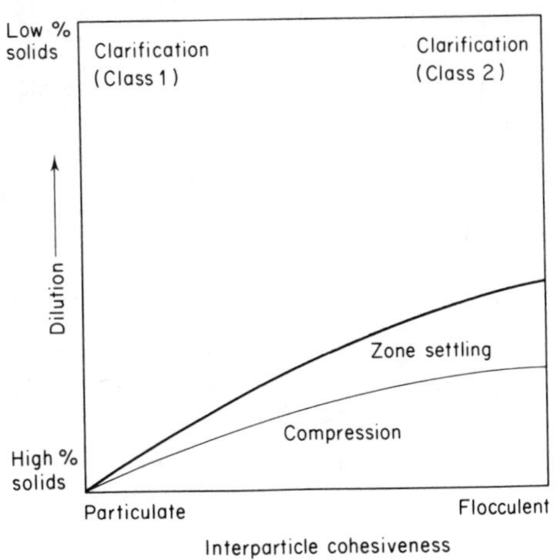

FIG. 4. Four zones of settling in a sedimentation tank (from ref. 11).

If the dilution is decreased and particles exhibit any cohesive force, they enter a regime of zone settling, as shown in Fig. 4. In zone settling the solids are considered close enough to cohere into a plastic structure. The forces between the particles are strong enough to drag each particle along in the same relative position.

As settled particles begin piling up from the bottom of the vessel, each layer of sludge solids provides some mechanical support to the layers above. The solids are no longer completely in suspension; instead, each layer of floc is compacted or thickened. This regime is called compression settling.

In Class 1 clarification, particles settle through the suspending fluid at a rate which remains constant. This rate is related to the size and density of the particle, the viscosity and density of the fluid medium, and the force of gravity acting on the particle.

Clarification tanks are designed on the basis of overflow rates. Typical overflow rates vary from 8000 to 30 000 litres/day/m^2 of surface area (8 to

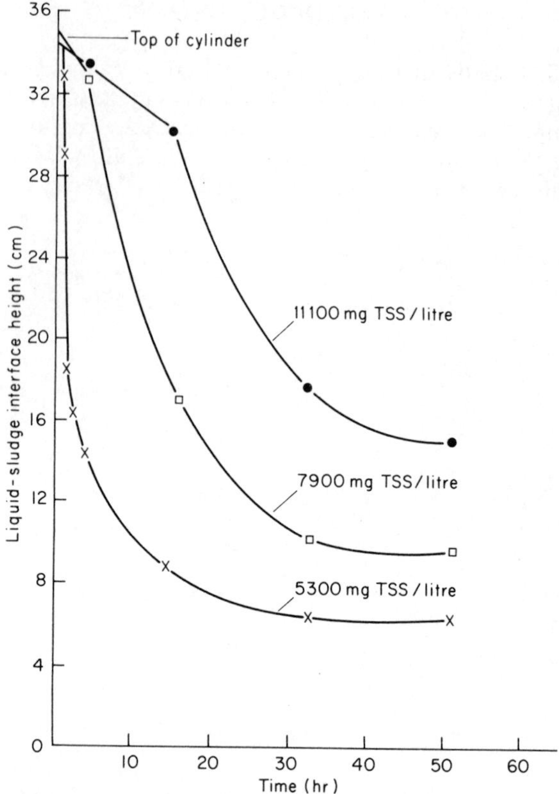

FIG. 5. Settling curves for oxidation ditch mixed liquor with different TSS concentrations.

30 m^3/day/m^2) for primary sedimentation and from 40 000 to 120 000 litres/day/m^2 for biologically treated wastewaters in final clarification operations.[12]

The required surface area for clarification for a wastewater whose sedimentation characteristics are not well established can be determined theoretically by using Stokes' Law[12] or by determining the settling velocity of the solids. The latter can be accomplished by developing settling curves similar to those shown in Fig. 5. The settling velocity for the wastewater with a TSS content of 7900 mg/litre is 2 cm/hr, while for the liquid with 5300 mg TSS/litre, the settling velocity is 9 cm/hr.[13] If the flow rate through the clarifier is 20 000 cm^3, then the surface area required for the 7900 mg TSS/litre liquid would be 10 m^2, and for the 5300 mg TSS/litre swine wastewater, 2 m^2.

According to Stokes' Law, for discrete spherical particles, for Reynolds number less than 1 and at constant temperature, the settling velocity of the particles in a quiescent viscous fluid would be given by the following formula:[14]

$$v_s = \frac{g}{18} \frac{(\rho_s - \rho)}{\mu} d^2$$

where v_s = settling velocity, g = acceleration of gravity, ρ_s = particle density, ρ = liquid density, μ = dynamic viscosity of liquid and d = particle diameter.

Figures 5, 6 and 7 show the effect of detention time, overflow rate and total suspended solids (TSS) content of the swine feedlot wastewater on the rate of solids separation and/or removal of solids or COD. Sedimentation tanks

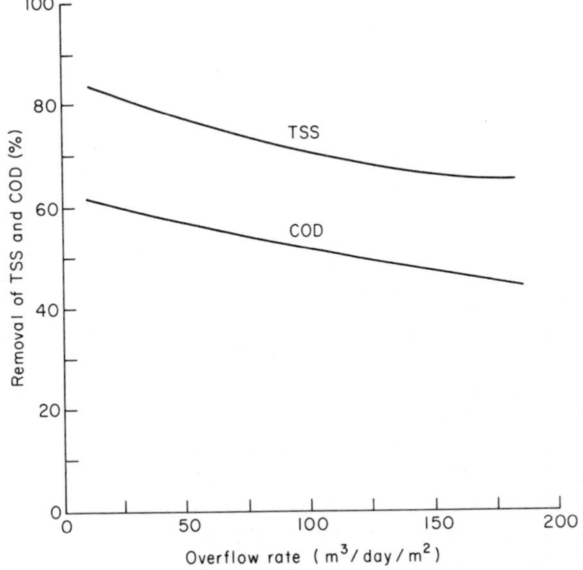

FIG. 6. TSS and COD removal efficiencies at various overflow rates for a slurry with initial TSS concentration of 5000 mg/litre (from ref. 11).

are not effective in clarification settling when TSS concentration of the influent exceeds 8000 mg/litre, because, as is shown in Fig. 5, initial settling is very slow. The clarifier must have a detention time equal to or greater than the time of the initial settling phase shown in Fig. 5.

As is shown in Fig. 6, removal of TSS and COD from swine feedlot wastewaters decreases as the overflow rate is increased beyond

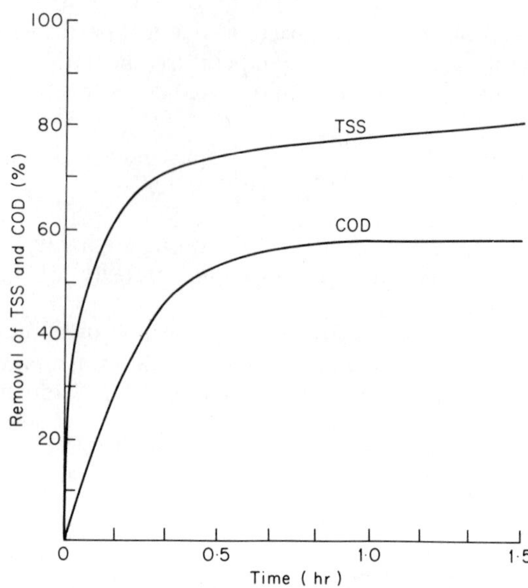

FIG. 7. TSS and COD removal efficiencies at various detention times for a slurry with initial TSS concentration of 5000 mg/litre (from ref. 11).

20 $m^3/day/m^2$ of surface area of the settling tanks. However, Fig. 7 indicates that detention times longer than 30 min do not result in significantly greater removals of TSS or COD.

Solids settling in slurries with an initial concentration of TSS greater than 1% (10 000 mg/litre) occurs as a zone settling process.[11]

For open feedlots exposed to rainfall, runoff contains solids which may be separated from runoff flow by the use of settling basins. These settling basins are made of concrete with a depth of at least 1 m. However, the cross-sectional area of these basins should be such that the velocity of flow at time of maximum runoff should not exceed 20 m/min so that coarse solids may settle out in the basin. Average flow rate into and through the basin should not exceed 30 $m^3/day/m^2$ of basin bottom area.[15] The basin should be equipped with a scum baffle and a weir outlet designed on the basis of 20 to 200 litres/min/m of weir length.[15,16] Maximum runoff flow is based on the runoff from a storm of 1 hr duration and frequency of occurrence of once in 10 years.[16]

Earth basins or channels may be used to settle out solids by slowing down the runoff velocity or by the use of porous dams or perforated pipe.[16] Basins and channels should be so designed so as to permit the removal of the settled-out solids by scraping with bulldozers or with tractors equipped with buckets or scrapers. Side slopes of 3:1 with a minimum of 1 m of berm width and grassed slopes are recommended for these earthen structures. The porous dams may be constructed with wooden boards spaced 2–3 cm apart, with welded wire fabric or expanded metal mesh.[16] Settling channels and basins should be cleaned at least once a year. Provision should be made for the proper disposal of the solids removed.

REFERENCES

1. Miner, R. J. and Smith, R. J. (eds.), *Livestock Waste Management with Pollution Control*. Midwest Plan Service, Ames, Iowa, 1975.
2. Aronsson, G., Decanters for sludge dewatering. *Process Biochemistry*, Jan. 1970.
3. Blaha, K., *Separation of Organic Components from Agricultural Waste*. Agricultural Engineering Research Institute, Prague, ČSSR, 1975.
4. Glerum, J. C., Klomp, G. and Paelma, H. R., The separation of solid and liquid parts of pig slurry. In: *Livestock Waste Management and Pollution Abatement*, pp. 345–7. ASAE, St Joseph, Mich., 1971.
5. Ross, I. J., Begin, J. J. and Midden, T. M., Dewatering poultry manure by centrifugation. In: *Livestock Waste Management and Pollution Abatement*, pp. 348–50. ASAE, St Joseph, Mich., 1971.
6. Shutt, J. W., Liquid cyclone for separation and concentration of swine manure solids. MS thesis, Ohio State University, Columbus, Ohio, 1974.
7. Shutt, J. W., White, R. K., Taiganides, E. P. and Mote, C. R., Evaluation of solids separation devices. In: *Proceedings of International Symposium on Livestock Wastes* (Urbana, Ill., Apr. 1975). ASAE, St Joseph, Mich., 1975.
8. Taiganides, E. P. and White, R. K., Automated treatment and recycling of swine feedlot wastewaters. OSU Research Foundation Terminal Report, EPA Grant No. R-801125, Columbus, Ohio, 1974.
9. Ngoddy, P. O. *et al.*, *Closed System Waste Management for Livestock*, EPA-RS 13040 DKP 06/71. Washington, DC, 1971.
10. Cassel, E. A., Warner, A. F. and Jacobs, G. B., Dewatering chicken manures by vacuum filtration. In: *Management of Farm Animal Wastes*, pp. 85–90. ASAE, St Joseph, Mich., 1966.
11. Thompson, A. V., The separation of solids from swine waste slurries. MS thesis, Ohio State University, Columbus, Ohio, 1971.
12. Nemerow, N. L., *Liquid Waste of Industry*. Addison-Wesley, London, 1971.
13. Mote, C. R., A computer simulation of biological treatment, storage, and land disposal of swine wastes. Ph.D. thesis, Ohio State University, Columbus, Ohio, 1974.
14. Clark, J. W. and Viessman, W., *Water Supply and Pollution Control*. International Textbook Co., Scranton, Pa., 1965.
15. Taiganides, E. P. and White, R. K., Automated handling, treatment and recycling of wastewater from an animal confinement production unit. In: *Livestock Waste Management and Pollution Abatement*, pp. 146–8. ASAE, St Joseph, Mich., 1971.
16. Midwest Plan Service, *Livestock Waste Facilities Handbook* (MWPS-18). Iowa State University, Ames, Iowa, 1975.

17

Dehydration Systems for Feedlot Wastes

MERLE L. ESMAY

Professor of Agricultural Engineering, Michigan State University, East Lansing, Michigan, USA

INTRODUCTION

Dehydration of feedlot manure is defined as the reduction of moisture content. The term 'manure' refers to faeces and urine. If other products are added to manure, such as bedding (litter), rain or other water, soil, etc., then the term 'waste' is used.

Moisture content will be referred to as percentage wet basis. For example, 100 kg of manure at 80 % moisture content contains 80 kg of water and 20 kg of dry matter. Table I indicates the relationship between percentage moisture content wet basis and the portion of water removed. When 50 % of the water content of 80 % moisture manure is removed, the wet basis moisture content becomes 67 %, not half of 80 %. The majority of water must be removed before moisture content is significantly reduced. Three-quarters of the water must be removed to lower the moisture content to 50 %, etc.

The dehydration of livestock manure can serve as a pollution control device and greatly improve the handling characteristics of the product. The

TABLE I

COMPUTED WATER QUANTITIES REMOVED WITH MOISTURE CONTENT REDUCED FROM 80 % TO 10 % (WET BASIS)

Wet basis moisture content reduction (%)	*Dry matter* (*kg*)	*Water* (*kg*)	*Total weight* (*kg*)	*Water removed* (%)
80	20	80	100	0
67	20	40	60	50
50	20	20	40	75
33	20	10	30	88
20	20	5	25	94
10	20	2	22	98

TABLE II
EXCRETA PRODUCTION FROM VARIOUS ANIMALS
(from ref. 1)

Animal	*Manure (kg/day/ 1 000 kg l.w.)*[a]	*Total solids*		*Moisture*
		(kg/day/ 1 000 kg l.w.)	*(% wet basis)*	*(% wet basis)*
Dairy cow	82	10·4	13	87
Beef feeder	60	6·9	12	88
Swine feeder	65	6·0	9	91
Sheep feeder	40	10·0	25	75
Poultry layer	53	13·4	25	75
Poultry broiler	71	17·1	25	75
Horse	45	9·4	21	79

[a] l.w. = live weight of animals.

reduction of manure moisture to the 10–15 % level produces a stable odour-free product that can be stored, transported and conveniently used as fertilizer or as feed supplement.

Freshly voided excreta does not have a highly offensive, penetrating odour. If the water reduction process can be started immediately, the possibility of the manure reverting to an anaerobic process is eliminated and the production of undesirable gases minimized. The high moisture content of voided excreta excludes oxygen and thus promotes the development of the uncontrolled anaerobic process which will cause air pollution. Moisture content of manure as produced by various animals is given in Table II. Moisture content ranges from a low of 75 % to a high of 91 %.

The most common moisture reduction processes are:

1. Evaporation.
2. Physical separation of liquid.
3. Increase of dry matter portion by addition of moisture-absorbing materials.

Dehydration by the evaporation process is the main subject of this chapter and will be discussed in considerable detail. First, a brief description of the latter two processes is given.

PHYSICAL SEPARATION

Water can be separated as a liquid from the dry matter of livestock manure by various natural and mechanical means. The most common natural means is to let the excess liquid drain away from the solids. Also, natural gravity settling and flotation of solids with subsequent separation from the liquid provides a means of obtaining a lower moisture content solid material.

Various mechanical means are being used to squeeze the water from manure mixtures. A wide range of such equipment is now available from commercial companies. Fairly high energy requirements are necessary for the mechanical process. Neither the resulting solids nor the liquids are generally completely stable. The solids may still have a 40–50 % moisture content and the liquid portion would still contain some solids. Further treatment of both materials may then be necessary depending on conditions and intended use for the products.

Water may also be separated from solids through a screening or filtering process. This method can be at least partially successful for the handling of manure from ruminant animals which contains a high proportion of fairly fibrous solids. Solid and liquid separation processes are generally quite difficult to keep operating on a continuous basis due to clogging of the screens.

Centrifugal separation is also possible, but has not found extensive acceptance on a continuous economic basis.

DRY MATTER ADDITIONS

The traditional way of reducing the level of moisture content of manure in livestock shelters has been to add dry matter to the manure in the form of

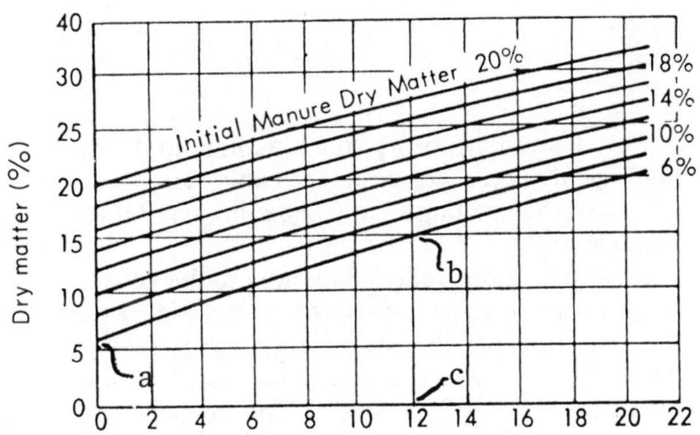

FIG. 1. Changing manure dry matter by adding bedding. Bedding is assumed to have 10 % moisture. Each sloping line across the chart is for one initial dry matter from 6 % to 20 %.

TABLE III
WATER ABSORPTION OF BEDDING
(from ref. 1)

Material	*kg water absorbed per kg bedding*
Wood	
Tanning bark	4·0
Dry fine bark	2·5
Pine chips	3·0
sawdust	2·5
shavings	2·0
needles	1·0
Hardwood chips, shavings or sawdust	1·5
Corn	
Shredded stover	2·5
Ground cobs	2·1
Straw	
Flax	2·6
Oats, threshed	2·8
combined	2·5
chopped	2·4
Wheat, combined	2·2
chopped	2·1
Hay, chopped mature	3·0
Shells, hulls	
Cocoa	2·7
Peanut, cottonseed	2·5
Oats	2·0

bedding or litter. The water absorption capability and density of various common bedding materials are given in Tables III and IV. The initial moisture content of the bedding as calculated in these tables is assumed to be approximately 10%.

Bedding has typically been added to manure to thicken it for handling as a solid. This mainly involves soaking up the urine. Figure 1 can be used for estimating the amount of bedding to add to manure to raise its dry matter or lower its moisture content. For example, if the initial dry matter content of the manure is assumed to be 6% (a) and it is desired to increase it to 15% (b), it would require the addition of 12 kg (c) of bedding per 100 kg of manure. It should be remembered that this process of lowering the moisture level by adding dry matter does not remove any water from the total mass of the manure/bedding mixture. In fact it adds dry matter weight and bulk to the mixture so there is more to handle than initially. This reason, along with

TABLE IV
BEDDING MATERIAL DENSITY
(from ref. 1)

Form	*Material*	*Density* (kg/m^3)
Loose	Alfalfa	4–4·4
	Non-legume hay	3·3–4·4
	Straw	2–3
	Shavings	9
	Sawdust	12
Baled	Alfalfa	6–10
	Non-legume hay	6–8
	Straw	4–5
	Shavings	20
Chopped	Alfalfa	5·5–7
	Non-legume hay	5–6·7
	Straw	5·7–8

problems of management and labour requirements for handling bedding, have caused it to be used less and less in modern livestock enterprises.

EVAPORATION

Dehydration of manure by evaporation involves the change of phase of the water from a liquid to a gaseous vapour and removal of the vapour by air acting as a carrier medium. To bring about this phase change, called evaporation, an input of energy is required. Table V shows the heat requirement for the change of water from a liquid to a vapour at various temperature levels. It takes 10 % more heat to vaporize water at 0 °C than at 100 °C. The required heat for vaporization is called latent heat. As air receives water vapour, the total enthalpy or heat content of the air is increased by both the latent and sensible heat of the vapour added.

Dehydration by evaporation can be accomplished in various ways and with different sources of energy:

1. Natural air dehydration by wind.
2. Forced air dehydration by mechanically controlled ventilation with fans.
3. Heated air dehydration with mechanical driers having heat sources of (a) fuel oils, (b) natural or liquefied petroleum gas, (c) biomass materials, such as wood or other plant waste products, (d) coal, (e) solar radiation.

TABLE V
HEAT OF VAPORIZATION FOR WATER
(from ref. 2)

Temperature (°C)	*Latent Heat (kcal/kg)*
0	596·5
4·5	593·8
15·5	588·2
26·5	581·0
38·0	574·4
49·0	568·3
71·0	555·0
93·0	542·2
100	537·8

MANURE DEHYDRATION BY VENTILATION AIR

Either natural or forced air evaporation of water requires the energy input as indicated by Table V. When evaporation takes place without a supplemental source of energy, the heat of vaporization must be supplied from the air itself. A reduction of sensible heat of air or of any matter lowers the temperature of the mass. Thus, as air absorbs water vapour, sensible heat is taken from the air mass and converted to the latent heat of the water vapour. The total energy is unchanged in this adiabatic process, but the air temperature is lowered. This process, then, is often referred to as evaporative cooling. Where dry, low-humidity climatic conditions prevail, evaporative cooling can be used as an economic means of cooling during high-temperature summer conditions. Evaporative cooling does not, however, reduce the heat stress on animals or birds under both hot and humid conditions (temperatures near 35 °C and relative humidities of 75 % and higher).

A controlled environment livestock building functions essentially as a large evaporator or dehydrator. Outside climatic air is brought into the building to provide fresh air for the animals. This ventilation air then serves also as a medium for picking up water vapour along with the undesirable metabolic gases. In order to maintain stable temperature and humidity conditions in the building, the heat and mass transfer for the building as a whole must be in balance. The heat balance equation in the building is as follows:

$$q_a + q_s + M_a h_1 = q_c + M_a h_2$$

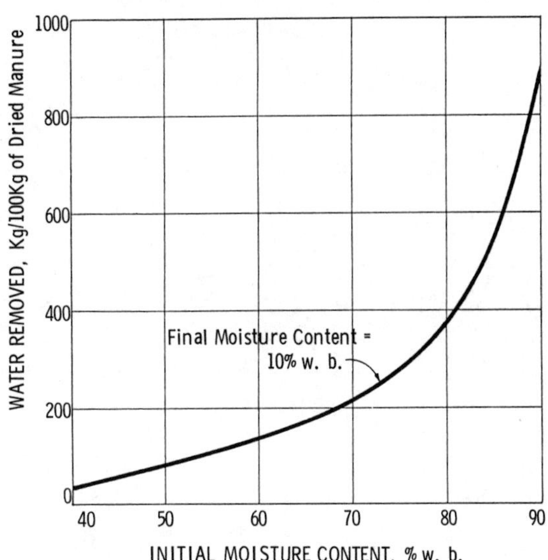

FIG. 2. Water that must be removed to produce 100 kg of dried excreta from wet excreta.

where q_a = heat from the animals or birds, q_s = heat added supplementally, $M_a h_1$ = mass of incoming air multiplied by enthalpy per unit of mass per hour, q_c = heat conducted through the exterior surfaces of the building, and $M_a h_2$ = mass of ventilation air exchange multiplied by the outgoing enthalpy per unit of mass per hour. The heat and moisture pick-up of the ventilation air is accounted for by the change of enthalpy of the incoming and outgoing ventilation air.

Livestock produce considerable water vapour directly from their respiratory systems besides the liquid water produced in the faeces and urine. The ventilation air must absorb the directly produced water vapour to prevent condensation within the livestock structure.

Manure dehydration then takes place within the housing structure to the extent that the ventilation exchange air can be induced to absorb additional water vapour by evaporation from the faeces and urine mixture. The housed animals or birds also produce heat which can be used to provide the necessary heat of evaporation for dehydrating manure. During hot climatic weather the conversion of the heat from the animals or birds from sensible to latent heat through the evaporation process is advantageous in preventing the sensible heat from increasing the air temperature, and thus possible heat stress conditions in the building. During the colder winter conditions more of the sensible heat from the animals must be used to keep the in-house environment at a comfortably warm temperature, so less is

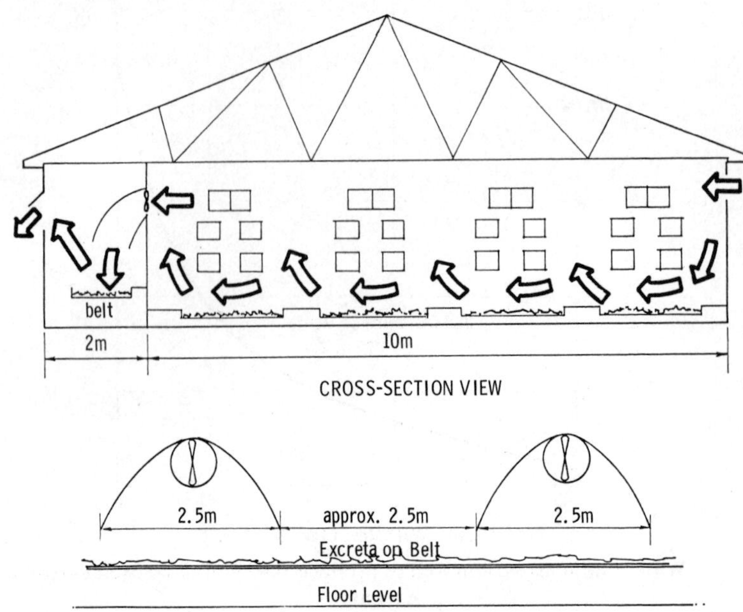

FIG. 3. Three-quarters of the water can be removed from manure in a cage-type poultry house with ventilation air. Exhaust fans direct air on to manure spread on the belt in the drying tunnel to provide maximum turbulence for drying (from ref. 3).

available for evaporating water from manure. Artificial supplemental heat might of course be used for providing the desirable environment and for evaporating additional water, but economics rules this out at least for the housing of adult animals.

Ventilation air in environmentally controlled structures can be used to remove three-quarters of the water (from 80% to 50% wet basis, see Fig. 2) from the manure by evaporation.[2] For example, in a cage-type poultry house the ventilation air should be controlled as shown in Fig. 3 for maximization of water dehydration. The cross-flow, exhaust-type ventilation system is standard for a cage-type poultry house. The exhaust air from the house is, however, used in the drying tunnel adjacent to the house for additional manure drying. The manure from the dropping pits below the cages is removed daily and conveyed on to the belt in the tunnel where it remains for one extra day of drying.

An alternate way of maximizing in-house dehydration by use of the ventilation air is to suspend small 35 cm fans every 20 m apart below the cage rows. The fans are centred over the dropping pits and oriented to move the air lengthwise of the pit. The air turbulence over the manure maximizes evaporation. Once a day stirring and the use of the fans will dry the manure to below 50 % wet basis during all seasons. Stirring can be accomplished by modifying the pit scraper to function as a stirring device as needed. With this arrangement the manure need only be removed every two to three months.

The electric energy for operating the small 35 cm fans on a continuous basis is much less costly than fuel for a mechanical drier. For a cost of 1 cent, 2·75 kg of water can be removed with the fans if electricity is assumed to cost 3·5 cents/kWh. Only 0·5 kg of water may be removed by a mechanical drier with afterburner, if fuel oil is assumed to cost 9 cents/litre.

DEHYDRATION WITH MECHANICAL DRIERS

Mechanical driers operate in the range of 50 % or less thermal efficiency.[3] Theoretical energy requirements to evaporate 1 kg of water range from 538 to 600 kcal (see Table V). Figure 4 provides a graphical means of calculating drier heat energy requirements for various water evaporation capacities. The bottom curve labelled 550 kcal/kg is for a theoretical 100 % thermal efficient operation. For the typical mechanical manure drier the

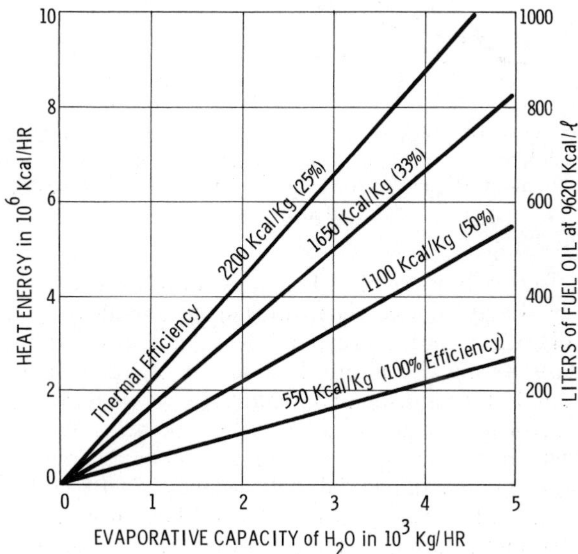

FIG. 4. Heat energy requirements for manure driers with various rates of efficiencies.

1100 kcal/kg curve should be used.[3] For example, a drier with an evaporative capacity of 3000 kg of water per hour would then require about $3{\cdot}3 \times 10^6$ kcal of energy per hour. The energy content of fuel oil is approximately 10 000 kcal/litre, thus this example of drier would require 300 litres of fuel oil per hour.

Afterburners for driers, if required for control of air pollution from exhaust gases, may double fuel requirements per kg of water evaporated,

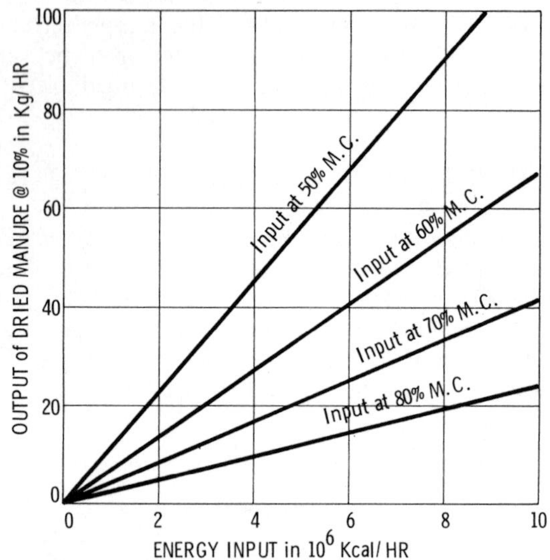

FIG. 5. Production of 10 % moisture content dried animal or poultry waste with driers having different energy inputs and with input of wet manure at various levels of moisture content. Drier performance efficiency of 1100 kcal/kg of water evaporated is assumed.

particularly for small driers of 100 kg/hr of water or less. The upper two curves of Fig. 4 may be used for making such estimates.

The rate of water evaporation capability of a manure drier should remain quite constant during operation. The output of dried manure is then inversely related to the moisture content of the wet manure going into the drier. Figure 5 provides a graphical means of estimating the production of dried product for various sizes of driers with known energy inputs (fuel requirements) and an assumed thermal efficiency of 50 % (1100 kcal/kg of water evaporated). For example, for a 4×10^6 kcal/hr energy input drier, the dry matter outputs can be obtained on the vertical scale for the input of wet manure at various moisture levels. It will be noted that the output of dried manure goes from about 900 kg/hr with an input of 80 % moisture content manure to about 4500 kg/hr with 50 % moisture content manure. This is an increase of dry matter output of approximately five times.

The conventional and most reliable type of manure drier consists of the rotary drum type that is charged with wet manure internally and fired internally. Drum-type driers are used traditionally for dehydration of many different biological products. Drum driers can be designed and constructed for a wide range of sizes and capacities. The rotating drum provides effective mixing of the manure and thus good exposure to the combustion energy for evaporation of water. The drum requires a minimum of power for the rotation movement and also has a minimum of moving parts (see Table VI). Figure 6 illustrates a typical drum drier.

A much less common and less reliable type of manure drier is the shaker type.[3] They are charged with wet manure at the top and fired at the bottom. The wet manure moves downwards across a series of inclined, shaking screens through the hot air stream. A rotating flail-type beater intercepts the manure at the bottom end of the top inclined screen and keeps throwing the wet manure back up the incline until it is dry and fine enough to drop through the screen mesh. The flail requires considerable power and the drying unit as a whole has many moving parts which amplifies maintenance and repair requirements.

Many commercially manufactured manure driers are being produced in various countries around the world. Drier performance data from the manufacturing company or test measurement data from an independent evaluator should be obtained before purchasing a commercial manure drier. Data similar to those included in Table VI should be used for an economic analysis. The water removal capability of the mechanical drier is the most critical criterion. The size and drying capability of the unit should be determined by the amount of manure drying to be accomplished. It is noted from the example data of Table VI, which pertain to one manufactured drier, that there is some increased efficiency and initial cost savings with the larger-sized driers. For the smallest drier of Table VI the fuel cost is above \$10 per 1000 kg of water removed, while for larger units the cost drops to \$8.50 per 1000 kg.

Depreciation costs can be determined by assuming some length of life and hours per year of operating time. If 40 hr per week for 50 weeks for 8 years is assumed (a working life of 16 000 hr), the depreciation cost is less than \$1/hr, or about 10 % of the fuel cost.

Installation cost, shelter and accessories may cost as much as the initial price of the drier. This, then, would be another 10 % cost of fuel for drying. Afterburner and pollution control devices might add another 5 % of the fuel cost to drying. Also, the fuel for the afterburner can equal that for the drier itself, particularly for the small driers of 500–1000 kg of water evaporating capacity per hour.

For example, using the data of Table VI and the above assumptions, a

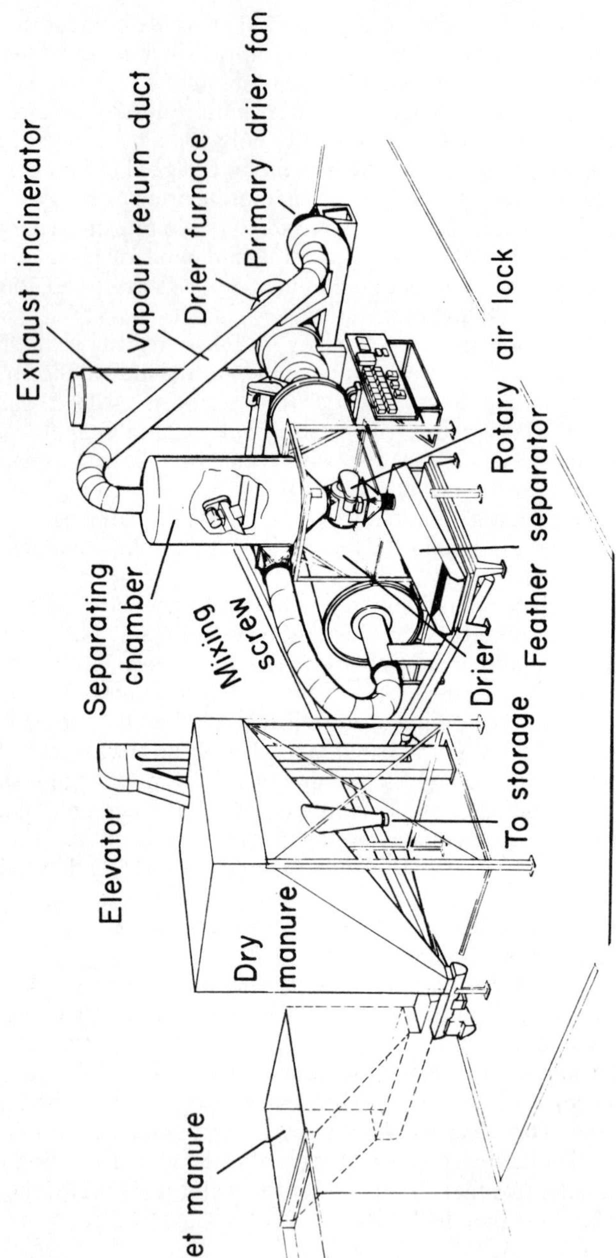

FIG. 6. Drum drier.

TABLE VI
OPERATING DATA AND ESTIMATED PRICES FOR VARIOUS SIZES OF ONE DRUM TYPE DRIER (HEIL CO.)
(from ref. 4)

Water removal (*kg/hr*)	*Initial price*		*Energy*			*Power*		*Fuel cost at 10¢/litre*	
	Estimated 1975 *price* (*US*\$)	\$/1 000 *kg* H_2O	10^6 *kcal/hr*	*Fuel/hr* (*litres*)	*Thermal efficiency* (%)	*h.p./hr*	*h.p./*1 000 *kg* H_2O*/hr*	*Cost* (\$/*hr*)	*Cost* (\$/1 000 *kg* H_2O)
455	23 000	50 000	0·5	50	50	11	24	5.00	11.00
2 275	33 000	15 000	2·3	230	55	50	22	23.00	10.00
2 730	37 000	14 000	2·8	280	55	54	20	28.00	10.00
4 100	56 000	13 000	3·5	350	65	60	15	35.00	8.50
5 455	67 000	12 000	5·5	550	65	80	15	55.00	8.50
8 180	84 000	11 000	7·0	700	65	122	15	70.00	8.50

cost analysis for the next to smallest drier (2275 kg H_2O/hr) shows the following:

Fuel cost per 1000 kg H_2O	$10.00
Initial cost per 1000 kg H_2O	1.00
Installation cost per 1000 kg H_2O	1.00
Pollution control per 1000 kg H_2O	0.50
Afterburner fuel	10.00
Total	$22.50/ton

Labour cost, interest and insurance must be added to this.

The drying cost per ton of dried manure product varies with the moisture content of the ingoing wet manure as shown by Fig. 5. For the drier of the above example, about 1 ton of dried manure can be produced per hour if the ingoing wet manure is at about 70 % moisture content wet basis. Thus, the dried manure would cost $22.50 per ton for drying. If the initial moisture content was as high as 80 % the drying cost would be nearly doubled, and if as low as 55 % the drying cost would be one-half.

DESIGN EXAMPLE

A 200 000 laying hen enterprise is used here as an example for manure dehydration. Table II may be used for determination of excreta dry matter production. The daily excreta dry matter production of 200 000 hens is indicated as about 5000 kg/day. The weight of dried manure product at 10 % moisture would be 5500 kg/day. The total production of manure from the 200 000 hens as voided at 80 % moisture would amount to 25 000 kg/day. If the worst condition of absolutely no pre-hydration in the poultry house is assumed, the mechanical drier must remove 19 500 kg of water per day. If a 10 hr drying day is assumed, the drier must have a water removal capacity of about 2000 kg/hr. If the typical 50 % thermal efficiency is assumed from Fig. 4, it can be determined that the drier needs a heat energy input of about $2{\cdot}2 \times 10^6$ kcal/hr. At 10 000 kcal/litre of fuel oil, this means a fuel input of 220 litres/hr.

If in-house dehydration of manure to 50 % moisture is assumed, the daily amount of water to be removed by the drier will be only 5000 kg. For a 10 hr drying day the evaporation capacity of the drier will need to be about 500 kg/hr. From Fig. 4 the heat energy requirement would then be only about 500 000 kcal/hr or about 50 litres per hour of fuel oil.

Similar examples for dehydration of manure from other animals may be carried out following the same procedure. For a dairy herd of 2400 milking cows at 500 kg each there would be a dry matter excreta production of $10{\cdot}4 \times 0{\cdot}5 \times 2400 = 12\,480$ kg/day (Table II). The excreta is produced at

87 % moisture content wet basis. Assuming it is dried to 80 % by ventilation air in the dairy barn, then, from Fig. 2, 375 kg of water must be removed to produce each 100 kg of dried manure at 10 % wet basis. Total daily water to be removed would be 125 × 375 = 46 875 kg/day. For 10 hr/day operation a 5000 kg water/hr drier would be required. Figure 4 and Table VI may be referred to in determining the drier size and fuel requirements.

REFERENCES

1. Midwest Plan Service, *Livestock Waste Facilities Handbook* (MWPS-19). Iowa State University, Ames, Iowa, July 1975.
2. Esmay, M. L., *Principles of Animal Environment*. AVI Publishing Co. Inc., Westport, Conn., 1969.
3. Esmay, M. L., Flegal, C. J., Gerrish, J. B., Dixon, J. E., Sheppard, C. C., Zindel, H. C. and Chang, T. S., In-house handling and dehydration of poultry manure from a caged layer operation: a project overview. In: *Proceedings of International Symposium on Livestock Wastes*. ASAE, St Joseph, Mich., 1975.
4. Taiganides, E. P., *Design of a Dehydrator*. Agricultural Engineering Dept, Ohio State University, Columbus, Ohio, 1974.

18

Lagoon Systems for Animal Wastes

RICHARD K. WHITE

Associate Professor, Department of Agricultural Engineering, Ohio State University, Columbus, Ohio, USA

INTRODUCTION

The current trend towards confined livestock feeding requires the development of integrated, mechanized and automated waste handling systems. Lagoons are often used in modern livestock feedlots since they fit into an automated liquid waste handling system.

Lagoons are the most common method of biological treatment of organic wastes where sufficient land is available. They are commonly used for the treatment of wastewaters from rural communities and industries.[11] Lagoons are defined as earthen structures which are designed, constructed and operated to provide storage and/or treatment to wastes and wastewaters.

Lagoons may be classified into impounding lagoons or storage ponds and flow-through lagoons.

In the first type, either there is no overflow, or there is intermittent discharge into a stream during periods of high rainfall and high river flows. The volumetric capacity of impounding lagoons is equal to the total waste flow times the days of detention, less volume lost through evaporation and percolation. In view of the extremely large area requirements, impounding lagoons are usually limited to areas where evaporation exceeds annual precipitation, or are suited for sources with low daily volumes of wastewater, for seasonal operations such as in the vegetable canning industry or for impounding feedlot runoff which is subsequently pumped out and spread by irrigation on cropland.

Flow-through lagoons, in contrast to impounding lagoons, are designed to provide, in addition to storage, solids reduction, stabilization of organic matter through mineralization or growth of biomass, odour management and waste disposal. Flow-through lagoons will be discussed in detail because they require more engineering design. The term 'lagoon' will henceforth be used to refer exclusively to flow-through lagoons.

The design and operation of livestock waste lagoons must include the disposal of effluent and solids. The main disposal method is to spread lagoon contents on cropland for utilization of the fertilizer elements and for irrigation water. Where discharge into a stream occurs, additional, extensive treatment is necessary. In many cases the effluent is recycled and used as flushing water to transport the manure from the feedlot to the lagoon. If the lagoon evaporation exceeds rainfall, water will need to be added to prevent accumulation of salts to levels toxic to the microflora of the lagoon.

LAGOON CLASSIFICATION

Removal of pollutants and stabilization occurs in lagoons mainly through two processes. One of the processes is the settling out of solids through sedimentation and/or chemical precipitation. The second process involves the stabilization of organics through biological transformations.

The mode of the biological transformation determines the type of lagoon. Usually four types of lagoons are recognized.

Aerobic-Algal Lagoons

Figure 1 presents the processes and biological transformations which occur in an aerobic lagoon where bacteria and algae, in a symbiotic mode of operation, combine to stabilize organic wastewaters. The effluent from the aerobic-algal lagoon will contain algal cells and though the BOD may be low, the solids content may be significant.

Since sunlight is essential for oxygen production by algae, two factors are extremely critical in this type of lagoon: depth of water and turbidity. Depth where algae grow profusely is limited to about 50 cm, which is the maximum light penetration for most lagoon waters. Lagoons are made 1–1·5 m deep which allows mixing of water by wind action, restricts growth of aquatic plants and provides for the deposition of sludges.

Feedlot wastewaters are considerably more turbid than standard wastewaters. As a result, light penetration is restricted, making aerobic-algal lagoons not well suited for feedlot wastes. Swine feedlot wastewaters may contain copper which is toxic to algae and would also restrict this type of lagoon.

Aerated Lagoons

Aerated lagoons are 3–4 m deep, and their contents are mixed and aerated by mechanical surface or diffused air equipment. The oxygen transfer into the water is increased for a given surface area by providing greater air/water contact through pumping water into the air or pumping air into the water. Also, the mixing action of the aerator will permit greater depths. Most floating aerators and platform-mounted aerators will pump the water into the air and to the side, providing aeration and mixing as shown in Fig. 2(a).

LIGHT
WASTE
NUTRIENTS
CO_2
BACTERIA
ALGAE
O_2
ALGAE, BACTERIA,
DISSOLVED AND
SUSPENDED SOLIDS
SLUDGE

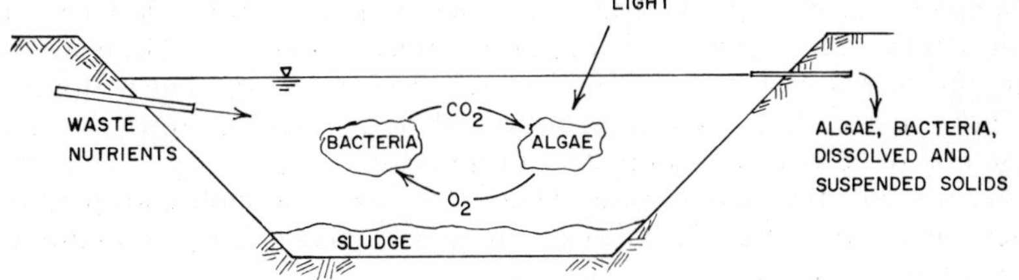

FIG. 1. Biochemical processes occurring in an aerobic-algal lagoon.

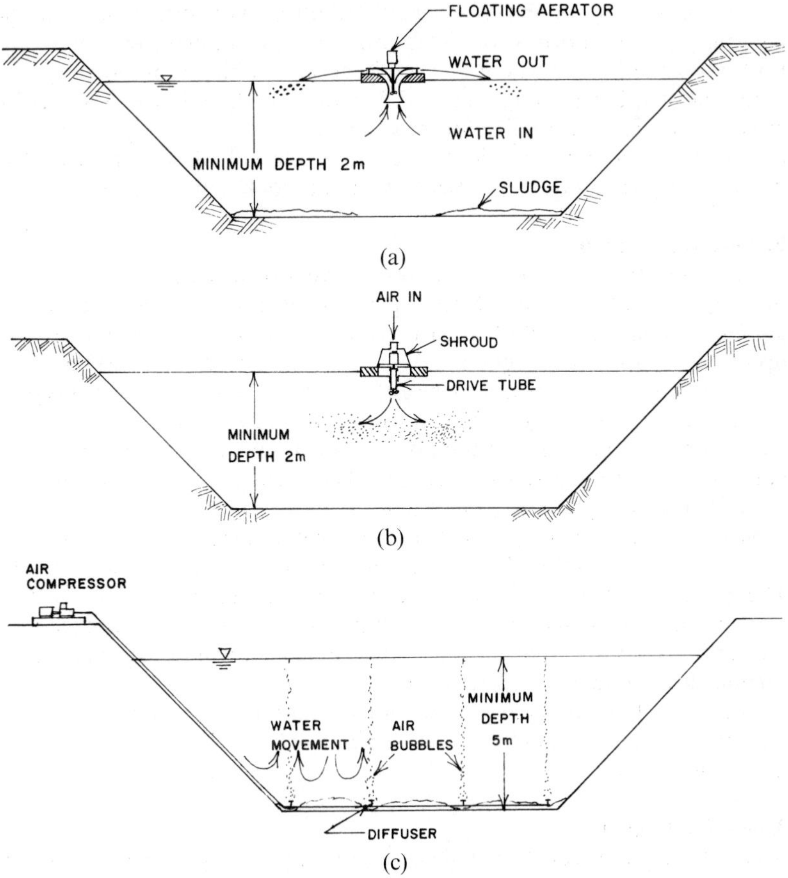

FLOATING AERATOR
WATER OUT
WATER IN
MINIMUM DEPTH 2m
SLUDGE
(a)
AIR IN
SHROUD
DRIVE TUBE
MINIMUM
DEPTH 2m
(b)
AIR
COMPRESSOR
MINIMUM
DEPTH
5m
WATER
MOVEMENT
AIR
BUBBLES
DIFFUSER
(c)

FIG. 2. Typical types of mechanical equipment used in aerated lagoons. (a) Aerated lagoon with most common type of surface aerator equipment. (b) Aerated lagoon with downdraught, induced-aspiration aerator. (c) Aerated lagoon with diffused aeration equipment.

Floating downdraught, induced-aspiration devices are available where the air is drawn into the vortex created at the impeller (Fig. 2(b)). The blades of the impeller break the air into fine bubbles. Pumping air through perforated plastic pipes or diffusers near the bottom of the lagoon (Fig. 2(c)) is the third method of aerating a lagoon. The advantage of the pump-type aerator is that it mixes the lagoon water while it aerates. The induced-aspiration aerator and the diffused aeration are more suitable for freezing conditions than the pump type.

Aerated lagoons differ from activated-sludge basins in that the degree of agitation is not sufficient to maintain all solids in suspension at all times. Mixing the lagoon contents provides distribution of oxygen throughout the lagoon. A major portion of the heavy, discrete solids settle out to the bottom of the lagoon where they undergo anaerobic decomposition.

Of the aerobic lagoons, aerated lagoons are best suited for feedlot wastewaters. Their use in feedlots has increased in the last few years, mainly due to the development of floating aerators specially designed for feedlot wastes and due to odour problems with anaerobic lagoons.

Facultative Lagoon

As shown in Fig. 3, facultative lagoons have a surface aerobic zone and a bottom anaerobic zone. During periods of non-freezing conditions, lagoons may have a well-defined thermocline dividing the water into an upper layer of warmer, lighter water and a lower layer of cooler, heavier water. A barrier to vertical mixing is formed by the thermocline. Oxygen content in the aerobic zone ranges from zero to saturation from night to daylight hours following diurnal variations. At times odours may emanate, particularly during periods when the thermocline is disturbed and the aerobic surface layer is not maintained.

Most of the biomass formed in the aerobic zone will settle into the anaerobic zone and undergo decomposition with the influent solids. Therefore, less algal and bacterial cell will be expected in the effluent than from aerobic-algal lagoons. The facultative lagoon has an advantage over the aerobic lagoon in that the lower anaerobic zone will remove additional carbon through methane formation.

Minimum depth of facultative lagoons is 1 m; usual depth is 2 m. Many feedlot waste lagoons designed as aerobic-algal lagoons usually function as facultative lagoons.

Anaerobic Lagoons

Anaerobic lagoons function without atmospheric oxygen. Decomposition of the organic matter is a reduction process carried out by anaerobic bacteria in two stages.

In the first stage, hydrolytic, acid bacteria reduce the organic compounds into volatile acids which are further reduced by methane fermentation

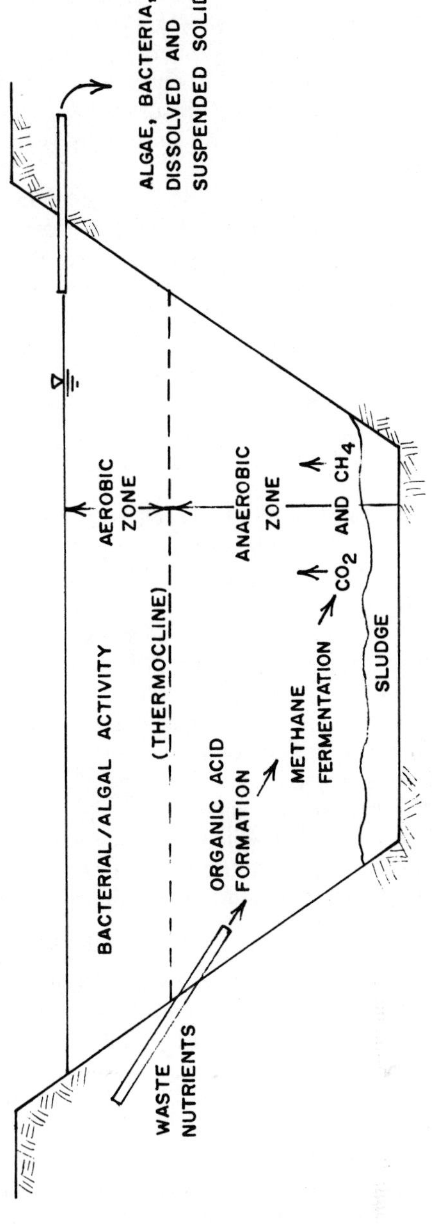

FIG. 3. Zones of biological reactions in a facultative lagoon.

which produces carbon dioxide and methane gas. The release of the carbon dioxide and methane from the lagoon is the principal mechanism for removing carbon from the treatment system.

In general, the purpose of anaerobic lagoons is not water purification but stabilization of organic matter. As such, anaerobic lagoons differ greatly from aerobic-algal lagoons which are wastewater purification schemes. Anaerobic lagoons may be used as sedimentation units to reduce the load on subsequent aerobic treatment units.

Anaerobic lagoons can be made as deep as groundwater levels and temperature profiles permit. It is best to avoid penetrating into the water table even though anaerobic lagoons seal up quickly. There is no need to provide large surface areas since neither oxygen diffusion from the atmosphere nor light penetration is important.

LAGOON DESIGN CRITERIA

Design of lagoons is based on waste properties and on environmental factors.

Waste properties which are used in lagoon design are mainly: volume of waste, biochemical oxygen demand (BOD) for aerobic lagoons, and total volatile solids (TVS) for anaerobic lagoons. Although only two parameters might be used in sizing lagoons, it is assumed that the lagoon designer is a professional engineer who is in a position to consider all the other bio-engineering waste properties in the design and would be in a position to judge the applicability of the general formulae to the local conditions.

Environmental factors which must be considered and used in the design of lagoons are: temperature, volume or surface area, solids concentration, and detention time. For aerobic systems, the design must include the transfer of oxygen to the biomass operating in the lagoon. There are many other factors and considerations which must enter into the process of lagoon design by the engineer. For example, pH, alkalinity, acidity, etc., are important factors which are considered but are not necessarily included in a quantitative way in design formulae.

Lagoons are designed on the basis of loading rates. These loading rates are expressed as units of an appropriate waste parameter per unit of time for each unit of the relevant environmental parameter (see Table I). The environmental parameter is usually surface area for aerobic-algal lagoons and volume for anaerobic systems.

Aerobic-Algal Lagoons

Aerobic-algal lagoons are aerated through natural diffusion of oxygen from the air and from the photosynthetic production of oxygen. The oxygen diffusion rate is low and is affected significantly by many factors such as

TABLE I

LIVESTOCK WASTE PROPERTIES RELATED TO LAGOON DESIGN AND PERFORMANCE. VALUES WILL VARY DUE TO ANIMAL AGE, FEED RATION, BEDDING, ETC. ALL VALUES ARE PER 1000 kg OF LIVESTOCK WASTE (from ref. 10)

Parameter	*Units*	*Swine feeder*	*Dairy cow*	*Beef feeder*	*Poultry layer*
Raw manure					
Weight	kg/day	65	82	60	53
Volume	m^3/day	0·069	0·081	0·062	0·054
TS	kg/day	6·0	10·4	6·9	13·4
TVS	kg/day	4·8	8·6	5·9	9·4
TSS[a]	kg/day	2·3	3·1	2·4	5·4
BOD_5	kg/day	2·0	1·7	1·6	3·5
COD	kg/day	5·7	9·1	6·6	12·0
N_{total}	kg/day	0·45	0·41	0·34	0·72
P_{total}	kg/day	0·15	0·073	0·11	0·28

[a] Calculated as follows: swine, 0·39 × TS; dairy, 0·3 × 75; beef, 0·35 × TS; poultry, 0·40 × TS.

temperature, wind action, wastewater properties, etc. Since there is no way of controlling such factors, loading rates must be extremely low.

Loadings range from 20 kg BOD_5/ha-day in cold climates where ice cover forms over lagoons during winter, to 50 kg BOD_5/ha-day in warm climates where no ice forms during winter.[11] These loading ranges should ensure free oxygen concentration of a least 1 mg/litre. There is little malodour emanating from lagoons of this type. Depth should be 1–1·5 m. Loading rates as high as 200–400 kg BOD_5/ha-day can be used for shallow depths, not deeper than 50 cm.[9] Detention time varies from 5–7 days to several weeks. Solids concentration must be below 0·5% and preferably 0·1–0·2%, in order to keep turbidity low and permit light penetration.

A 10 000 swine feedlot with 500 000 kg of live weight would have a daily BOD production of 1000 kg. In a cold climate this 10 000 head feedlot would require a surface area of 50 ha for an aerobic-algal lagoon. In addition to this surface area, approximately 600 000 litres of water would be needed each day to dilute the waste to about 0·5% solids. Both the excessively large surface area requirements and the dilution water needed, make this type of lagoon impracticable for livestock waste treatment.

Aerated Lagoons

The amount of oxygen required for a given wastewater is related to the amount and type of BOD and the total suspended solids (TSS) of the wastewater. In general, the oxygen required is related to the BOD removed according to eqn. (1):

$$Y = ax \tag{1}$$

where Y = kg of O_2 required per unit of time, x = kg of BOD_5 removed from wastewater per unit of time, and a = coefficient of oxygen transfer (kg O_2/kg BOD_5 specific to the aeration equipment and wastewater characteristics).

Because of the difficulty of measuring the coefficient of oxygen transfer in wastewaters, and because aeration equipment is usually rated in clear water, it is best to determine oxygen requirements for a given wastewater in a pilot-plant study conducted under conditions expected in the field. Based on pilot-plant data, the design engineer is then in a position to compute the capacity of the aerator, the dilution of the wastewater needed to bring the TSS to an optimum level, and the detention time needed for a specific BOD reduction.

For livestock wastes, the oxygen transfer coefficient, a of eqn. (1), varies from 1·5 to 2·0. The value usually used is 2·0.[19] Therefore, for feedlot wastes:

$$\text{kg } O_2 \text{ required/day-animal} = 2{\cdot}0 \text{ (kg } BOD_5\text{/day-animal)} \tag{2}$$

Equation (2) gives the oxygenation capacity needed to be provided by the aeration equipment. The oxygen transfer capacity of each type of aerator is related to its design and is given in terms of its horsepower. The oxygen transfer capacity of aeration equipment ranges between 1·0 and 2·0 kg O_2/h.p.-hr (1·4–2·7 kg O_2/kWh), with 1·4 kg O_2/h.p.-hr commonly used.

In addition to oxygenation capacity, the mixing capacity of the aerator must be considered. Mixing capacity is affected by the density of suspended solids (SS) and the pumping characteristics of the aerator. Mixing requirements for lagoons are normally between 3 and 18 h.p./1000 m^3, with higher SS requiring higher values. The mixing requirement would be even higher if the aerated lagoon were designed to keep all solids in suspension. A value of 10 h.p./1000 m^3 is recommended for livestock waste lagoons which will give adequate mixing to maintain aerobic conditions throughout the liquid of a properly sized lagoon.

Surface area is not a design parameter in aerated lagoons and should be minimized to reduce heat loss due to convection, evaporation and radiation. Depth of aeration lagoon should be at least 3 m and may be deeper, which generally improves mixing. The design detention time is 10 days in non-freezing areas to 30 days where ice cover is experienced for several weeks. The SS concentration should be maintained at 0·3–0·5 %, although satisfactory performance will occur up to 1·0 %. With provision for sedimentation of suspended solids in the effluent, BOD reductions of

90% or more may be expected. Figure 4 gives expected BOD reduction for different detention times and temperatures.

Once the h.p. requirements for oxygenation and mixing are determined, the engineer must then determine the size and number of aerators to use. It is preferable to select several smaller units to provide agitation over the entire surface area than to select a few large units, for example four 20 h.p. units as compared to one 80 h.p. unit.

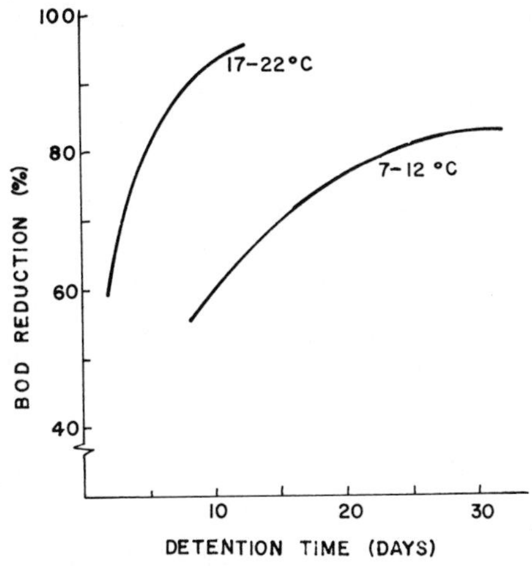

FIG. 4. Expected BOD reductions in aerated lagoons for summer and winter operation with sedimentation of effluent suspended solids.

Facultative Lagoons

In that facultative lagoons have a surface aerobic layer, their design with respect to oxygenation is on the same basis as the aerobic lagoon. The difference is that their depth is increased to allow a bottom anaerobic layer of water to form. For aerobic-algal lagoons to function as facultative lagoons, the depth should be greater than 2 m. However, the same factors making aerobic-algal lagoons unsuitable for livestock wastes would apply. Aerated lagoons may be designed so that only the surface layer of about 2 m is oxygenated. A total depth of 4 m or more would be required.

The advantage of the facultative lagoon occurs when the h.p. for mixing, as controlled by volume/detention times, is greater than that required for oxygenation. Less h.p. may be used by providing only the needed oxygenation capacity. The facultative lagoon surface area should be designed on the basis of 30 m^2/h.p.[7] A detention time of 20–30 days should be provided, the longer period for areas where ice occurs in the winter.

Anaerobic Lagoons

The design of anaerobic lagoons should be such as to favour the development of methane bacteria. The primary factors are temperature, pH, detention time, loading frequency and loading rate. The presence of free oxygen is inhibitory to methane bacteria. Methane bacteria will grow below 15 °C but higher temperatures are preferred, as indicated by the curve

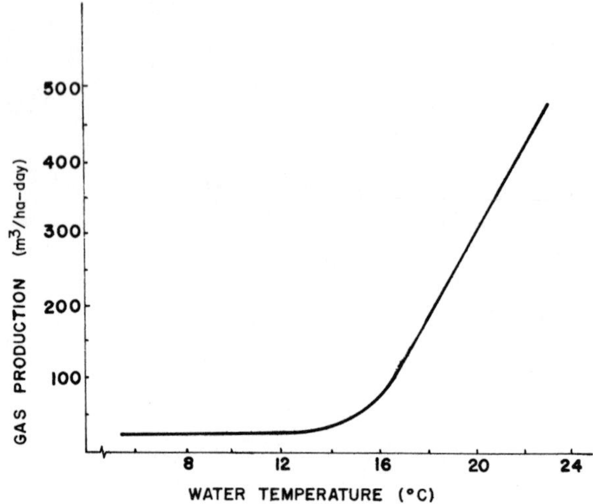

FIG. 5. Lagoon gas production as a function of water temperature at a constant loading (from ref. 16).

in Fig. 5 on lagoon gas production versus temperature. Therefore, during winter operation in cold climates, methane bacteria are virtually inactive. In warmer climates anaerobiosis increases, allowing larger loading rates.

Detention times of 30–60 days should be employed during periods of active decomposition, *i.e.* lagoon temperatures above 15 °C. With most livestock systems, the hydraulic loading is too small to provide these detention times unless dilution water is added or effluent is recycled. During cold periods the detention time should be lengthened by decreasing the amount of water added.

The pH range for active methane fermentation is 6·8–7·4. Experience has indicated that anaerobic livestock waste lagoons normally operate with a pH of 7·0 and above. If the pH drops below 6·0, offensive odours will be produced.

Daily or 'continuous' loading is recommended. Weekly or even bi-weekly loadings have been shown to allow satisfactory performance in an area where freezing does not occur.[5]

The sizing of anaerobic lagoons is based on organic matter loading for a unit volume, g VS/m^3 for each day. Loadings of 6 to 166 g VS/m^3-day have been reported for lagoons treating a variety of livestock wastes.[9] Based on the VS production in Table I, the recommended lagoon design volume for various livestock at different loading rates and average January temperatures is presented in Fig. 6. These loading rates are based upon data

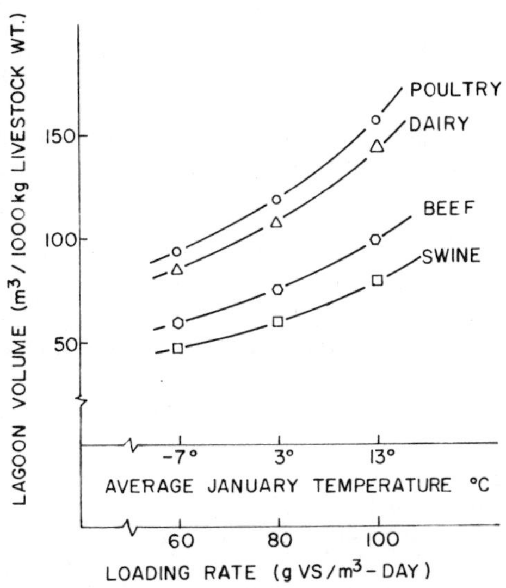

FIG. 6. Anaerobic lagoon design volume for various loading conditions or temperatures, and for different livestock.

reported for existing lagoons that appeared to be working satisfactorily, *i.e.* with minimum odour nuisance and good gas production (frequent gas boils). The presence of purple sulphur bacteria also indicates a satisfactorily functioning lagoon.

Figure 7 shows a map of recommended loading rates for the northern hemisphere related to average January temperatures. The average January temperature for the 100 g VS/m^3-day curve is about 13 °C and for the 60 g VS/m^3-day curve, about −7 °C.

For small livestock production units the lagoon depth should be at least 1·5 m, but greater depths are desirable. Depths up to 6 m may be used. The lowest operating level of the lagoon should be above the highest level of the groundwater table.

Loading of anaerobic lagoons should begin in the spring or early summer so that a balanced biomass may be developed before cold conditions set in.

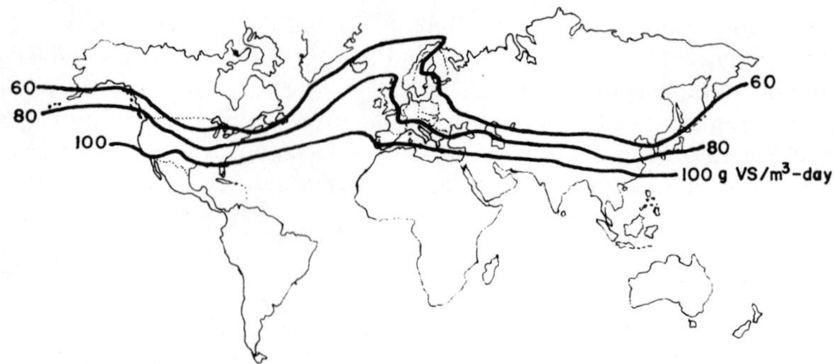

FIG. 7. Recommended maximum loading rates in grammes of volatile solids (VS) per cubic metre each day for feedlot anaerobic lagoons in the northern hemisphere.

Seeding the lagoon with digester sewage sludge or sludge from a functioning anaerobic lagoon will speed up its functioning.

Sludge accumulation will vary with loading rate and type of livestock waste. For cattle on a high-roughage feed, the annual sludge accumulation rate will be 15–25 % of the design volume. Sludge accumulation for swine and poultry lagoons will be at a lower, annual rate of 5–10 % of the lagoon volume per year.

Multiple Cell Lagoons

A lagoon system may have multiple cells in series or in parallel. They may all be anaerobic, aerobic, facultative or a combination. The lagoon receiving the feedlot waste should be loaded at the design level discussed previously. Later stages can be sized based on the performance characteristics of the preceding lagoon(s).

MANAGEMENT OF LAGOONS

Satisfactory operation of lagoons and ponds depends upon good management in addition to correct design.

Odour Control

Spring warm-up for both aerobic and anaerobic lagoons may produce higher odour levels. A period of four to six weeks may occur when odours are emitted. Mechanical mixing can shorten the odorous period in the spring warm-up. Research has indicated that mechanical aeration of the surface layer at a lower aeration rate than that required for treatment can provide odour control, which is providing a condition similar to that of a facultative lagoon.

Effluent Disposal
Disposal of effluent will be needed either because of excess liquid or because of the need for dilution when evaporation increases the salts content in a lagoon, as in the case of recycling lagoon effluent for flushing wastes from buildings. Also, feedlot runoff collected in a detention basin will need to be field spread. Normally, irrigation by sprinklers, ridge and furrow or overland flow is used for effluent disposal. For small livestock facilities a tank wagon may be used to haul the liquid (usually with sludge) to cropland.

A maximum inorganic salts content of 5000 mg/litre can be irrigated without plant damage. In regions with low rainfall (less than 40–50 cm) and high evaporation rates, a maximum salts content of 3000 mg/litre is recommended; otherwise plant toxicity and saline soil conditions may result.

Solids Disposal
Some form of solids disposal must be utilized to maintain a functioning treatment lagoon. In the case of mechanically aerated lagoons the mixing action of the aerator will keep solids in suspension and disposal of effluent/solids may be performed at the same time through irrigation.

In a functioning anaerobic swine lagoon (frequent gas boils), solids will be resuspended and irrigating supernatant should remove enough of the solids to maintain the required volume. Additional solids may be removed by pumping from the bottom sludge layer.

In anaerobic lagoons for cattle waste, sludge will accumulate and need to be removed. The type of sludge removal scheme—batch (once every four to six years), periodic (once or twice a year), or continuous (monthly or more frequent)—will depend on farm management preference, cropping practices, equipment availability and non-farm markets for the sludge.[13]

Pumps and Pipes for Recycling Flush Water
In facilities recycling supernatant from anaerobic and aerobic lagoons, a white crystalline material has been found to deposit on metal surfaces in the pump and pipes, causing plugging and equipment malfunction. This material is reported to be $MgNH_4PO_4$.[2] It is recommended that plastic pipe and a pump with a plastic housing and impeller be used for recycling flush water as the deposition of $MgNH_4PO_4$ is greatly reduced. Flushing the recycling system with a 1:50 dilution of glacial acetic acid as a regular maintenance procedure will prevent the crystalline build-up.

PERFORMANCE OF LAGOONS

The performance of livestock waste lagoons must be evaluated on the basis of public health, aesthetics, ecology and economics in addition to pollutant removal efficiency.

Generally speaking, both aerobic and anaerobic lagoons have been found to be a safe practice from the standpoint of public health. Under good management, fly and mosquito breeding is inhibited and most pathogens and viruses are killed or greatly reduced in number.[1,8] Sanitation around a feedlot facility is improved with lagoon treatment of waste as compared to conventional handling of manures.

In the area of aesthetics, the principal problem is odours. Aerobic lagoons meet this condition completely. Anaerobic lagoons can cause an odour nuisance. However, with proper design, operation and management the odour level can be reduced to an acceptable level and then only for a brief period in the spring warm-up.

Lagoons reduce the amount and impact of organic wastes from livestock facilities reaching streams. The lagoon effluents and sludges that are spread on cropland in a designed irrigation system do not affect the environment adversely.

With regard to economics, the construction and operation of ponds and lagoons are usually the most cost-effective methods of livestock waste treatment. With the high cost of energy the facultative and anaerobic lagoons are being looked at more favourably than aerated lagoons except when odour is potentially a problem.

Expected reductions of wastewater parameters for aerated and anaerobic single-stage lagoons are shown in Table II. The higher reduction values can be expected for properly designed and managed lagoons during non-freezing conditions. During frozen conditions, reduction of the various parameters is less. Greater reductions can be obtained if a lagoon of more than one stage is used.

The quality of lagoon effluent is too poor for direct discharge to a waterway. Controlled irrigation of effluent on cropland can effectively remove and utilize the organic material, nitrogen and phosphorus.

EXAMPLES OF LAGOONS FOR LARGE FEEDLOTS

If the siting of a livestock facility can allow for some odour, the use of an anaerobic lagoon is recommended as the first stage of a lagoon system. If odour control is needed, then an aerobic lagoon system is needed. Storage capacity for wastewater loading should be provided for the period that effluent irrigation may be restricted due to freezing conditions or cropping.

Dairy

Assume 2400 cow equivalents and associated milking facility wastes to be treated in a two-stage, series lagoon system, anaerobic and unaerated facultative. Mean January temperature is 3°C. Cows are in total confinement. Odours are not expected to be a major problem. What will be

TABLE II

EXPECTED PERFORMANCE OF SINGLE-STAGE FEEDLOT WASTE LAGOONS
(from refs. 6, 9, 12, 14, 15, 17, 18, 20)

Parameter	*Aerated lagoon*				*Anaerobic lagoon*		
	Influent range (ppm)	*Effluent range (ppm)*	*Reduction range*		*Influent range (ppm)*	*Effluent range (ppm)*	*Reduction range (%)*
			Unclarified (%)	*Clarified (%)*			
TS	1 000–10 000	300–3 500	40–75	60–85	2 000–15 000	1 000–3 000	50–75
TVS	700–7 500	200–2 400	40–75	60–85	1 200–10 000	300–4 000	60–90
TSS	300–6 000	150–3 000	50–80	70–95	500–8 000	100–3 000	60–90
BOD_5	400–3 000	150–1 200	60–80	70–95	600–3 500	200–1 500	60–90
COD	1 200–8 000	400–3 000	60–75	70–95	1 500–9 000	450–3 000	60–90
N_{total}	400–5 000	300–2 400	40–70	50–80	500–6 500	200–1 200	60–85
P_{total}	80–1 500	60–1 200	20–30	70–85	100–1 800	20–200	70–95

the size of the lagoons? What treatment may be expected? What management features must be incorporated?

Table III presents a summary of the design and operation for the lagoon system. The loading rate for the anaerobic lagoon is the recommended value for mean January temperatures of 3 °C. The unaerated facultative lagoon is based on the loading rate for aerobic-algal lagoons.

The milking facility wastewater production was calculated on the basis of 20 litres/cow/day. It was assumed that the TVS, TSS, BOD_5, COD, N and P of the milking facility wastewater was negligible as compared to that of manure.

The detention time for the wastewater hydraulic loading is too long, 800 days. Effluent from the final lagoon needs to be recycled, either via water for flushing wastes to the lagoon or directly to the anaerobic lagoon. Sixty days detention time is needed: total hydraulic flow = 1600 m^3/day. The recycled effluent volume will need to be 1480 m^3/day or about 1·0 m^3/min. The detention time of the second lagoon will be 75 days.

With recycling of effluent, the amount of evaporation is increased. A sample of lagoon mixed liquor should be tested for its inorganic dissolved salts concentration at least once a year. When this concentration exceeds 3000 mg/litre, add fresh water. Surface or roof water may be directed to the lagoon to provide dilution.

An effluent and sludge disposal programme is needed. If sludge and mixed liquor from the anaerobic and/or facultative lagoon is irrigated, the TS and inorganic salts concentrations may be controlled. If only effluent is irrigated, a periodic (every third or fourth year) sludge removal programme must be used for the anaerobic lagoon.

Particularly for the design of the anaerobic lagoons, it would be desirable to provide parallel lagoons to handle equal amounts of manure and wastewater production. This would provide some flexibility in operation and facilitate sludge removal.

An alternative waste system to the lagoons would be a storage pond. The design volume would be based on the production of manure, wastewater and direct rainfall on the pond surface. All rainfall runoff and roof water should be diverted. Commercially available piston-type pumps can be used to transport the manure and wastewater to the pond. The waste should enter the pond beneath the surface, as a floating crust will form which controls odour release. Provision must be made to agitate (mix) the contents of the pond before application to cropland either by tank wagon or irrigation. The volume of storage pond needed for 365 days is 44 000 m^3. Assuming a depth of 4·0 m, the surface area needed would be a little more than 1·1 ha.

Swine

Assume a 100 000 head, total confinement, finishing operation with an

TABLE III

DESIGN AND EXPECTED OPERATION OF A LAGOON SYSTEM FOR 2400 DAIRY COWS

		Anaerobic		*Facultative*	*Effluent*[a]
Loading rate		80 g VS/m^3-day		50 kg BOD/ha-day	—
Hydraulic loading		120 m^3/day		120 m^3/day	—
Volume of lagoon		96 000 m^3		120 000 m^3	
Depth		5·0 m		2·0 m	—
Surface area (2:1 side slope)		2·25 ha		6·25 ha	—
Suspended solids		5 000 ppm		750 ppm	—
Detention time		800 days		1 000 days	—
Dilution (recycling) water		1 480 m^3/day		1 480 m^3/day	—
Detention time (with recycling)		60 days		75 days	—
Effluent parameter (% removal in parentheses)					
TVS	(80)	1 540 kg/day	(70)	460 kg/day	3 800 ppm
TSS	(85)	420 kg/day	(70)	130 kg/day	1 080 ppm
BOD_5	(85)	230 kg/day	(70)	70 kg/day	580 ppm
COD	(80)	1 630 kg/day	(70)	490 kg/day	4 080 ppm
N_{total}	(70)	110 kg/day	(60)	45 kg/day	380 ppm
P_{total}	(85)	10 kg/day	(75)	3 kg/day	25 ppm
Sludge removal		3–4 years[b]		10–12 years[b]	

[a] Effluent from second lagoon. Based on total recycling for required detention time, *i.e.* no additional dilution.
[b] Period can be lengthened by irrigating liquid and sludge for effluent disposal.

TABLE IV

DESIGN OF AN AERATED LAGOON AND CLARIFIER FOR 100 000 FEEDER PIGS

Lagoon:	
Manure volume	410 m^3/day
Dilution water (influent 1·0 % TS)	3 200 m^3/day
Detention time	30 days
Lagoon volume	108 000 m^3
Oxygenation (2·0 × BOD_5)	24 000 kg O_2/day
h.p. (based on 1·4 kg O_2/h.p.-hr)	720 h.p.
h.p. (based on 10 h.p./1 000 m^3)	1 080 h.p.
Depth	5·0 m
Surface area (2:1 side slope)	2·3 ha
Clarifier:	
Surface area (40 m^3/day-m^2)	90 m^2
Volume (3 hr detention)	450 m^3

average weight of 60 kg per pig and with the waste to be treated with an aerated lagoon to provide odour control. Clarified water from the aerated lagoon is recycled for flushing wastes from buildings. Average January temperature is −7 °C. What size of lagoon will be required? How much aeration is required for the lagoon?

Table IV presents a summary of the design and Table V a summary of the expected operation of the aerated lagoon. In order to maintain a satisfactory MLSS, the influent waste needs to be diluted to 1·0 % TS by the addition of 3200 m^3/day. The h.p. required for mixing the volume of 108 000 m^3 controls the design. Therefore, 11 aerators of 100 h.p. or 22 aerators of 50 h.p. are needed. The lagoon geometry should be selected to provide an equal area of influence for each aerator, *e.g.* a 65 × 355 m lagoon with two rows of 11 aerators each.

TABLE V

EXPECTED OPERATION OF AN AERATED LAGOON FOR 100 000 FEEDER PIGS

		Clarified effluent		
Parameter	*Influent (kg/day)*	*Removal (%)*	*Total (kg/day)*	*Concentration (ppm)*
TS	36 000	80	7 200	2 000
TVS	28 800	80	5 800	1 600
TSS	13 800	90	1 400	390
BOD_5	12 000	90	1 200	330
COD	34 200	85	5 100	1 400
N_{total}	2 700	60	1 100	310
P_{total}	900	80	180	50

The settled sludge should be recycled to the influent wastewater. Provision is needed to dispose of effluent and solids to control the inorganic salts concentration.

REFERENCES

1. Axtell, R. C., Rutz, D. A., Overcash, M. R. and Humenik, F. J., Mosquito production and control in animal waste lagoons. In: *Proceedings of International Symposium on Livestock Wastes*. ASAE, St Joseph, Mich., 1975.
2. Booram, C. V. and Smith, R. J., Manure management in a 700-head swine finishing unit in the American Midwest. *Water Research*, **8** (1974) 1089–97.
3. Booram, C. V., Hazen, T. E. and Smith, R. J., Trends and variations in an anaerobic lagoon with recycling. In: *Proceedings of International Symposium on Livestock Wastes*. ASAE, St Joseph, Mich., 1975.
4. Eckenfelder, W. W., *Industrial Water Pollution Control*. McGraw-Hill, New York, 1966.
5. Howell, E. S., Overcash, M. R. and Humenik, F. J., Unaerated lagoon response to loading intensity and frequency. Paper No. 74–4514, ASAE, St Joseph, Mich., 1974.
6. Humenik, F. J., Overcash, M. R. and Driggers, L. B., *Swine Production Industry Waste Characterization and Management*. Agricultural Engineering Dept., North Carolina State University, Raleigh, NC, n.d.
7. Humenik, F. J., Overcash, M. R. and Milter, T., Surface aeration: design and performance for swine lagoons. In: *Proceedings of International Symposium on Livestock Wastes*. ASAE, St Joseph, Mich., 1975.
8. Krieger, D. J., Bond, J. H. and Barth, C. L., Survival of salmonellae and fecal coliforms in swine waste lagoon effluents. In: *Proceedings of International Symposium on Livestock Wastes*. ASAE, St Joseph, Mich., 1975.
9. Loehr, R. C. *Agricultural Waste Management*. Academic Press, New York, 1974.
10. Midwest Plan Service, *Livestock Waste Facilities Handbook* (MWPS-18). Iowa State University, Agricultural Engineering Building, Ames, Iowa, 1975.
11. Missouri Basin Engineering Health Council, Waste Treatment Lagoons, State of the Art Publication 17090 EHX 07/71. US Environmental Protection Agency, Washington, DC, 1971.
12. Nordstedt, R. A., Baldwin, L. B. and Hortenstine, C. C., Multistage lagoon systems for treatment of dairy farm waste. In: *Livestock Waste Management and Pollution Abatement*. ASAE, St Joseph, Mich., 1971.
13. Nordstedt, R. A. and Baldwin, L. B., Sludge accumulation in anaerobic dairy waste lagoons. Paper No. 74–4039. ASAE, St Joseph, Mich., 1974.
14. Nordstedt, R. A. and Baldwin, L. B., Sludge management for anaerobic dairy waste lagoons. In: *Proceedings of International Symposium on Livestock Wastes*. ASAE, St Joseph, Mich., 1975.
15. North Central Regional Committee 93, *Livestock Waste Management with Pollution Control* (MWPS-19), ed. R. J. Smith and R. J. Miner. Midwest Plan Service, Iowa State University, Agricultural Engineering Building, Ames, Iowa, 1975.

16. Oswald, W. J., Advances in anaerobic pond systems design. In: *Advances in Water Quality Improvements*, ed. E. F. Gloyna and W. E. Eckenfelder, Jr. University of Texas Press, Austin, Tex., 1968.
17. Smith, R. J., Hazen, T. E. and Miner, J. R., Manure management in a 700-head swine building: two approaches using renovated wastewater. In: *Livestock Waste Management and Pollution Abatement*. ASAE, St Joseph, Mich., 1971.
18. Taiganides, E. P., Theory and practice of anaerobic digesters and lagoons. *2nd National Poultry Litter and Waste Management Seminar*, College Station, Texas A & M University, Tex., 1968.
19. Taiganides, E. P., Aeration of animal wastes. Paper presented at 1968 Farm Power Conference, Pittsburgh, Pa., 1968.
20. Willrich, T. L., Primary treatment of swine wastes by lagooning. In: *Management of Farm Animal Wastes*. ASAE, St Joseph, Mich., 1966.

19

Aerobic Stabilization of Pig Feedlot Wastewaters in Czechoslovakia

J. JONAS

Hydroprojekt, Prague, ČSSR

INTRODUCTION

In the Czechoslovak Socialist Republic (ČSSR), pig wastes from large-scale pig feedlots are disposed of on fields after a period of storage under anaerobic conditions. Difficulties arising are odours and occasionally surface-water pollution. To alleviate these problems, a series of pig waste stabilization systems were tested.

Two main stabilization and disposal systems for hog wastes are being considered in ČSSR. The first system only improves the properties of the hog wastes before application as a fertilizer, *i.e.* odour removal, improvement of pumping and hygienic properties. The second system provides sufficient treatment to allow for the liquid effluent to be discharged into streams and to diminish substantially the volume of solid residue while improving its value as fertilizer. In the first method the aim is to preserve to the maximum the fertilizing components of the hog wastes, while in the second method the aim is to remove the fertilizing components and/or to convert them into insoluble residues.

THERMOPHILIC ACTION

The use of thermophilic aerobic stabilization is considered to be the most convenient method of stabilization of hog wastes which are to be used for fertilization on agricultural land.[1–4] Several such facilities have been designed and should be in operation in ČSSR in 1976. The main difficulty of this system appears to be foaming, which greatly impedes the system's performance. Furthermore, storage for 90 days must be provided in the system. Stabilization of hog wastes on an operational scale was tested in a waste channel designed with rounded ends, similar to an oxidation ditch (Fig. 1). The channel was aerated by vertically mounted aeration mixers set approximately 30 m apart. The aeration mixers are designed both for

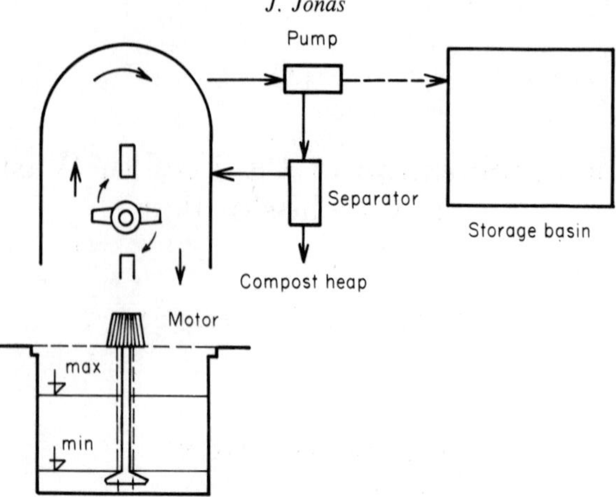

FIG. 1. Schematic layout of waste channel stabilization of hog wastes.

aeration and for the propelling of wastes. Evaporation of 8 mm of water per day occurred, which translates into a reduction of 2 litres per head of the daily waste volume. The channel was designed on the basis of 0·25 m^2/pig. The initial liquid height was 20 cm (water was used at first). When a height of 100 cm in the channel was reached from the addition of pig wastes, part of the stabilized hog wastes were pumped into storage. The temperature and pH of the mixture began to rise slowly only after 20 days of operation. After 30 days of operation the temperature reached 35 °C and the pH 8·1; the maxima attained were 46·5 °C and a pH of about 9 (Fig. 2).

This method of aerobic stabilization in the lower thermophilic range furnishes an odour-free product, but with only partially improved hygienic

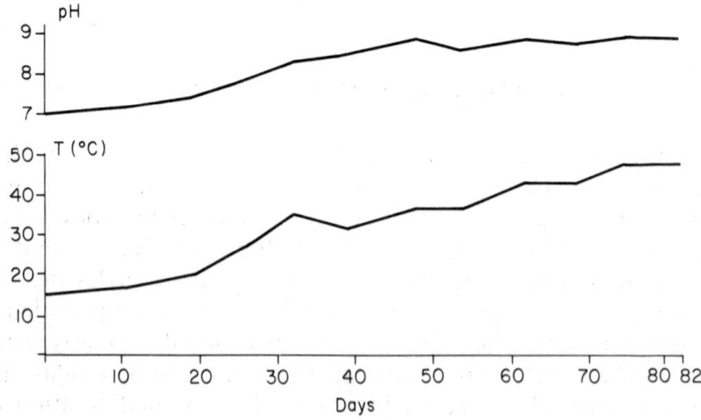

FIG. 2. The course of pH and temperature (experiment in Tošov).

qualities. During storage the stabilized wastes do not settle. They can easily be pumped and spread on fields. Nitrogen compounds appear to be bound in the form of humic acids. The organic loading rate in the process is very low, which may be responsible for the relatively low heat production. The loading of organic matter should exceed 1 kg/m^2/day to reach the lower limit of the thermophilic range, *i.e.* temperatures over 40 °C. Power requirements for the aeration devices are in the range of 20–30 W/pig.

Reduction of volatile solids was in the range of 40–50 %. The total amount of nitrogen was reduced by 20–30 %, while the number of mesophilic and psychrophilic bacteria was reduced by 99–99·9 %. The stabilized wastes smelled of pond mud with a slight ammonia odour.

ACTIVATED SLUDGE AERATION SYSTEMS

A two-stage activated sludge plant was designed for 1400 pigs.[5] Raw hog wastes are conveyed mechanically twice daily into a pit, and coarse solids are removed on a vibrating screen. The liquid effluent is fed into the first stage of the aeration basin which has a volume of 63 m^3 (45 litres/pig capacity) and a surface aerator powered by a 11 kW motor. The wastewater then flows directly into the second-stage aeration basin which has a volume of 200 litres/pig, including the secondary clarifier volume. The second-stage aeration basin is also equipped with a surface aerator powered by a 30 kW motor. Mixed liquor or return sludge from the clarifier is returned into the first aeration basin or it can be returned to the inflow of the vibrating screen. Also, the plant liquid effluent may be recycled through the plant. Influent liquid from the screen is diluted with returned mixed liquor or plant effluent in the ratio of 1:5 to achieve a 7000–8000 mg/litre BOD concentration in the diluted wastes entering the first stage of aeration.

In the second stage, aeration is carried to the point where denitrification of wastes takes place. As a result, the effluent contains 100–150 ppm of total nitrogen in the summer, which increases to 600–800 ppm in the winter, when the nitrification processes are inhibited by low temperatures in the aeration basin (0·5–7 °C). The plant effluent has a BOD of 40–120 ppm and a COD of 400–1000 ppm. The mechanical screen removes 42 % of solids on average; solids content of the separated solids was 22 %. The separated solids were put on a field compost heap where spontaneous rise of temperature was observed.

Another activated sludge treatment plant tested was the Bioclar system. It is similar to the first system, the only difference being that the mixed liquor from the first stage flows through a sedimentation tank before going into the second aeration basin. The settled-out sludge is returned into the first-stage aeration tank. The same process is used for the second stage of aeration. The tanks have a circular cross-section of 3 m diameter and are cylindrical in shape. They are aerated by compressed air. The first and second stages each provide a volume of 135 litres/pig. The plant was designed to treat wastewaters from 1200 pigs.

During the operation, excessive foaming occurred which was controlled by dosing the liquid with defoaming agents such as oleic acid, at the rate of 1 kg/100 hogs/day. The effluent in winter had 1200–2400 ppm of total nitrogen and 1000–3000 ppm COD. The degree of nitrification and denitrification in this plant was very low. The foaming difficulties were attributed to inadequate supplies of oxygen. Surface mechanical aerators are more suitable for the aeration of the hog wastes in the activated sludge processes because they help to destroy the resulting foam.

COMBINED MUNICIPAL–FEEDLOT WASTEWATER TREATMENT

A treatment plant designed to treat the wastewaters from the city of Třeboň and from a 25 000 pig feedlot went into operation at the beginning of 1974. Since there is no other similar plant in the world, a detailed description of the treatment plant and the results obtained during the first year of operation may be in order.

The treatment plant is designed for 29 900 population equivalent (PE) and dry weather flow of 3186 m^3/day.

Municipal wastewaters are conveyed from the town into a pumping station through a combined sewer system and from there pumped through a 3 km long pressure pipeline of 50 cm diameter into the treatment plant. The pumping station is equipped with two working pumps with a capacity of 150 litres/sec each and a third pump of the same capacity as a standby reserve.

A flow chart of the plant is given in Fig. 3. The treatment plant unit operations, in their order of sequence, are: five screens, 100 cm wide with a 2 cm bar spacing; one Parshall flume; an aerated grit chamber of 60 m^3; two rectangular sedimentation basins, 18 m long, 3 m wide, 3·10 m deep, with an area of 116 m^2 and useful capacity of 360 m^3; a storm weather flow by-pass; six aeration basins provided with surface aerators of 1250 mm diameter with 7·6 kW motors (the basins are 6 × 6 m with a water depth of 2·6 m and total useful capacity of 536 m^3; the retention time is approximately 6 hr for average flow); four secondary sedimentation basins, each 18 m long and 3 m wide, with a water depth of 2·4 m, an area of 216 m^2 and useful volume of 517 m^3; activated sludge is pumped from the sludge sumps at the rate of 40 litres/sec. Surplus sludge is returned to the inflow of the sedimentation basins. The sludge from the sedimentation basins is pumped into the mixing and pumping sump adjoining the digester. Pig wastes from the feedlot are pumped into the same sump. Both types of sludge are pumped through sludge heaters into the first-stage digester of 15 m diameter with a useful

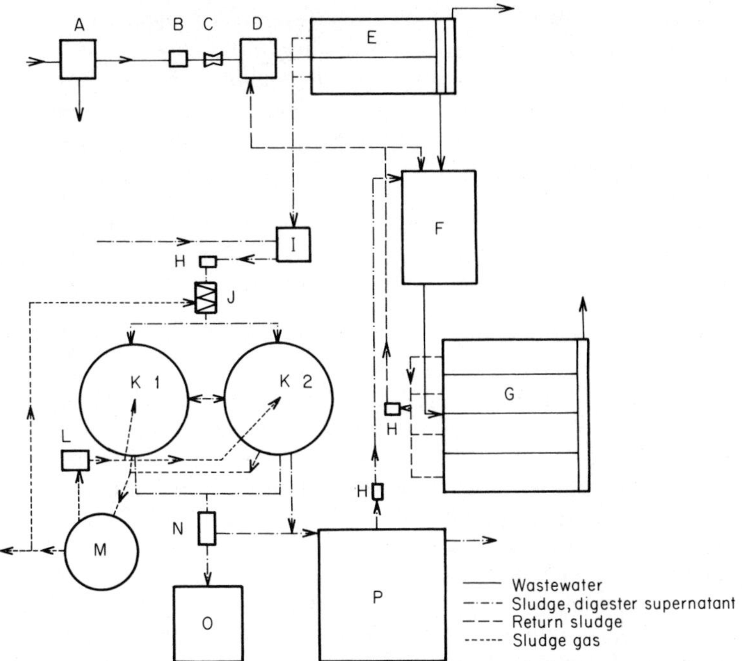

FIG. 3. Flow diagram of the Třeboň waste treatment plant. A, pumping station; B, screens; C, measuring flume; D, aerated grit chamber; E, sedimentation basin; F, aeration basin; G, secondary sedimentation basin; H, pump; I, sludge sump; J, heat exchanger; K, digester tank; L, gas compressor; M, gas holder; N, decanter centrifuge; O, composting plant; P, storage basin.

volume of 3200 m^3. The digested sludge is then pumped into the second-stage digester of 15 m diameter and useful volume of 3000 m^3. The digesters are heated to 33 °C and mixed by compressed sludge gas.

The digested sludge from the second stage was to be dewatered by a decanter centrifuge after thickening through sedimentation. The digester supernatant should have been released together with the centrifugate into a storage basin with a capacity of 13 000 m^3 and spread on land. However, in the course of construction of the treatment plant, the possibility of common treatment of digester supernatant from the digestion of the hog wastes and the municipal sewage in the aeration basins was tested. It appeared that it was possible to treat digester supernatant to a high degree provided a dilution ratio of 1:15 or more was attained, the concentration of BOD in the mixed wastewater did not exceed 400 ppm BOD, nitrogen content was below 200 ppm, and 5–6 hr of aeration time was achieved.

The digester supernatant contained 3000–4000 mg/litre total nitrogen,

3000–4000 mg/litre BOD and 4000–5000 mg/litre suspended solids. After treatment with municipal wastes the quality of the effluent was 25–30 ppm BOD, 30–40 ppm suspended solids and 30 ppm total nitrogen. It is possible, therefore, to treat digester supernatant together with municipal sewage during the ten months of the year when it is not possible to apply this material on agricultural land.

During the start-up of the plant it appeared that the flow of the municipal wastewater was only 1000–2000 m^3/day instead of 3186 m^3/day and its concentration was 150 mg/litre BOD instead of 400 mg/litre. The volume of hog wastes was greater than presumed in the design—150 m^3/day instead of 93 m^3/day.

The inoculation of raw hog wastes with warm sludge from the digester resulted in obnoxious odours and had to be stopped. The second stage of digestion is superfluous because the mixture of the municipal sludge and raw hog wastes is digested almost completely in the first stage. The digested sludge compacts in the second stage to such a degree that after a few hours of settling it has the tendency to clog the withdrawal pipelines. It is therefore necessary to agitate continuously the contents of the second-stage digester.

The concentration of the raw sludge is 8 %. The raw organic loading of the primary digester is 10 tonnes of pig wastes and 0·6 tonnes of sewage sludge. Detention time is 20 days. The volatile solids reduction in the first stage is 40 %. The daily production of sludge gas is 4000 m^3 containing 67 % of methane. This amounts to about 250 litres of methane gas per day per kg of wet waste (or 3 m^3 of CH_4/kg of dry solids/day). The digested sludge has a total solids content of 5·5 %, of which 75 % is volatile. The solids content, after dewatering, is 35 %. The dewatered digested sludge is composted and, after spontaneous heating up to 50–60 °C, forms a humus containing fertilizer which is quite safe from a hygienic point of view. The centrifugate has a solids content of 1·5 %. The centrifugate is pumped from the storage basins at the rate of 1·5 litres/sec into the aeration basin. The concentration of BOD of the centrifugate in the storage basin was measured to be less than 1500 mg/litre and the nitrogen content was 1000–1500 mg/litre.

Proposed changes in the plant include revision of the sludge withdrawal method from the secondary settling tanks, the elimination of the secondary digestion stage, and an increase in the capacity of digestion to handle wastes from 50 000 pigs. The supply of municipal sewage will be adapted to attain an even flow of wastewaters into the treatment plant.

Capital costs of a treatment plant for 30 000 PE will increase by 40–50 % to cover the treatment of waste from 60 000 pigs. Operational costs will increase by 20–25 % provided the volume of pig wastes is kept at a minimum by eliminating leakage of water from waterers, washing pens, etc. The increase in the power for aeration will not exceed 2 W/pig. However, the increase in the production of sludge gas would be 150 litres/day/pig with a heat value of 850 kcal.

It is necessary to design the treatment plant for nitrification followed by denitrification so as to minimize the nitrogen content of the effluent. Digester supernatant and centrifugate contain, during the start-up of the plant in Třeboň, 10 000–15 000 mg/litre total solids, 3000 mg/litre nitrogen and 2000–4000 mg/litre BOD. These concentrations could be decreased to about half of their original value by treatment in activated sludge basins after preliminary reduction of BOD and nitrogen content in the storage basin, and dilution with municipal sewage.

COMBINED INDUSTRIAL–FEEDLOT WASTEWATER TREATMENT

In the sewage treatment plant of the town of Štětí, 120 m^3 of raw pig wastes were added daily to a flow of 60 000 m^3 of industrial wastewaters, mainly from a kraft pulp plant. The quality of the plant effluent was not affected significantly by the addition of pig wastes. Total phosphorus in the effluent remained below 1 mg/litre, while nitrogen remained below 6 mg/litre on average. The addition of raw hog wastes can replace the necessary dosing with nutrients such as phosphorus and nitrogen salts, which is a common practice in the treatment of industrial wastes, deficient in nutrients essential to the activated sludge process. In combined treatment, raw pig wastes require dilution by municipal wastewaters equal to the volume of 5 PE. Furthermore, if the BOD concentration of mixed wastes is around 400 mg/litre, filamentous sludge may be produced resulting in sludge bulking. This requires special arrangements for the aeration basins.[6]

Combining pig wastes with the treatment of municipal or industrial wastewaters requires less capital and costs less than if each were treated separately. Furthermore, better performance is attained during winter when pig wastewaters are warmed up by the relatively warmer sewage wastewaters.

On the basis of studies such as these, it has been decided that future construction of large-scale pig feedlots (over 10 000 head) should be abandoned in cases where the wastes are to be disposed of by application on land after thermophilic stabilization. Feedlots with a capacity higher than 25 000 pigs should always be connected to a municipal or industrial wastewater treatment plant, using one of the methods of treatment described above.

REFERENCES

1. Pöpel, F., Selbsterwärmung bei der aeroben Reinigung hoch konzentrierter Substrate mit Hilfe von Umwälzbelüftern. *Landtechnische Forschung*, **18**(5) (1970).

2. Rincke, G., Auftreten und Einfluss thermophiler Prozesse bei der aeroben Stabilisation. *Aachen Gewässerschutz-Wasser-Abwasser*, **1**(6) (1971) 25–35.
3. Rüprich, W., Einsatz der Umwälzbelüfter für die Flüssigmistaufbereitung. *Landtechnische Forschung*, **18**(5) (1970).
4. Hotař, Z., Tekuté kompostování—nový způsob úpravy kejdy prasat. *Vodní hospodářství B*, **24**(8) (1974) 212–17.
5. Mach, M., Pavlík, M., Rešetka, D. and Nový, J., Výsledky zkušebního provozu zpracování tekutého podílu výkalů na farmě Rýcholka. *Vodní hospodářství B*, **24**(8) (1974) 207–11.
6. Grau, P., Maděra, V. and Ottová V., *Aktivační systémy s různým stupněm podélné disperze*. VŠCHT-KTVP, Prague, 1973.

20

Composting of Feedlot Wastes*

E. Paul Taiganides

Professor of Environmental Engineering, Department of Agricultural Engineering, Ohio State University, Columbus, Ohio, USA

INTRODUCTION

Composting is the aerobic, thermophilic decomposition of organic wastes to a relatively stable humus.[1] The resulting humus may contain up to 25% dead or living organisms and is subject to further slower decay, but is sufficiently stable so as not to reheat significantly, nor create odour nuisance, nor breed flies. The high temperatures achieved during the composting process kill pathogenic organisms, plant seeds and insect larvae. Furthermore, composting conserves the chemical nutrients in the raw wastes and enhances their availability and value as plant nutrients. In composting, mixing and aeration are provided to maintain aerobic conditions and permit adequate heat development, without excessive convective cooling occurring during mixing and aeration. The decomposition is done by aerobic organisms, primarily bacteria, actinomycetes, fungi and protozoa.

THE COMPOSTING PROCESS

Composting being mainly a biological process, environmental factors affecting the microorganisms involved in the process must be controlled so as to attain high decomposition speeds and, at the same time, produce a stable end-product. The main factors of strategic importance in composting are: moisture content, aeration, carbon to nitrogen (C/N) ratio, phosphorus and potassium content, temperature, pH, particle size, plus micronutrients and microenvironmental conditions conducive to the growth of aerobic organisms.

Moisture Content

One of the main advantages of the composting process is that it occurs at

* Prepared by Professor Taiganides specially for this publication.

low moisture contents. While most biological treatments of organic wastes are carried out at moisture contents well above 90 %, optimum moisture content of the composting process is 40–60 %. Thus, composting is most suitable for solid wastes in the range of 40–60 % total solids content. It is important, particularly during the initial phase of composting, that moisture content be maintained between 40 % and 60 % so as to ensure adequate water supply for the proper growth of the biological organisms involved in the process and for the biochemical reactions occurring during composting. Furthermore, high temperatures attained during the early stages of composting would dry up the pile, unless there is adequate water.

Excessive moisture tends to create anaerobic conditions which could result in the release of offensive malodours. Furthermore, essential nutrients and intermediate products of decomposition could leach out.

Moisture content can be maintained by spraying water over the compost mixture if moisture content is low. If moisture is too high, moisture-absorbing materials such as straw, sawdust, bedding materials, etc., are added. Quite often it is difficult to maintain the proper moisture content mainly because of the size, shape and nature of the particles in the compost mixture.[2] The situation may be remedied by shredding the particles if they are too large. On the other hand, some materials such as mature tree leaves, rye and wheat straw, sawdust, and others with high cellulose content tend to shed water and are difficult to wet. To remedy such cases, a small quantity of detergent may be added to the water which is being sprayed over the pile to enhance the wetting of the surface of the solid particles. Once decomposition is initiated and the composting process is in progress, moisture content decreases and may even reach 10–15 % at the end of the composting process.

Dairy manure from stanchion barns where bedding is used makes an ideal composting mixture, provided the moisture content is in the range of 40–55 %.[3] At moisture contents above 75 %, thermophilic temperatures may never be reached. High moisture content would increase the energy requirements for mixing and/or aeration of the pile and may create mechanical difficulties such as stickiness to machinery, plugging of air inlets, etc. When rotary drums are used and the moisture content is excessive, balls may be formed whose centre could remain anaerobic.[4,9] Excessive or low moisture contents also tend to slow the decomposition rate and thus prolong the composting process. Furthermore, under such conditions the composting process is incomplete.

Aeration

Aeration is required to provide free oxygen for the aerobic organisms involved in the composting process. Both the quantity of air supplied and the rate at which oxygen is supplied are critical. Natural or mechanical aeration of the compost mixture is essential. Mechanical aeration combined with uniform air distribution tends to speed up the composting process. Naturally aerated piles which are not turned over frequently would compost in 8–12 weeks, provided moisture and other parameters remain within the acceptable ranges.[5] Mechanical aeration reduces the time required for composting to 2–3 weeks.[3] After active composting, it is usually recommended that the material be allowed to age for a period of several weeks to several months, depending on the requirements for the stability of the end-product which, in turn, are determined by the intended use of the compost humus. For agricultural or farm uses, compost does not need to age. However, if the compost is to be marketed for greenhouses or for home garden use, ageing for several months is advisable. During the ageing period the compost mixture decomposes further at a very slow rate.

Rate of aeration and quantity of air per unit of composting materials must be controlled. Aeration rate depends not only on the nature of the composting materials, environmental factors and moisture content, but also on the stage of the composting process. The aeration should be varied during the composting process.[6] Oxygen use is highest during the first two or three days when anaerobic conditions are being neutralized. High aeration rates should be confirmed when thermophilic temperatures are reached, usually after the first two days. Once the thermophilic decomposition stage is completed, usually within a few days to one week, and as the temperature begins to fall, aeration rate must be decreased so as to prevent fast cooling of the compost pile.

Oxygen consumptions for dairy manure during the first few days of operation may be in the range of 9·7 to 35 litres/min/ton of manure compost.[3] Because of the large number of variables affecting the aeration requirements in composting, aeration rates applicable for all situations cannot be specified. Aeration should be managed during the process by an experienced person, so that oxygen saturation of not less than 10 % is maintained throughout the pile, while temperatures are kept at 50–70 °C.[8] Another way to specify aeration rate is to supply air at such a rate that the gases exhausted from the compost pile contain at least 5 % residual oxygen by volume.[9]

Mechanical aeration may be accomplished by forcing air through the compost mixture or by mechanically mixing and rearranging the compost pile so that air reaches the inner core of the pile. The latter is usually windrow composting and requires less capital investment. Forced-air aeration is expensive, but results in better control of the composting process. Forced aeration is usually accomplished with the use of perforated floors. Compressed air is forced up through the floors, or air is sucked out of the compost pile through the bottom. The sucked-out air must go through the pile to reach the bottom. This latter system allows better dust and odour control.

C/N Ratio

Nitrogen content and C/N ratio are critical in composting. Generally, if total nitrogen content of the compost exceeds 1·5–2·5%, additional quantities of nitrogen would not be needed for proper composting. However, the nature of the nitrogen and how readily it may become available to the microorganisms are also important. For example, aquatic plants do not compost readily even though their total nitrogen content may exceed 2–3%. In that case, additional nitrogen in the form of urea, ammonium sulphate or ammonium nitrate needs to be added. The same holds true for certain sewage sludges, which may contain 3·5–4·5% total nitrogen, but would not compost readily because of high concentrations of heavy metals. The inhibitory effect of these heavy metals may be overcome by the addition of 50–100 kg of nitrogen per tonne of dry solids in the compost mixture.[2] Additions of nitrogen might need to be supplemented with equal amounts of limestone because of the acidity which may result from the addition of nitrogen. If acidity develops and is not buffered with the addition of appropriate quantities of lime, microbial activity may be curbed to the point where the compost process ceases to proceed satisfactorily. When composting is completed, the resulting humus should have a 2·5–3·5% nitrogen content whose availability value should range from 50% to 70% of that of ammonium sulphate.[2]

The most widely used parameter in the design and operation of composting systems is the carbon to nitrogen (C/N) ratio. Microorganisms involved in the decomposition process of the carbon in the compost mixture require nitrogen for their growth and metabolic activities. These requirements vary with the type of organism. Thermophilic fungi, for example, require 1 part of nitrogen to 30 parts of carbon.[2] The optimum initial C/N ratio ranges from 30 to 50. If the C/N ratio is below 30, ammonification may occur resulting in loss of nitrogen. Mixtures with low C/N ratio have higher aeration requirements both in total oxygen demand and aeration rate. If C/N ratio exceeds 50, the initiation of the composting process may be delayed. Composting might take 50% more time than it would at the optimum C/N ratio.

Carbon and nitrogen are utilized by the bacteria during composting. A portion of the carbon is utilized to growing bacterial cells, another portion provides the energy needed for bacterial growth, and the remainder stays in the humus. The portion of carbon utilized for energy is converted to CO_2 and is released. Generally the reduction of the organics in the raw waste is in the range of 7% to 99·9%.[10] The carbon of low molecular weight carbohydrates is highly soluble and may be readily utilized by a variety of microorganisms. Thus lipids, starch and sugar may be reduced during composting by 77%, 88% and 99·9%, respectively. Crude fibre containing cellulose and lignin is not easily decomposed and thus may only be reduced by 7%. Decomposition of cellulose may take several weeks. Weight reduction which may range from 47 % to 80 % results from carbon loss as CO_2 and from moisture loss due to evaporation.[3] Moisture reduction may be 35 %.[10]

During the transformation of carbon, nitrogen is immobilized and stored in the bodies of the microorganisms. Thus, while carbon is lost from the compost mass, nitrogen is concentrated. As the C/N ratio decreases with the evolution of CO_2 and nitrogen is no longer needed for cellular growth, excess nitrogen may be liberated as ammonia, resulting in nitrogen loss. In sawdust–poultry manure compost mixtures, nitrogen losses may range between 3 % and 8 % of the initial nitrogen content.[9] In dairy manure composting, nitrogen losses are high during periods of anaerobic activity in pockets of the compost mixture. Adding superphosphates seems to reduce nitrogen losses.[11]

Initial C/N ratio is attained by combining wastes with high carbon content with wastes high in nitrogen. Sawdust, for example, has a C/N ratio of 390:1 while that of poultry manure may be 5:1.[9] However, when formulating the C/N ratio of the mixture of the two wastes, the nature of the carbon and nitrogen in each of the wastes being combined must also be considered. If the entire carbon content consists of lignin or cellulose, a 30:1 C/N ratio in the mixture would not necessarily be optimum for composting.

Final C/N ratio may be in the range of 12:1 to 27:1. If the final C/N ratio is high, *i.e.* higher than 20:1, and the carbon of the compost is not stable, then when this compost is added to the soil, soil bacteria may use soil nitrogen to decompose the carbon, thus affecting plant growth by reducing soil nitrogen levels. The nitrogen content of the final compost should be in the range of 1·2–1·5 % so as to avoid unfavourable effects on crops.[10]

Essential Chemicals

Besides carbon and nitrogen, phosphorus, potassium and small quantities of other nutrients such as calcium, iron, zinc, magnesium, etc., are also essential. Usually most or all of these chemicals are present in sufficient quantities for good microbiological decomposition. The usual problem is excessive concentrations which may reach toxic levels in the case of heavy metals. In some instances, particularly in industrial wastes, phosphorus or potassium might be present in too low concentrations. In such cases, superphosphates may be added at the rate of 7–10 kg/tonne and potassium chloride at rates of 2–5 kg/tonne.[2]

Ideally the compost should have P_2O_5 and K_2O contents of 1–1·5 %. The cation exchange capacity (CEC) should be in the range of 75–100 ml/100 g of compost. Its mineral ash content should be in the 10–20 % range.[2]

Temperature and pH

Both temperature and pH vary according to the stage of the composting process. Both are affected by the process itself and are hard to manipulate if the other process parameters are not within the optimum range.

During the first stage of composting, temperature begins to rise slowly while pH decreases to acidic level in the range of 5–6. As soon as the pH downward trend is reversed, the temperature begins to climb rapidly. Temperature climbs rapidly from the mesophilic (20–40 °C) to the thermophilic range (50–70 °C) soon after the pH exceeds 7 and reaches a level of 8–9. A high temperature of 65–70 °C may be maintained for a few hours to two days depending on the C/N ratio and the decomposition rate of carbon. While pH remains in the alkaline range of 7–9, temperature begins to decrease, reaching ambient temperature levels after several days or several weeks depending on the composting method used.[9,12,13]

Particle Size

Both the size and the distribution of particle sizes are important in composting. Particles varying from microns to centimetres should be used and mixed in such a manner as to permit good distribution of airflow and wetting of all parts of the composting mixture. The total mixture must remain aerobic at all times to prevent putrefaction or malodour production. Because it is expensive or technically difficult to produce a compost pile with ideal distribution of particles, mixing and frequent rearranging of the compost pile must be practised. The mode of mixing and its frequency are specified by the composting system being adopted, which in turn is based on experience and/or patented equipment. A good indicator of the need to turn over or mix a compost pile is the pile's temperature. If the temperature levels off and/or begins to decline slowly during the early stages of composting, then it is time to mix the compost mixture. Another indicator is when oxygen level in the gases within the composting manure drops below 2 %.[3]

Microorganisms

The composting process is initiated and maintained by microorganisms which begin to grow once optimum environmental and substrate conditions develop. There is no need to add bacterial cultures to the pile to get the process going.

What type of microorganism predominates depends on the environmental conditions, the nature of the materials being composted, and the stage of the process. In general, bacteria predominate, but their species changes as the process develops and temperatures increase. Fungus (moulds) predominate when the temperature is in the 20–30 °C range, but can grow well in the 50–60 °C range. Actinomycetes are most active in the mesophilic range (30–40 °C). It takes one to two weeks for actinomycetes to reach high growth rates. One genus of actinomycetes is thermophilic and is thus predominant in composting. Thermophilic bacteria predominate above 40 °C. Their optimal growth is in the 50–70 °C range. However, the thermophilic bacterial decomposition process is very sensitive to environmental factors and availability of substrate. Protozoa are bacterial predators. They are active at temperatures below 40 °C.

End-Product

Many of the characteristics expected in the end-product have already been given. Additionally, well-composted materials are brown in colour, have an earthy aroma which may be slightly musty, and have good water-holding capacity, in the range of 150–200 %.[2] However, compost ripeness cannot be judged on outer appearance, fragrance or age alone.[14] Chemical tests should be used to determine the ripeness of composting,[14,15] along with growth tests on an index plant such as garden cress.[14] Qualitative tests should show the absence of sulphides and free ammonia in a ripe compost. An 85 % reduction of the COD of the original materials is another good indicator of compost ripeness.[10] All composts tend to reheat to some degree when wetted to a moisture content of around 40 %. No test, however, has been developed indicating compost ripeness on the basis of the rate and degree of reheating of a compost.[16]

COMPOSTING METHODS

There are two major methods of composting: natural and mechanical. In both instances, pretreatment is necessary to optimize all the essential environmental parameters of the compost process. Pretreatment has already been discussed. Composting methods vary only in the techniques used for the aeration of the compost mixture.

Natural Aeration Methods

Natural methods of composting are variations of the traditional schemes originally developed by Sir Albert Howard in India. They are being practised by compost/garden enthusiasts in all parts of the world. The usual system consists of alternate layers of bulky trash, manures, grass cuttings, garbage, leaves, etc. The process depends on natural ventilation. It is therefore a slow process requiring a minimum of six months to be completed. Even then the end-product is not uniform in quality.

This natural aeration process is not suitable for large-scale composting. In small dairy cattle operations where building is used to absorb urine, compost piles may be formed. However, the compost process rarely produces a high-quality end-product unless the pile is turned over periodically and allowed to develop thermophilic temperatures. Poultry units using litter are in the same situation.

Modern feedlots where large quantities of wastes are generated and excessive odours cannot be tolerated must use mechanical aeration systems to get a good compost process started.

Mechanical Systems

Windrow Composting

The most common and easiest type of composting is the windrow process. Composting occurs when the materials are shaped into windrows which are shaped to shed rain water and to preserve temperature and humidity. Periodically the heaps are turned over manually or with specially designed mechanical equipment. Windrow composting requires little engineering but considerable experience gained through trial and error for the particular mixture of materials. The size of windrow composting operations may vary from small farm operations to operations consolidating thousands of cattle, as is being done in the dairy farms in the Chino, California, milkshed, or large beef feedlot operations in arid regions, to the huge operations composting solid wastes from a major part of a whole country such as is being done in the Netherlands. It is interesting to note that composted manure from the Chino, California, dairy corrals is shipped through the Panama Canal and the Atlantic Ocean and is applied to vineyards along the Rhine.[17]

Windrow composting takes time and also space, for the process itself and for storage of the materials. It is mainly a batch process. The quality of the end-product is not uniform. Odour problems may arise during material storage and during the process itself.

Composting of swine wastes in windrows 3 m wide by 1 m deep[13,17] can be accomplished by mixing the pile every 4–7 days. Composting may take 1–7 months.

Continuous Flow Composters

Both open and enclosed continuous flow composters are in operation in many parts of the world. The main component of these systems is the biological reactor or, as it is usually referred to, the bioreactor. The system takes its name from the inventor of the system or from its configuration or is known by a trade name. There are no standard designs, which is one of the reasons why composting has not been such an economic success as other systems which appeal less to the public.

In the bioreactor, or as it is sometimes called the digester, the environmental conditions are mechanically controlled to provide uniform airflow distribution and maintain optimum moisture. The detention time in the bioreactor may be reduced to as little as two weeks followed by four weeks of ageing in open windrows.[14]

The bioreactor might be an open pit or bins with a perforated floor from which air is sucked through the pile or air is forced into the pile.[19] In other cases, frequent mechanical agitation may provide the needed aeration.[7] Both systems have been successful with animal wastes.

Other bioreactor configurations are vertical silos with decks at various levels on which the materials are raked and then dropped from one level to the next, or by vertical augers mixing the silo contents, or by conveyor belts moving and dropping the mixtures, etc. Aeration or agitation may be carried out with screws, ploughs, rakes, lifts, etc. A common bioreactor configuration is the horizontally rotating drum. It has been used successfully in poultry farms of several million birds capacity in Southern California. The most common drum bioreactors are of steel drums set at a horizontal slope of 5° and rotating at 1·5–5 rev/min. The material enters from one end of the drum and works its way to the other end in 3–5 days. In the meantime, forced draught aeration is used to keep the mixture in suspension. The end-product is of high quality. Several experimental rotating drum bioreactors have been tested for poultry[20] and for beef cattle.[4]

Liquid Composting

A recent development in manure treatment is aeration of mixtures high in moisture content but facilitating the growth of thermophilic bacteria so as to reduce the pollution load and to kill pathogen organisms.[21–24] However, in order to reach high temperatures in the range of 50–70 °C, the bioreactor must be thoroughly insulated to prevent heat losses.

Liquid composting consists of submerged aeration with downdraught-type aerators equipped with propellers to agitate and aerate the liquid. Solids concentrations in the range of 2–7% may be aerated. High temperatures of the mixture of 45–55 °C are reached only at high solids concentrations of 8%.[24] The usual temperature reached by the process is around 35–42 °C.[23,25] The BOD of the effluent from this process is too high for discharge into streams and must therefore be contained under aerobic storage conditions because otherwise hydrogen sulphide formation may occur, making it malodorous.[24] The operating costs of the process to give complete treatment, which may include vacuum filtration, will be in the range of US$230/dairy cow/year.[22]

USES OF COMPOST

Despite the great public appeal of compost, it is not a widely used process because only a few large-scale composting operations have been successful. Thanks to the interest in ecology and environmental issues, there is a renewed interest in the use of compost in commercial gardens, around the home and in vineyards. It is not likely that farmers will use high-quality compost for field crop production. Therefore, a well-defined market for the compost product should be defined and developed before undertaking expensive composting operations. The only successful composting systems are those which succeed financially by identifying a market area and developing a network for marketing the end-product, before any composting equipment is installed. To maintain the market, however, a good end-product with uniform quality is essential. The composting system should therefore be engineered to meet the intended use of the end-product.

REFERENCES

1. American Society of Agricultural Engineers, *Agricultural Engineers Yearbook*, pp. 515–17, ASAE, St Joseph, Mich., 1975.
2. Toth, S. J., Composting agricultural and industrial wastes. In: *Symposium on Processing Agricultural and Municipal Wastes* (ed. G. E. Inglett), pp. 172–82. AVI Publishing Co., Westport, Conn., 1973.
3. Wilson, G. B., Composting dairy cow wastes. In: *Livestock Waste Management and Pollution Abatement*, pp. 163–5. ASAE, St Joseph, Mich., 1971.
4. Wells, D. M., Albin, R. C., Grub, W. and Wheaton, R. Z., Aerobic decomposition of solid wastes from cattle feedlots. In: *Animal Waste Management* (Proceedings, Cornell Agricultural Waste Conference), pp. 58–62, Ithaca, NY, 1969.
5. Toth, S. J. and Gold, B., Composting. In: *Agricultural Wastes: Principles and Guidelines for Practical Solutions* (Proceedings, Cornell Agricultural Waste Management Conference), pp. 115–20, Ithaca, NY, 1971.
6. Wilson, G. B. and Hummel, I. W., Aeration roles for rapid decomposition of dairy manure. In: *Waste Management Research* (Proceedings, Cornell Agricultural Waste Management Conference), pp. 145–58. Ithaca, NY, 1972.
7. Stombough, D. P. and White, R. K., Aerobic composting: new built-up technique. In: *Managing Livestock Wastes*, pp. 485–9. ASAE, St Joseph, Mich., 1975.
8. Miner, J. R. and Smith, R. S. (eds), *Livestock Waste Management and Pollution Control* (MWPS-19). Midwest Plan Service, Ames, Iowa, 1975.
9. Galler, W. S. and Davey, C. B., High rate poultry manure composting with sawdust. In: *Livestock Waste Management and Pollution Abatement*, pp. 159–62. ASAE, St Joseph, Mich., 1971.
10. Hays, S. T., Composting of municipal refuse. In: *Symposium on Processing Agricultural and Municipal Wastes* (ed. G. E. Inglett), pp. 205–15. AVI Publishing Co., Westport, Conn., 1973.
11. Wilson, G. B. and Hummel, J. W., Conservation of nitrogen in dairy manure during composting. In: *Managing Livestock Wastes*, pp. 490–1, 496. ASAE, St Joseph, Mich., 1975.
12. Cardenas, R. R. and Varro, S., Disposal of urban solid wastes by composting. In: *Symposium on Processing Agricultural and Municipal Wastes* (edited G. E. Inglett), pp. 183–204. AVI Publishing Co., Westport, Conn., 1973.
13. Singley, M. E., Decker, M. and Toth, S. J., Composting of swine wastes. In: *Managing Livestock Wastes*, pp. 492–6. ASAE, St Joseph, Mich., 1975.
14. Spohn, E., 100% waste recycling with the Renova-method. *Compost Science*, Mar.–Apr. 1972, pp. 8–11.
15. Knuth, D. T., Composting of solid organic waste. *Battelle Technical Review*, **17**(3) (1968) 14–20.

16. American Public Works Association, *Municipal Refuse Disposal*, 3rd ed. AWPA, Chicago, Ill., 1970.
17. Taiganides, E. P., *Livestock Waste Management* (mimeo). Ohio State University, Columbus, Ohio, 1973.
18. Martin, J., Decker, M. and Das, K. C., Windrow composting of swine wastes. In: *Waste Management Research* (Proceedings, Cornell Agricultural Waste Management Conference), pp. 159–72. Ithaca, NY, 1972.
19. Hummel, J. W. and Wilson, G. B., High rate mechanized composting of dairy manure. In: *Managing Livestock Wastes*, pp. 481–4. ASAE, St Joseph, Mich., 1975.
20. Bell, R. G. and Pos, J., Design and operation of a pilot plant for composting poultry manure. *Trans. ASAE*, **14** (1971) 1020–3.
21. Taiganides, E. P. and White, R. K., Automated handling, treatment and recycling of waste water from an animal confinement production unit. In: *Livestock Waste Management and Pollution Abatement*, pp. 146–8. ASAE, St Joseph, Mich., 1971.
22. Grant, F. A., Liquid composting of dairy manure. In: *Managing Livestock Wastes*, pp. 497–500, 505. ASAE, St Joseph, Mich., 1975.
23. Terwilleger, A. R. and Crawer, L. S., Liquid composting applied to agricultural wastes. In: *Managing Livestock Wastes*, pp. 501–5. ASAE, St Joseph, Mich., 1975.
24. Grabbe, K., Thaer, R. and Ahlers, R., Investigations on the procedure and the turn-over of organic matter by hot fermentation of liquid cattle manure. In: *Managing Livestock Wastes*, pp. 506–9. ASAE, St Joseph, Mich., 1975.
25. Taigainides, E. P. and White, R. K., *Automated Treatment and Recycle of Swine Feedlot Wastewaters*, EPA-R2-74. US Environmental Protection Agency, Washington, DC, 1974, 136 pp.

21

Nutrient Control Applicable to Animal Wastes

R. C. LOEHR

Director, Environmental Studies Program, Cornell University, Ithaca, New York, USA

INTRODUCTION

The amount of animal waste that causes environmental problems is not well documented. It is incorrect to use the amount of wastes defaecated by an animal as a direct indication of the surface and groundwater pollution that may result. Only a small portion of such waste may find its way into surface waters or groundwaters. However, under conditions of intensified animal production, large quantities of wastes must be disposed of under circumstances that can result in environmental problems. The most significant problems are odours, contaminated surface runoff, and soil percolate containing soluble waste constituents.

Historically, animal wastes have been recycled through the soil with a minimum of direct release to surface water. This practice, especially when it is part of crop production, continues to be the most appropriate approach for disposal of animal wastes.

Major emphasis in water pollution control has been on the reduction of carbonaceous oxygen-demanding substances released to surface waters. Now consideration is also being given to control of the nitrogen content of a waste or treated effluent. The reasons why nitrogen compounds are of concern include: the resultant nitrogenous oxygen demand in receiving waters, ammonia toxicity to fish, increased chlorine demand due to ammonia if the water is chlorinated, the role of nitrogen in eutrophication, and health problems in humans and animals. Regulatory authorities are beginning to set standards for nitrogenous materials in streams and effluents and treatment facilities are being required to produce effluents low in both unoxidized and total nitrogen.

Phosphorus control is also being emphasized due to its role in eutrophication. However, since land disposal is an integral part of animal waste management, control of nitrogen in the wastes is likely to be of more concern than that of phosphorus.

The waste management needs at a livestock production facility will be some combination of odour control, waste stabilization, nitrogen control consistent with land disposal constraints, and ease of waste handling and disposal. The best alternatives to accomplish these needs appear to be (a) some form of aerobic biological treatment followed by land disposal, or (b) direct land disposal in a prudent, environmentally sound manner.

The desirable level of nutrient control is related to the utilization of the nutrients when the wastes are ultimately disposed of on the land. If the wastes are intended to provide a source of nutrients for crop production, then the obvious objective is to conserve as great a quantity of nutrients as possible. On the other hand, if the supply of nutrients from disposal is potentially excessive, then it is desirable to reduce the nutrient content of the wastes prior to disposal. To utilize suitable means of nutrient control. it is necessary to understand the factors that affect the control processes and their applicability to livestock production systems.

PHOSPHORUS CONTROL

General

The general movement of phosphorus in the environment is noted in Fig. 1. Inorganic phosphorus is the most important form for plant growth. As plant roots remove phosphorus from the soil solution, phosphorus adsorbed to soil particles enters the soil-water solution to help replenish that removed. The soluble, available phosphorus constitutes only a small fraction of the total phosphorus in the soil. Organic phosphorus, converted

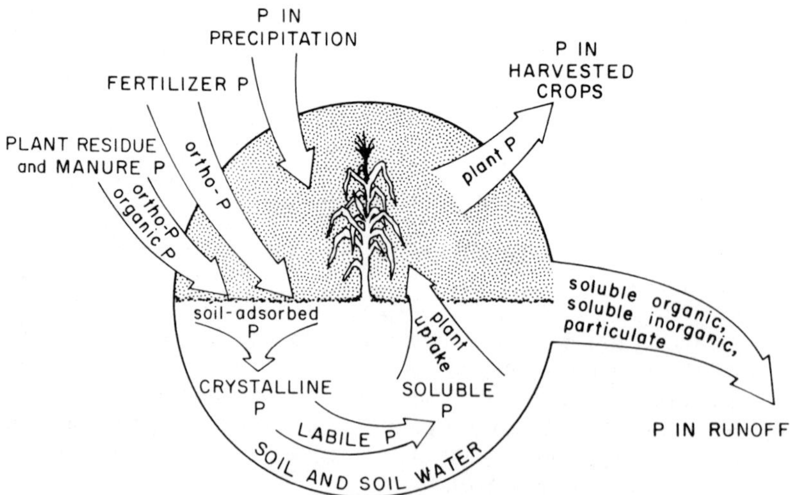

FIG. 1. General movement of phosphorus in the environment.

to the inorganic form by microbial action, also acts as a source of supply for the soluble phosphorus.

The quantity of soluble phosphorus in the soil is determined by (a) soil pH, (b) iron, aluminium, manganese and calcium content of the soil, (c) decomposition of organic matter, and (d) microbial activity. These factors are interrelated because their effects are dependent upon the conditions of the soil environment. In alkaline soils, soluble phosphate from fertilizers and manures reacts with exchangeable calcium ions and salts to form slightly soluble calcium phosphates.

The iron and aluminium phosphates have a minimum solubility near pH 3–4. At higher pH values these phosphates are more soluble. At a pH about 6, phosphorus precipitation as calcium compounds begins. The greatest phosphate availability to plants is when the soil pH is maintained in the range of 6–7.

The removal of phosphorus from the soil is almost entirely due to plant uptake, with some losses occurring in land runoff. Gaseous losses of phosphorus do not occur naturally. Some phosphorus becomes airborne in dust and is removed from the local site.

Water in contact with soils contains relatively low concentrations of phosphorus. Applications of manure and fertilizer to soil can increase the concentration of nitrogen and phosphorus in soil water. Surface runoff from manured and fertilized soils will contain higher concentrations of phosphorus than subsurface waters.

TABLE I
METHODS FOR REDUCING THE PHOSPHORUS CONTENT OF WASTEWATERS

Classification	*Method*
Physical and chemical methods	Land application
	Electrochemical
	Ion exchange
	Precipitation
	Sorption
Biological methods	Activated sludge
	Algal utilization

The pH–phosphorus solubility relationships are equally important in the removal of phosphorus from wastes and wastewaters, since chemical precipitation is an important technology in such removal. The possible methods of phosphorus removal from wastes are noted in Table I.

The possible alternatives for control of phosphorus in animal wastes or wastewaters are (a) land disposal and (b) chemical precipitation. Chemical precipitation is more applicable to liquid wastes that are intended to be discharged to surface waters. The effluent limitations in the United States do not permit discharge of animal wastes to surface waters except due to extreme storms, and in the special case of duck wastewaters, which must receive a specified degree of treatment before such discharge.[1] This no-discharge requirement means that the wastes must be disposed of on the land.

Technology

Although land disposal is a proper approach for phosphorus control with animal wastes, chemical precipitation of animal wastewaters has been evaluated.[2] This evaluation dealt primarily with liquid poultry and duck wastes. However, the results have indicated chemical demand relationships, type of sludge production and relative costs that can evaluate the practicality of chemical precipitation for other animal wastewaters.

The chemicals which may be used to precipitate phosphorus are lime, alum and ferric salts. Lime reacts with orthophosphate in solution to precipitate hydroxylapatite. When alum is used to remove phosphates, the removal mechanism is by incorporation in a complex with aluminium or by adsorption on aluminium hydroxide floc. Ferric ions and phosphates react to form insoluble ferric phosphate precipitates. Each of the reactions has a specific pH optimum range.

The quantity of chemical to achieve specific phosphorus removals depends upon the characteristics of the wastewater, such as pH, alkalinity, phosphate concentration and related factors that affect coagulant demand. These factors vary from wastewater to wastewater. Empirical relationships are used to estimate the feasible type and quantity of chemical.

To evaluate the removal of phosphates from animal wastewaters by the above chemicals, laboratory jar test experiments should first be conducted. Such tests may determine the appropriate chemical, the effectiveness of the chemicals with wastewaters of varying characteristics, and predictive relationships that could be used for potential design purposes.

Each type of wastewater will have its own chemical demand relationship. The chemical demand will be in proportion to parameters such as the initial phosphate, alkalinity or hardness concentrations.

The most sensitive predictive relationships are likely to be:

(a) chemical dosage per remaining total or orthophosphate concentration versus percentage total or orthophosphate removal (mg/litre of chemical per mg/litre PO_4 remaining versus percentage PO_4 removal); and

(b) chemical dosage per initial calcium hardness versus total or orthophosphate removal (mg/litre of chemical per mg/litre of initial hardness versus percentage PO_4 removal).

An example of such a relationship with specific animal wastewaters is noted in Fig. 2. These relationships have been found to be reasonably sensitive for the different wastewaters, for different phosphate concentrations and for the chemical used.[2,5] Other relationships involving pH and alkalinity of the solution may also prove to be useful.

Required chemical concentrations are in proportion to the characteristics of the wastewater. Ratios of chemical dosages per initial orthophosphate concentration may range up to 8–10 for alum and lime to achieve low residual orthophosphate concentrations (less than 5–10 mg/litre) and high orthophosphate removals (greater than 90 %). Sludge production may average between 0·5 and 1·0 mg/litre suspended solids increase per mg/litre of chemical used.

To achieve low residual phosphate concentrations, a wastewater containing 100 mg/litre of orthophosphate may require about 800–1000 mg/litre of chemicals which may produce an additional 400–1000 mg/litre of suspended solids for ultimate disposal. The large chemical demand and sludge production are decided disadvantages to this method of phosphate control for concentrated animal wastewaters.

Cost Relationships

Estimates of costs were obtained by (a) using the noted predictive relationships with average animal waste parameters, and (b) using the predictive relationships and results obtained for duck wastewaters.[3]

There are no typical amounts of water that dairy or poultry operators add to their waste to handle them in a liquid manner. Enough water is added to make the resultant slurry pumpable by available equipment. For the purposes of these cost estimates, three slurries were assumed, *i.e.* approximately 1 %, 5 % and 10 % total solids. The 5 % and 10 % slurries represent conditions such as might exist in actual animal production operations and the 1 % slurry represents a dilute waste. The chemical costs used for the estimates were: alum, \$0·127/kg as $Al_2(SO_4)_3$; lime, \$0·022/kg. These were the actual costs of the available chemicals.

The chemical costs for dairy cattle and poultry layers ranged from \$0.10 to \$1·10 per day for 100 dairy cattle and \$0·05 to \$0·40 per day for 1000 birds, depending upon the chemical used. Sizeable costs, *i.e.* up to \$120/day, for chemical costs for a 300 000 bird operation to obtain 90 % total phosphate removal would be involved. Even at the lowest dilution used in the estimate, 1 % total solids, the chemical costs are at least an order of magnitude larger than chemical costs quoted for phosphate removal from municipal wastewaters.

These cost estimates should not be viewed as totally realistic. It would be folly to treat waste solutions chemically containing 5 % and 10 % or even 1 % solids. There are more applicable methods of phosphate control for these slurries, namely controlled land disposal. The estimates were made only to provide some measure of the costs that could be involved when

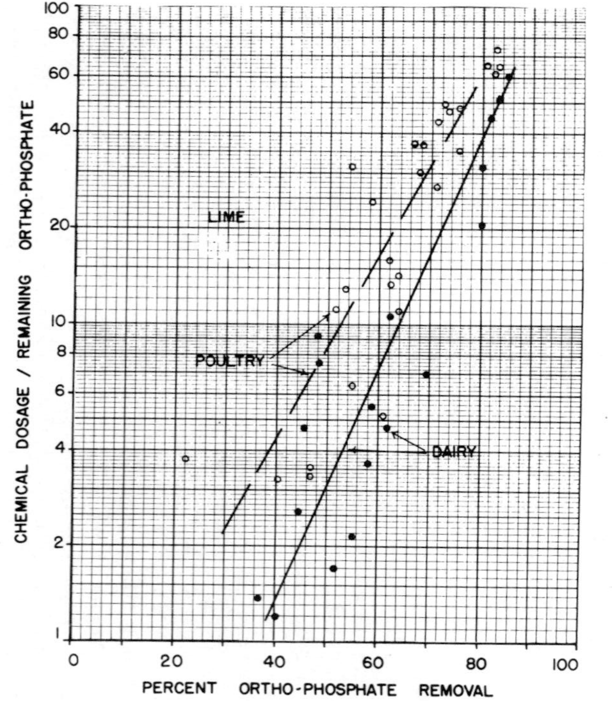

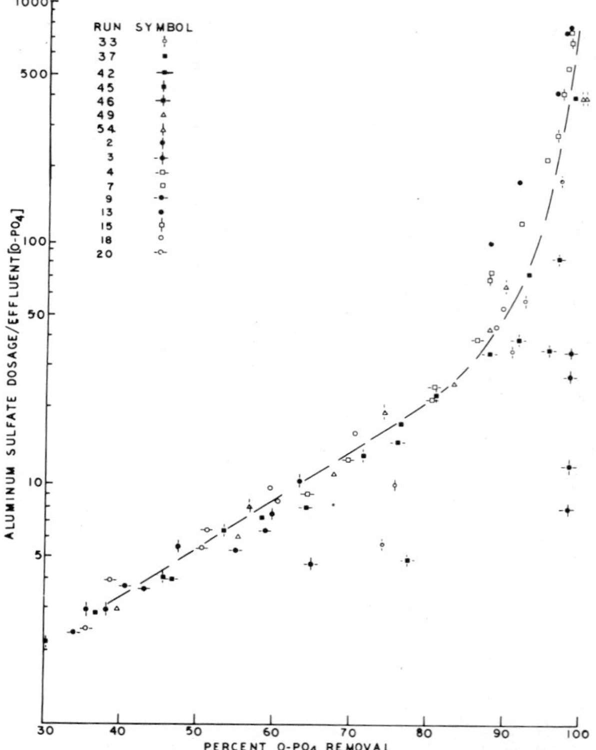

FIG. 2. Chemical dosage relationships to remove phosphorus from liquid animal wastes.

considering chemical precipitation for phosphate control from such wastes. Chemical precipitation is more applicable to dilute wastes.

More realistic cost relationships for phosphate removal can be obtained by analysis of the removals and associated costs that occurred at the full-scale, operating duck farms.[3] Lime was the chemical of choice for these wastewaters. Over the orthophosphate removal range of 50–90 %, the chemical costs of lime ranged from $0·002 to $0·011/1000 litres of wastewater/day. The chemical cost for alum was $0·006 to $0·011 and for ferric chloride $0·024 to $0·067/1000 litres of wastewater/day.

The alum and lime costs were similar to or slightly higher than the chemical costs observed with municipal wastewater, which were about $0·004 to $0·005/1000 litres for lime and $0·002/1000 litres for alum at the time of the study for 90 % phosphate removal and a phosphate residual of less than 0·5 mg/litre. Although chemical costs and orthophosphate percentage removals may be similar between duck wastewaters and municipal wastewaters, the 'average' duck wastewater still had a residual orthophosphate concentration of about 3·5 mg/litre at 90 % removal. To reduce this phosphate concentration further will require considerably more chemicals and greater costs.

Although more realistic than the previous costs for poultry and dairy manure wastes, the costs for phosphate removal from duck wastewater should be viewed as estimates rather than precise costs. When evaluating the overall costs of phosphate removal, the following should be added to the above chemical costs: (a) cost of energy for pumping and mixing, (b) costs of additional units and equipment, (c) costs of sludge handling and disposal, and (d) costs of additional qualified manpower.

Summary

The most effective method of phosphate control with animal wastewaters is not chemical precipitation. Controlled land disposal should be considered as a high-priority method for phosphorus control from agricultural wastewaters because it is more amenable to normal agricultural production operations, avoids the need for chemical control and treatment plant operation, and eliminates the additional problems of chemical costs and sludge production.

NITROGEN CONTROL

General

The general movement of nitrogen in the environment is noted in Fig. 3. Nitrogen in the soil consists of inorganic and organic fractions. The organic fraction is larger and contains components differing in biodegradability. At any time, most of the nitrogen is immobilized as organic nitrogen and is not available for plant growth, leaching or gaseous loss. Through the process of mineralization, microorganisms oxidize organic nitrogen to inorganic forms that can be utilized by plants. Plants absorb primarily inorganic nitrogen, and fertilizer is added to the existing nitrogen in the soil, especially when the rate of conversion of organic and inorganic nitrogen in the soil is insufficient for desired plant growth.

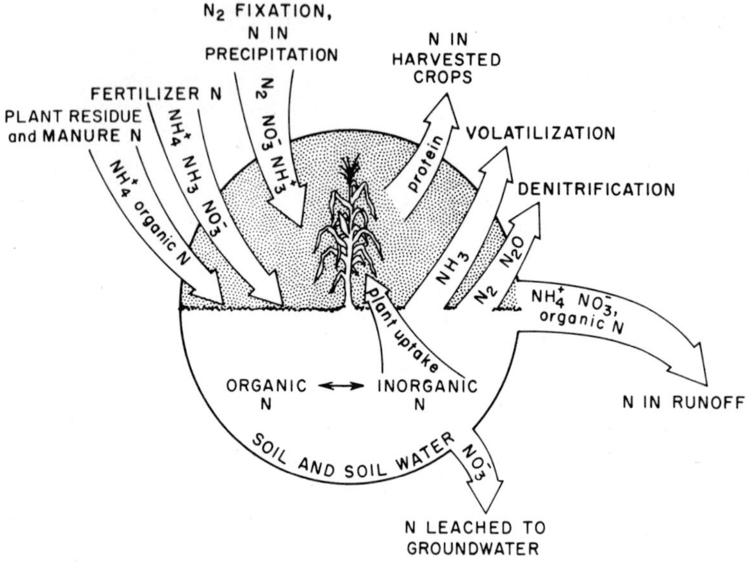

FIG. 3. General movement of nitrogen in the environment.

Some nitrogen can be lost from the soil before it is able to be assimilated by plants. The principal losses occur through denitrification, ammonia volatilization and leaching of nitrate. Denitrification is the microbial reduction of nitrate to nitrogen oxides or nitrogen gas which returns to the atmosphere. The denitrification process provides a means of removing nitrogen from waste in a non-pollutional manner.

Ammonia does not occur in large quantities in the soil. However, when fertilizers and manure are applied to the soil surface, free ammonia can escape to the atmosphere. Over 50% of the nitrogen excreted by animals can be lost in this manner. The fate of this ammonia is largely unknown, but it probably dissolves in rainwater and is returned to land or water surfaces.

A more undesirable loss is nitrate nitrogen which, being highly soluble, is readily removed by water flowing over or through the soil. This loss may contaminate water resources. Losses of organic nitrogen can be controlled by utilization of sound soil conservation practices.

Nitrogen occurring in precipitation is in the nitrate and ammoniacal forms and is readily available to plants. In a humid temperate climate, about 0·5–0·75 g of inorganic nitrogen annually falls on a square metre of land.

Where excess nitrogen in animal wastes is implicated in pollution problems, nitrogen control may be needed. A summary of possible methods is noted in Table II. The state of art of nutrient removal from agricultural wastes is in its infancy, although some of the methods have been tested on a laboratory and pilot-plant scale.

TABLE II

METHODS FOR REDUCING THE NITROGEN CONTENT OF WASTEWATER

Method	*Nitrogen compounds removed*
Physical and chemical:	
Land application	NH_3, NH_4^+, organic N
Ammonia stripping	NH_3
Ion exchange	NO_3^-, NH_4^+
Breakpoint chlorination	NH_4^+, organic N
Biological:	
Algal utilization	All forms
Microbial denitrification	NO_3^-, NO_2^-
Land application	All forms

Of the noted methods, only nitrification–denitrification appears feasible as a controllable method with current animal production operations. Controlled ammonia desorption from liquid and solid livestock wastes is a technical possibility. However, it requires a degree of pH control or aeration that is unlikely to be achieved or maintained by those operating animal waste management facilities. Some amount of uncontrolled ammonia desorption will result from any animal waste storage, handling and application to the land. The actual amount will be dependent on the degree of aeration or agitation, the time of storage and handling, and the pH of the slurry or mixture. A reasonable estimate is that between 10 % and 50 % of the ammonifiable nitrogen can be lost from the wastes under such uncontrolled conditions.

A high degree of nitrogen control may not be necessary in animal wastewater treatment facilities when the effluent is disposed of on land. The control that is necessary will be related to the level of conservation or loss that is desired. To provide a degree of nitrogen control by nitrification and denitrification, it is necessary to understand the processes which transform nitrogen in wastes from one form to another.

Nitrification and Denitrification

Details of these microbial processes are well described in many texts[4] and only the application of the process to the nutrient control of animal waste will be discussed. Nitrification can be defined as the biological conversion of nitrogen in inorganic or organic compounds from a reduced to a more oxidized state. In the field of water pollution control nitrification is generally referred to as a microbial process in which ammonium ions are oxidized initially to nitrite and then to nitrate.

Nitrogen in fresh animal excreta is essentially in organic form. Microbial manure stabilization systems produce a sequence of nitrogen transformations, the first step of which is the ammonification of the organic nitrogen:

$$\text{Organic nitrogen} \xrightarrow[\text{(heterotrophs)}]{\text{ammonification}} NH_3 \underset{\text{hydrolysis}}{\overset{H_2O}{\rightleftharpoons}} NH_4^+ + OH^-$$

This step can result in an increase in pH. If the ammonium concentration and pH are sufficiently high, significant ammonia volatilization can occur. Such ammonia losses have been documented at beef feedlots and from other manure concentrations.

Under aerobic conditions, ammonium nitrogen can be microbially oxidized to nitrate by the autotrophic organisms *Nitrosomonas* and *Nitrobacter*. This oxidation of NH_4^+ to NO_3^- is termed nitrification:

$$NH_4^+ \xrightarrow{(Nitrosomonas)} NO_2^- \xrightarrow{(Nitrobacter)} NO_3^-$$

Under anaerobic conditions, nitrite and nitrate can be reduced by denitrifying organisms. This process is termed denitrification and may be represented as:

$$\text{Reduced organic matter} + NO_3^- \xrightarrow{\text{bacteria}} NO_2^- + \text{Oxidized organic matter}$$

which illustrates the reduction of nitrate to nitrite, and

$$\text{Reduced organic matter} + NO_2^- \xrightarrow{\text{bacteria}} \text{Oxidized organic matter} + N_2$$

which illustrates the reduction of nitrite to nitrogen gas. During denitrification the pH will increase. The degree of pH change is related to the amount of denitrification and the buffer capacity of the liquid.

Wastewaters containing oxidized nitrogen can be denitrified readily if these is an adequate supply of hydrogen donors available. Microbial denitrification takes place under anoxic conditions where nitrites and nitrates are used as terminal hydrogen acceptors in place of molecular oxygen.

Denitrification is brought about by facultative bacteria such as

Pseudomonas, *Serratia*, *Achromobacter*, *Bacillus* and *Micrococcus*. Of the several genera of nitrifying organisms that have been reported, only *Nitrosomonas* and *Nitrobacter* are generally encountered and are undoubtedly the nitrifying autotrophs of importance. Although nitrification is predominantly autotrophic, bacteria, actinomycetes and fungi can bring about heterotrophic oxidation of nitrogen to nitrite and nitrate.

If it were possible to control nitrification–denitrification in liquid aeration units, it would be possible to conserve or to remove the nitrogen content of the waste, prior to land disposal. Waste management systems most likely to use liquid aeration units are those for poultry (layer) and swine wastes, although under certain conditions such units may be used for beef cattle and dairy cattle wastes. Liquid aeration and hence nitrification–denitrification are not limited by geographical location. Their use is primarily determined by land availability and waste management objectives.

Technology

Denitrification following nitrification has been successful in laboratories, pilot plants and large-scale plants to remove the oxidized forms of nitrogen from sewage effluents. Some information is available on the denitrification of wastewaters from livestock operations. The use of aerated units for storage and treatment of animal waste slurries and wastewaters will minimize the odour problem normally present with anaerobic storage of the wastes and can provide a degree of stabilization prior to ultimate disposal. Nitrogen losses caused by denitrification will occur in these units.

With long detention times and adequate oxygen, the opportunity exists for nitrogen control by a nitrification–denitrification cycle. Nitrogen losses are generally determined by nitrogen balances on the waste stabilization units. In the majority of cases, losses due to volatilization have been included in the balances or have been determined to be negligible. The losses noted in Table III represent those presumably due only to denitrification.

Uncontrolled nitrification–denitrification patterns have been observed in

TABLE III

NITROGEN LOSSES ATTRIBUTED TO DENITRIFICATION: AERATED POULTRY WASTE UNITS

Liquid temperature (°*C*)	*Dissolved oxygen* (*mg/litre*)	*Nitrogen loss* (%)	*Ref.*
12–23	2–7	36	5
5–21	2–6	66	6
12–23	0–2	50–60	7
11–18	0–6	70–80	8

oxidation ditches for swine and poultry and in aeration units for cattle feedlot wastes. Aerated units treating animal wastes will contain ample unoxidized organic matter to achieve denitrification.

In developing technical methods for nitrogen control with livestock wastes, the methods should be economical, convenient to manage, perform satisfactorily under varying environmental conditions, and fit into a livestock production operation. Experience in many countries has shown that oxidation ditches are a viable method to stabilize livestock wastes and meet the above objectives. Consequently, the following information relates to the use of an oxidation ditch as a nitrogen control method. The information was obtained from both pilot-plant and full-scale production unit systems.

Four different modes of operation can be considered (Fig. 4):

I. Continuous rotor operation with no intentional wasting of effluent, *i.e.* as an aerated holding tank with continuous addition of solids.
II. Maintenance of a solids equilibrium by intentionally wasting some mixed liquor and subjecting the remaining mixed liquor to intermittent denitrification.
III. Maintenance of solids equilibrium but followed by a tank for settling the mixed liquor suspended solids and achieving denitrification of the recycled effluent.
IV. Intermittent periods of rotor aeration in which the rotor is connected via a time switch to control the time of operation of the rotor.

The results that can be expected when an oxidation ditch is operated in each of these modes are noted in Table IV. The results indicate that the effective denitrification period can be controlled without seriously affecting the performance of the nitrification phase or the overall performance of the oxidation ditch.

TABLE IV
SUMMARY OF NITROGEN AND COD LOSSES IN THE FOUR MODES OF OXIDATION DITCH OPERATION
(from ref. 9)

Mode of operation	*Total nitrogen loss* (%)	*COD loss* (%)
I Aerated holding tank	30–35	62
II Continuous flow	30 or greater	50
III Continuous flow with separate solids removal and combined denitrification	about 75	55
IV Intermittent aeration	about 90	50–60

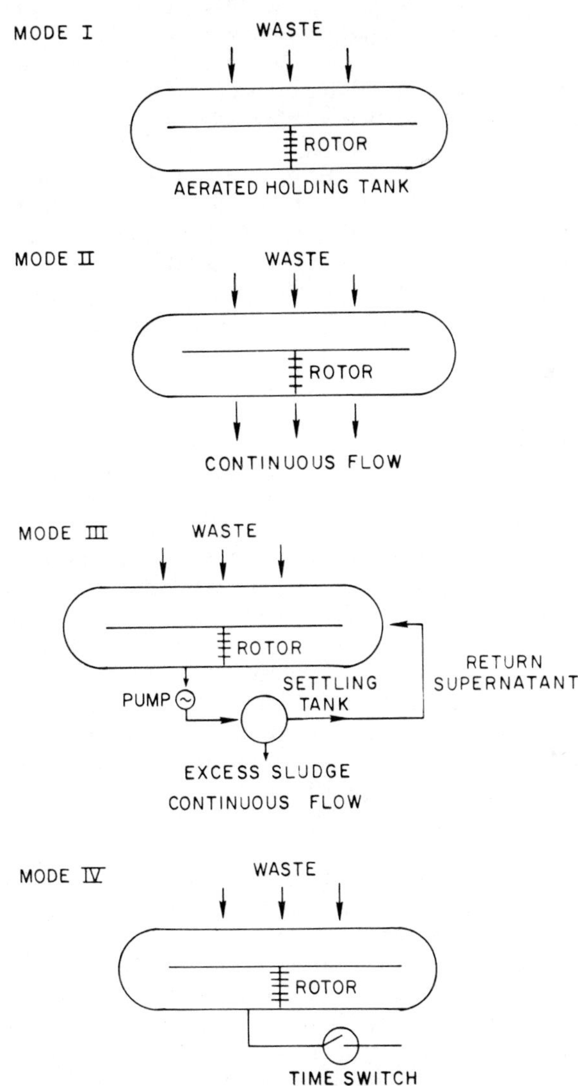

FIG. 4. Modes of nitrogen control with livestock wastes using an oxidation ditch.

Odour control can be achieved in all the modes of operation shown in Fig. 4. COD removals of 50–60 % result irrespective of the mode of operation. Although high COD removals are not required for disposal of stabilized livestock wastes on land, significant removals can be accomplished. Nitrogen control in oxidation ditches is consistent with the other objectives of livestock waste management, *i.e.* odour control and waste stabilization.

The data in Table IV have been corroborated by results obtained from monitoring oxidation ditches at two full-scale egg production facilities for over a year. All these oxidation ditches were designed as odour control, manure stabilization units and were not designed or operated to achieve a specified level of nitrogen control. The nitrogen losses at these facilities ranged from 29 % to 58 % and 47 % to 63 % respectively. These values indicate the level of nitrogen loss that is likely to occur in normal oxidation ditches stabilizing animal wastes.

Cost Relationships

Brief available information permits an estimate of the costs of operating an oxidation ditch to achieve some degree of nitrogen control. The following data represent the annual costs at one of the two full-scale units noted above. For the 300 000 eggs produced annually at this farm, the annual costs for depreciation, interest and repairs were about $0·007 per dozen eggs. The electrical power to operate the rotors amounted to about $0·01 per dozen eggs. Thus the total cost at this facility was about $0·017 per dozen eggs. These costs are those of operating the ditch to minimize odours, handle wastes in a liquid manner and result in a nitrogen loss of 29–58 %.

Recent data from a large egg production facility, of about 300 000 birds, indicate that the electrical power costs for operating an oxidation ditch can be less than $0·01 per dozen eggs.

Summary

Oxidation ditches can be managed in several ways to stabilize animal wastes and control nitrogen. The available information identifies approaches to minimize or enhance nitrogen loss depending upon the waste management objective. The decision on what nitrogen control approach is appropriate depends upon the economic, legal, manpower and other constraints imposed upon the producer. The basic decisions are (a) maximum nitrogen conservation and use in crop production as the wastes are disposed of on the land, or (b) nitrogen loss to the level that the nitrogen remaining in the wastes is consistent with the land application guidelines. In the latter case, excess nitrogen loss to the environment can be minimized. Depending upon the mode of oxidation ditch operation, it is possible either to conserve as much as 70 % or remove as much as 90 % of the input manurial nitrogen.

It is extremely difficult to conserve all the nitrogen in animal wastes.

Approximately 30% will be lost even in a well-mixed and aerated stabilization unit. The minimum amount of nitrogen loss will result from the use of completely mixed aeration units that consistently maintain a high dissolved oxygen concentration, *i.e.* above 2 mg/litre. The nitrogen in this mixed liquor will be in the form of organic, ammonia, nitrite and nitrate. The latter three forms are soluble in water, and care should be taken to minimize losses of these forms when the oxidation ditch mixed liquor is applied to the land. The greatest benefit will result when the soluble nitrogen can be utilized by the crops shortly after the mixed liquor is applied.

If the mixed liquor is stored before application to the land, the storage should be under well-aerated conditions. Otherwise the oxidized nitrogen, nitrite and nitrate will be denitrified, be lost from the mixed liquor, and negate the nitrogen conservation achieved in the initial oxidation ditch.

To achieve the minimum nitrogen loss, the oxidation ditch should be operated as a well-aerated holding tank with the contents only removed when it can be used effectively in crop production.

The maximum nitrogen loss will occur under conditions that favour ammonia volatilization and denitrification. These conditions occur when the oxidation ditch is poorly mixed, when aeration is intermittent, when nitrified wastes are held under non-aerated conditions, and when the dissolved oxygen concentration in the mixture remains low, *i.e.* below 1 mg/litre.

Achieving the desired nitrogen control can be done using the existing knowledge of oxidation ditch design and operation. The key factor is adequate rotor design to provide the necessary mixing and oxygen input for the ditch contents.

Even if the maximum amount of nitrogen conservation (70%) takes place in an oxidation ditch, additional losses will occur as the stabilized wastes are stored, transported and distributed on the land. Probably less than 50% of manurial nitrogen excreted by the animals remains after stabilization, storage and spreading the wastes. The amount actually available for plant growth will be even less and will depend upon the degree of nitrification, denitrification and leaching that takes place in the soil.

Nitrogen conservation, rather than loss, will be the goal of most livestock producers. The increased cost and possible uncertain availability of inorganic fertilizers are causing considerable interest in the nutrients in animal manures.

CONCLUSIONS

Proper management of livestock wastes includes control of whatever contaminants are likely to cause environmental problems. Such control can be for odours in livestock confinement structures and when anaerobic wastes are disposed of on the land, for discharge of unstabilized wastes to surface waters, for excess quantities of nutrients that can cause problems in surface and groundwaters, and for toxic materials.

When wastes are disposed of on land it is unnecessary to stabilize the wastes to the extent that would be necessary if the wastes were discharged to streams. Given adequate cropland, disposal of livestock wastes can be integrated with crop production, in which case odour control and conservation of nutrients prior to waste application to the land become the waste management objectives. Where land is limited, nutrients such as nitrogen may have to be removed so that the assimilative capacity of the available soil is not exceeded.

Since land disposal is an integral part of livestock waste management, phosphorus control, such as by chemical means, is not necessary. Chemical precipitation of phosphorus in dilute livestock waste requires additional equipment, incurs chemical costs and increased solids disposal, and requires a higher degree of waste management than exists or is needed.

It is difficult to operate a waste stabilization system without losing some nitrogen. Losses due to ammonia volatilization can be minimized by nitrifying the waste. Facultative conditions prevailing in a nitrifying mixed liquor can lead to denitrification of the oxidized nitrogen in the system. If conservation of nitrogen is desired, such losses should be minimized by maintaining adequate levels of oxygen throughout the system. It is possible to achieve different degrees of nitrogen removal (ranging from 30 % to 90 %) with different modes of oxidation ditch operation by controlling the time of effective nitrification and denitrification. Thus various approaches are available to meet the nitrogen content that is needed.

The nitrogen control that does occur in aerated liquid livestock waste stabilization systems, such as oxidation ditches, is consistent with other waste management needs of livestock producers: odour control, waste stabilization, and ease of mechanical handling.

REFERENCES

1. Loehr, R. C. and Denit, J. D., Effluent regulations for animal feedlots in the USA (this volume, pp. 77–89).
2. Loehr, R. C., *Agricultural Waste Management*. Academic Press, New York and London, 1974.
3. Loehr, R. C. and Johanson, K. J., Phosphate removal from duck farm wastes. *J. Water Poll. Control Fed.* (1974) 1692–1714.
4. Brady, N. C., *The Nature and Properties of Soils*, Chap. 16. Macmillan, New York, 1974.
5. Loehr, R. C., Anderson, D. F. and Anthonisen, A. C., An oxidation ditch for the handling and treatment of poultry wastes. In: *Livestock Waste Management and Pollution Abatement*, pp. 209–12. ASAE, St Joseph, Mich., 1971.

6. Stewart, T. A. and McIlwain, R., Aerobic storage of poultry manure. In: *Livestock Waste Management Pollution and Abatement*, pp. 261–3. ASAE, St Joseph, Mich., 1971.
7. Hashimoto, A. G., An analysis of a diffused air aeration system under caged laying hens. ASAE Paper NAR-41-428, St Joseph, Mich., 1971.
8. Dunn, G. G. and Robinson, J. B., Nitrogen losses through denitrification and other changes in continuously aerated poultry manure. In: *Proc. Agric. Waste Management Conference*, pp. 545–54. Cornell University, Ithaca, NY, 1972.
9. Prakasam, T. B. S., Srinath, E. G., Anthonisen, A. C., Martin, J. H. and Loehr, R. C., Approaches for the control of nitrogen with an oxidation ditch. In: *Proc. Agric. Waste Management Conference*, pp. 421–35. Cornell University, Ithaca, NY, 1974.

PART III

Utilization and Disposal of Animal Wastes

22

Energy Recovery from Animal Wastes: Anaerobic Digestion, Pyrolysis, Hydrogenation

W. J. JEWELL* and R. C. LOEHR†

*Associate Professor, †Director, Environmental Studies Program, Cornell University, Ithaca, New York, USA

INTRODUCTION

Numerous studies have emphasized the possibilities of producing useful fuel from various organic wastes such as animal manures. In animal production operations, waste management may require as much as one-fifth of both the manpower and the total investment in the operation, and systems that have potential to decrease these commitments are receiving increasing attention. This paper will review animal waste management processes that offer the potential of energy recovery. The specific objectives are:

(a) to examine the potential application of energy recovery from animal wastes;
(b) to review the status of pyrolysis, hydrogenation and anaerobic digestion and their application to animal wastes;
(c) to discuss the potential utility of the energy produced from the above processes; and
(d) to estimate the economics of using the technology in animal production systems.

BACKGROUND

Until recently there were no full-scale agricultural waste management energy recovery systems in use in the United States. Most reports in the public press were related to small farms or inventors who had made laboratory studies. One large anaerobic digester went into operation in September 1974 on a 350 head beef feedlot,[14] and a pyrolysis unit has been in operation since 1972.[11] The pyrolysis unit treats sawmill wastes at a rate of about 45 360 kg/day. Although most of the existing information continues to be from research studies, many people are now considering the possibility of using energy recovery systems in agricultural operations.

The feasibility of energy recovery from agricultural wastes depends on many factors, including: (a) the amount of recoverable energy remaining in the wastes and its relation to the total energy needed in the operation; (b) compatibility of the form of the energy to the uses for it; (c) the availability of equipment and the skills needed to maintain the process; and (d) the cost of using the system and of the resultant energy. Two points should be emphasized. First, detailed information relating to the above factors is not extensive since the technology remains in the early development stages. Sound, preliminary data indicate that some systems are potentially useful and can generate significant amounts of usable energy.

ENERGY FLOWS IN FOOD PRODUCTION OPERATIONS

The quantity of energy that is currently lost in animal manures and other agricultural wastes has not been defined. Recent detailed energy flows have provided greater insight into the energy needs and losses for US dairies and beef feedlots. Figure 1 indicates the annual energy sources and uses for a 100 milking cow dairy and a 1000 head beef feedlot, respectively. The following general conditions were utilized. Both operations were assumed to have adequate land to produce all the feed needed for the animals. Maintenance energy is that energy that does not exist in the live animal or manure and represents animal metabolism energy. The farm and home energy is that energy used to operate the farm, including the homes of the people on the farm, vehicles, pumps, scrapers and similar required equipment. Such energy is the equivalent of the net input of fuel and electricity. The data represent only direct farm energy input and do not include the off-farm energy input such as that involved in the production of fertilizer, machinery and other items that may be needed.

The petroleum and electricity are the energy inputs that have the potential of being replaced by an alternative source such as methane. In practice, not all this energy may be able to be replaced.

In both cases, the gross amount of energy available in the wastes (manure and crop residues) is more than the total consumed in the operation. In a beef feedlot of 1000 head the amount of energy in the manure alone is more than five times that consumed in the operation. If all the energy in the total organic wastes were recovered, it would be almost ten times that consumed in the feedlot. Obviously all the energy cannot be recovered by digestion. However, if even a reasonable fraction were recovered, it appears possible to meet the energy inputs.

Using a similar energy budgeting approach, the amount of energy remaining in the plant residues in wheat and soybean farming has been estimated at more than 12 times that consumed in the production of the

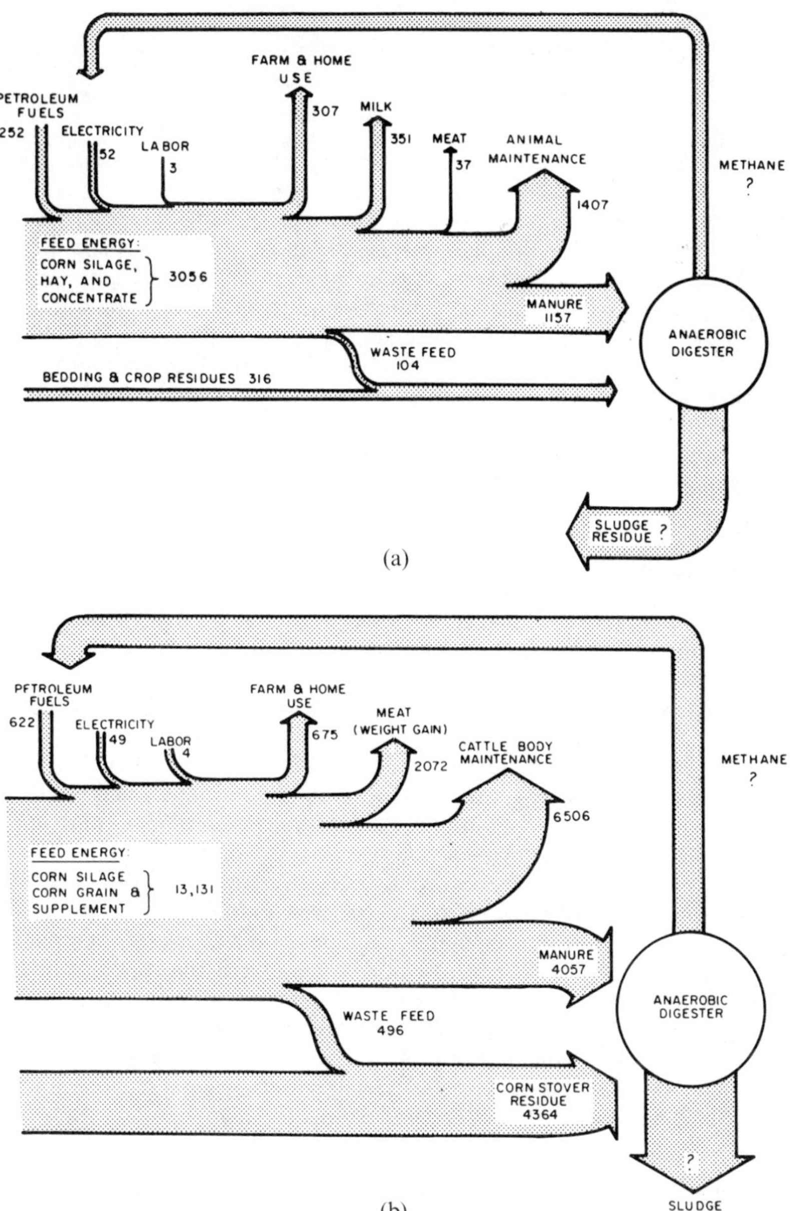

FIG. 1. Annual energy pathways in (a) US dairies (100 head) and (b) beeflots (1000 head) (from ref. 21). (All units in 10^6 kcal.)

crops.[2] Efforts to determine methods of extracting energy now unused in agricultural wastes have an interesting potential.

PROCESSES FOR RECLAMATION OF ENERGY FROM ANIMAL WASTES

There are many factors to consider when determining the potential for energy recovery from animal wastes. Some of these are: size of the installation, status of the technology, availability of full-scale equipment, degree of skilled operation required, operational time commitment, complexity of the process, and whether it is expected that the unit will be operated by agriculturally trained personnel. Comprehensive information on these factors is not yet available for any animal waste energy recovery process. However, consideration of the basic characteristics of several processes and their capability with agricultural wastes enables an estimate to be made of their applicability.

Pyrolysis

The pyrolysis process can be described as high-temperature decomposition of organics in the absence of oxygen. It is sometimes called destructive distillation. Reactor temperatures range above 900 °C and high pressures are common. This process has been investigated as a method of treating municipal solid wastes.[4, 6] Initial studies conducted in 1970 reported that this process was more than 75% efficient. Efficiency was defined as the net output of energy after subtracting the amount required to support the process. A subsequent review of pyrolysis indicated that in many applications, all the energy products formed were consumed in maintaining the extreme operating conditions.[5] In further developments and other applications, the efficiency of pyrolysis has been improved.

The art of pyrolysis is not new. An English patent was issued in AD 1620 to convert organics to charcoal using this process. Pyrolysis, as a commercial process, has been in use for many years in the production of methanol, acetic acid and turpentine from wood as well as in the recovery of the residual charcoal.

A characteristic of this process, which is a disadvantage under some circumstances, is that it generates three forms of energy-containing products: solids, liquids and gases. In trials with municipal refuse (which is highly cellulosic and therefore similar to cow manure), the following products were identified:

Solids: charcoal
Liquids: water, methyl alcohol, ethyl alcohol, methyl acetone, methyl acetate, oils and residual tars
Gases: CO_2, CO, O_2, H_2, CH_4, C_nH_m, N_2

A minimum of 23 constituents were identified in the liquid obtained with municipal refuse. The existence of many basic chemicals in the liquid products of pyrolysis has led to the proposal that by-products from pyrolysis of waste organics have the potential of supporting new chemical industries.[19]

In considering pyrolysis of animal wastes, attention should be given to the total energy balance. The conditions required for pyrolysis cause all moisture to be vaporized, and the heat required for drying or vaporization of the waste materials will indicate the applicability of the processes to animal wastes. In general, animal wastes are produced at a water content greater than 80 % by weight. Even when a significant amount of dry bedding material is used, the resulting mixture of manure and bedding will often have a moisture content greater than 50 %. The gross amount of energy that is available in various organics is summarized in Table I. At a 10 % solids

TABLE I

ENERGY VALUES OF COMMON FOSSIL FUELS AND AGRICULTURAL WASTES

	Energy content (dry weights)	
	kcal/gm	*BTU/lb*
Fossil fuels:		
Coal, bituminous	9·2	13 100
Gasoline	—	20 250
Methane	—	21 500
Organic materials:		
Wood, pine	4·5	6 200
Charcoal	8·1	11 500
Methanol	—	8 600
Cow manure	—	6 000–9 000
Leaves	5·2	7 000

(Btu/lb × 2·32 × 10^3 = J/kg)

content, the energy in organics such as cow manure is equivalent to that required to vaporize the water. However, considering the inefficient use of fuel to vaporize water (about 50 %), the minimum allowable moisture content which will result in operation of pyrolysis with no net energy production is 65 %. Thus, on the basis of moisture content alone, it is clear that high-temperature, high-pressure pyrolytic processes have limited application to high-moisture animal wastes. Only in cases where air drying produces animal wastes at greater than 50 % solids does pyrolysis appear to be a potential means of energy recovery.

Pyrolysis has been investigated as a potential process for several types of agricultural wastes. Low-temperature pyrolysis of egg farm wastes (hen and pullet undercage waste, dead birds, and wood shavings used as litter) was found to require a small energy input to convert the waste to a char.[17] This resulted in a 50–88 % weight reduction, and a similar volume reduction. The char contained 3 % nitrogen, 7 % phosphorus and 6 % potassium by weight. Land application of the char was limited due to its high salt concentration.

The pyrolysis of animal wastes indicated that the heating value of gases from dried beef cattle wastes, dried poultry wastes and dried swine wastes was about $4{\cdot}4 \times 10^6$, $3{\cdot}8 \times 10^6$ and $3{\cdot}2 \times 10^6$ J/kg, respectively.[23]

Pyrolysis temperatures of 400–500 °C, when used with beef cattle feedlot wastes, resulted in maximum yields of liquid organic compounds.[7] The char was composed of one-third ash and was difficult to burn in open air but had some properties as an absorbent. Estimated gas production had an energy equivalent of $2{\cdot}3$–$3{\cdot}5 \times 10^6$ J/kg (2–3 million Btu/ton) of dry cattle feedlot manure. It was concluded that the pyrolysis process applied to cattle feedlot wastes was uneconomical, since the cost of equipment to separate potentially valuable materials from the exhaust stream was not offset by the market value of these materials. In addition, some objectionable odours would require control.

One application of pyrolysis uses lower temperatures (300–400 °C) and a unit which is partially open to the atmosphere, allowing oxygen to enter the reactor.[11] This has several advantages besides simplifying the process. The oxygen supports combustion in the reactor, thus providing a high rate of internal heating. This combustion produces a significant amount of carbon monoxide which provides conditions favourable for hydrogenation and production of oil. The lower temperatures decrease the formation of volatile products and increase the relative quantities of char and oil. The generated gases are used to dry the input material and the char is mixed with the oil for subsequent use. This char–oil mixture is an easily stored, pumpable mixture with an energy content close to that of coal but with sulphur concentrations less than 0·3 %. A heat balance from experiments with pine bark at 37 % moisture resulted in obtaining 50 % of the input heat energy in the char–oil mixture (36 % in the char and 14 % in the condensed oil). About one-quarter of the energy in the gases was used to dry the wood waste to 5 % solids. The oil–char mixture had an energy content of about 28×10^6 J/kg (12 000 Btu/lb). The full-scale facility has operated since 1972 at a lumber operation without major problems.

Hydrogenation

In early developments of the pyrolysis process, it was noted that addition of carbon monoxide and a catalyst favoured the formation of oily end-products. For example, heating cellulose above 250 °C in the absence of oxygen results in the production of a char. If carbon monoxide is present, the production of the char through dehydration is inhibited. The effect of this change is to increase the hydrogen to carbon ratio and to produce a heavy oil end-product rather than a char. The formation of the oil is favoured by the presence of water and catalysts such as sodium carbonate. The term hydrogenation is applied to this process since the waste is essentially treated with hydrogen produced by reacting carbon monoxide and steam under high temperature and pressure.

In a short series of hydrogenation experiments with municipal solid waste (refuse), the oil in the experiment had the following composition (% by weight): carbon 79·6, hydrogen 9·5, nitrogen 1·9, sulphur 0·13 and oxygen 8·9.[1]

In considering the basic operating modes of both pyrolysis and hydrogenation, it is important to note the complexity of the products and methods of handling these products. For example, in hydrogenation the water is separated from the oil, and extraction procedures are used to separate various oil-base products. About 60 litres of highly contaminated water will be generated in treating 100 kg of 50 % moisture municipal refuse. Very little data have been published on alternative methods of handling or disposing of this liquid.

Anaerobic Digestion

The anaerobic digestion process is a microbial degradation of biodegradable organics which results in a gas containing methane and carbon dioxide. This method of energy production from organic wastes is a complex process. It is, however, the most developed of the three processes discussed.

Although this process has been used for animal wastes in many areas, limited data are available to define the details of its overall feasibility with animal wastes. After World War II, 17 large-scale complex systems were constructed on farms in Germany.[20] These units were constructed under unique circumstances and the resulting cost of the energy was high. These systems included gas compression and use as a tractor fuel.

A large amount of data is available on the anaerobic digestion of sewage sludge. Few in-depth research studies have been conducted on the digestion of animal wastes. No attempt will be made to provide a detailed review of this technology, since reviews of the process fundamentals are available.[10, 15] Instead, some of the operating limitations are noted.

Anaerobic digestion is a microbial process susceptible to failure (anaerobic microbial activity is reduced or ceases) when unfavourable conditions exist. This occurs because the process depends on two groups of organisms (volatile acid formers and the methane formers) to carry out the organic breakdown. The methane-forming bacteria grow at slower rates and are more sensitive than the acid-forming bacteria. Adverse environmental conditions can result in inhibition of the methane formers with the accumulation of volatile acids and depression of the pH.

Anaerobic digestion is usually designed to operate at 35 °C, although digesters will operate satisfactorily at temperatures as low as 20 °C and as high as 60 °C. The main variables in operating digesters are pH, volatile solids concentration, feed rate and mixing. With sewage sludge digestion, the desirable pH operating range is 6·8–7·2. It may not be possible to maintain this range for the more alkaline animal wastes such as cow manure or chicken manure. The fermenting solids concentration is usually maintained below 7 % solids to obtain good mixing and digestion. Recent data suggest that mixing may not be as critical with cow manure digestion as it may be with sewage sludge.[22]

Caution is advised in utilizing data obtained from sewage sludge digestion for the design of animal waste digesters. Animal wastes have different characteristics than sewage sludge and these may affect digester performance. Animal waste digesters appear to be more stable and less susceptible to operational difficulties.

The amount of energy which can be generated during anaerobic digestion is a function of the fraction of the total waste that is available to the anaerobic bacteria (*i.e.* the biodegradable fraction) and the amount of energy used to operate the process. Possible waste inputs at an animal production facility can include manure, bedding, waste feed, milk-house wastes and crop residues.

The total estimated gas production from various animals is shown in Table II. These incorporate average values obtained from the literature for waste production and biodegradability. The highest amount of gas

TABLE II

ESTIMATED MANURE AND BIO-GAS PRODUCTION FROM ANIMAL WASTES: ESTIMATED OUTPUT PER 1000 LB LIVE WEIGHT

(from ref. 16)

	Dairy cattle	*Beef cattle*	*Swine*	*Poultry*
Manure production (lb/day)	85	58	50	59
Total solids (lb/day)	10·6	7·4	7·2	17·4
Volatile solids (lb/day)	8·7	5·9	5·9	12·9
Digestive efficiency (% of VS)	35	50	55	65
Bio-gas production				
(ft^3/lb VS added)	4·7	6·7	7·3	8·6
(ft^3/1 000 lb animal/day)	40·9	39·5	43·1	110·9

(lb × 0·454 = kg; ft^3/lb × 0·062 = m^3/kg)

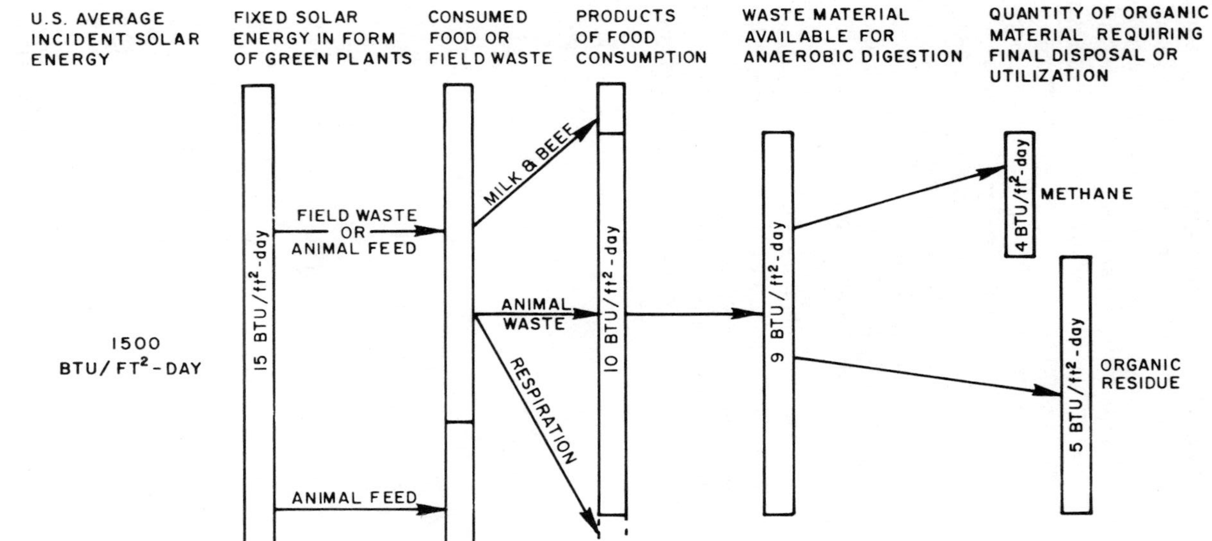

FIG. 2. Relative quantity and fate of plant material consumed by cows and processed by anaerobic digestion. Energy values are approximations related to average sunlight energy striking ground per square foot of surface area (from ref. 8). (Btu/ft^2 = 1·135 × 10^4 J/m^2.)

produced per 454 kg live weight was estimated to be from chickens, indicating the higher amount of biodegradable organics in that material. This information indicates that the minimum numbers of animals to produce the energy equivalent to 1 litre of gasoline would be: 2 beef cattle, 3·2 dairy cattle, 330 chickens and 16 swine. These values depend on the weight and feed ration of the animals and will vary over a wide range. The data in Table II is gross energy possible from digestion and does not include energy required to operate the process.

The fate of the organic material in plants and which is processed by animals and anaerobic digestion is illustrated in Fig. 2. An important aspect to be considered with the anaerobic digestion of animal wastes is that the total volume of residue that must be handled for final disposal is equal to or greater than the initial amount of animal wastes that are to be digested because of the liquid added to obtain a solution that can be mixed. Although considerable solids decomposition occurs in a digester (30–60 %), little volume reduction results.

Many studies of anaerobic digestion of animal wastes have been published[8, 13] which detail some of the design criteria when applied to animal wastes. The relationship between the volumetric loading rate of animal wastes and the efficiency of organic degradation is illustrated in Fig. 3. The loading rates for most animal waste digesters have been between 1·6

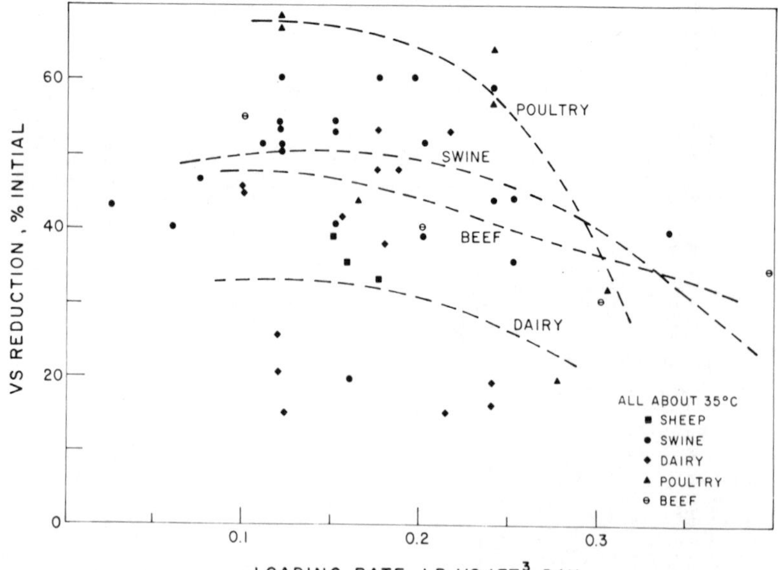

FIG. 3. Animal waste solids decomposition at various organic loading rates (from ref. 10). (lb/ft^3 × 16·02 = kg/m^3.)

and 4·8 kg/m^3 of reactor per day. The hydraulic detention period is usually 10–20 days. In general, the fraction of organic matter which is stabilized in poultry manure is higher, indicating that this material should produce the largest quantity of gas per quantity of solids introduced to the digester. The lower efficiency of the digesters operating on cow manure reflects the high content of non-biodegradable organics in ruminant manures.

DESIGN AND ECONOMIC CONSIDERATIONS

Pyrolysis and Hydrogenation

Pyrolysis and hydrogenation have not been used extensively in the treatment of animal wastes. These high-temperature energy recovery systems currently have limited application to wet animal manures. Because of their complexity they also cannot be constructed to handle small capacities. However, their high reaction rates (reaction times of 20 min or less) enable a relatively small unit to process large quantities of materials. This has led to the suggestion of mobile processes.[19] The economic feasibility of a mobile pyrolysis unit has been analyzed with wastes from a medium-sized sawmill. Waste handling costs exceed the income for capacities less than 13 600 kg/day. However, at feed rates above 68 000 kg/day the income derived from this energy recovery process could be significant.

At the present time the state of knowledge is not sufficiently advanced to identify feasible design and economic data for either of these processes when used with animal wastes. Considerably greater development and demonstration is needed to acquire and evaluate such data.

Anaerobic Digestion

Numerous design alternatives and combinations are available for anaerobic digestion (Fig. 4). The digesters identified in Fig. 5 can be used as part of the noted alternatives. Five types of digestion processes can be considered: (a) completely mixed digesters operated in the mesophilic range; (b) completely mixed digesters operated in the thermophilic range; (c) batch loaded digesters where there are two digesters in use with alternate loading, accumulation, digesting and emptying; (d) partially mixed digesters operated in the mesophilic range; and (e) plug flow reactors.

A detailed design of each of these five systems was developed to determine the economics of each possibility. The design criteria that were used are identified in Table III. These criteria are reasonable and, when a digester is operated and maintained properly, should result in successful digestion.

The costs were developed for a single animal operation, a 100 cow dairy, and using labour, equipment and types of construction available to an agricultural enterprise. The cost of using equipment and methods available

TABLE III

CRITERIA USED FOR DESIGN OF ANAEROBIC DIGESTION SYSTEMS[a]

Parameter	*Typical municipal digester*[d]	*Completely mixed (mesophilic)*	*Partially mixed (displacement)*	*Completely mixed (thermophilic)*	*Batch load*	*Plug flow*
Premix tank holding capacity (days)	2	2	—	2	2	—
Digester design criteria:						
Hydraulic retention time (days)	20	20	20	5	20[b]	40
Volumetric organic loading (lb VS/ft^3-day)	0·25	0·25	0·25	1·00	0·42	0·13
Temperature of operation (°C)	35	35	35	60	35	25
Gas production:						
% VS reduction expected	34	34	32	36	38	27
ft^3 gas produced/day[c]	5 500	5 500	5 180	5 830	6 150	4 370
10^6 Btu/year	1 200	1 200	1 130	1 270	1 340	960
Gas storage capacity (ft^3)	1 000	1 000	1 000	1 000	1 000	1 000
Residue storage capacity (months)	6	6	6	6	6	6

[a] Manure production from a 100 cow free-stall dairy taken from Table II and resulting in 11 400 lb of manure per day, 12·5% solids and 82% volatile solids. Dilution water of 260 gal/day was added to attain a total solids concentration of 10%. Resultant diluted waste volume was 240 ft^3/day.

[b] A 20 day detention time based upon adding waste for 10 days and using it as a batch digestion unit with no additional waste for an additional 10 days.

[c] A gas production rate of 13·3 ft^3/lb of volatile solids destroyed was assumed. Gas composition assumed to be 60% methane and 40% carbon dioxide.

[d] The municipal digester was designed using equipment and accessories immediately available from equipment manufacturers selling municipal sewage sludge digestion equipment.

($lb/ft^3 \times 16{\cdot}02 = kg/m^3$; $ft^3 \times 2{\cdot}83 \times 10^{-2} = m^3$; 10^6 Btu $= 1{\cdot}05 \times 10^9$ J; lb $\times$ 0·454 = kg; gal $\times 3{\cdot}78 \times 10^{-3} = m^3$)

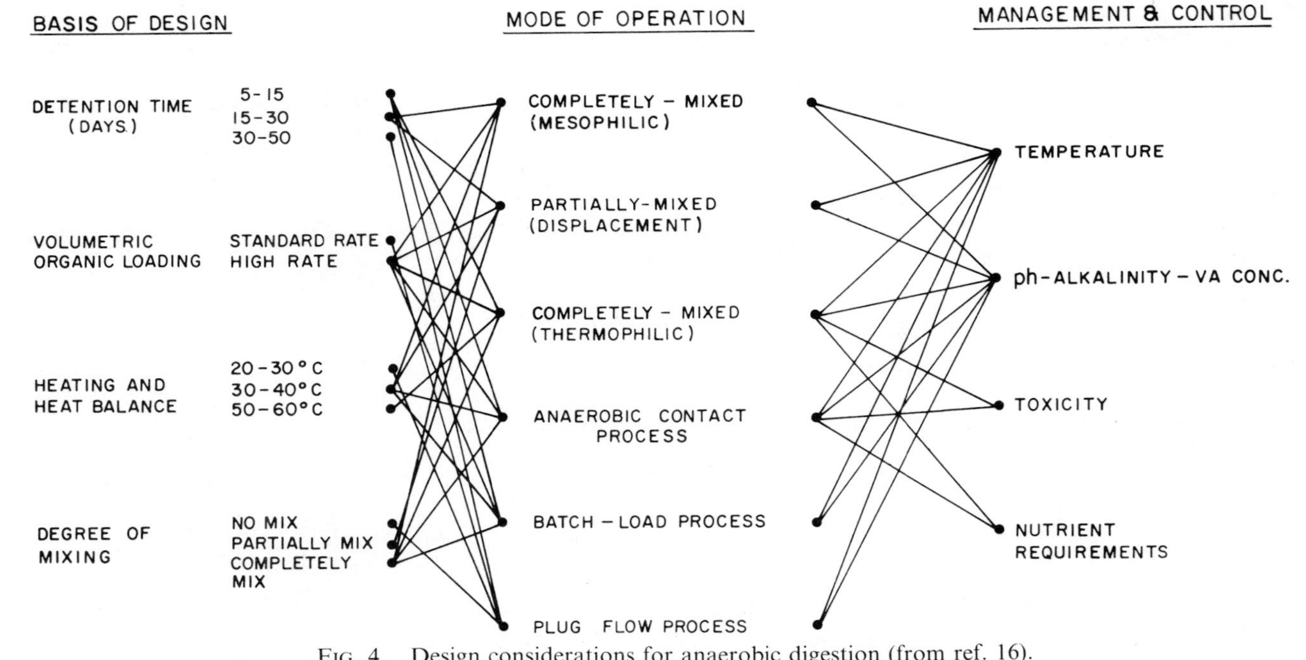

FIG. 4. Design considerations for anaerobic digestion (from ref. 16).

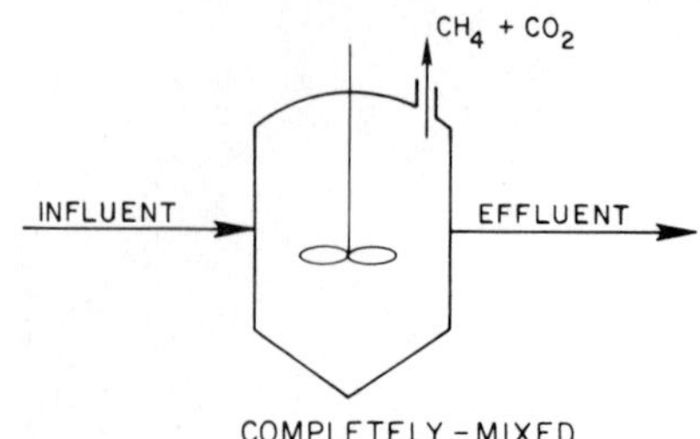

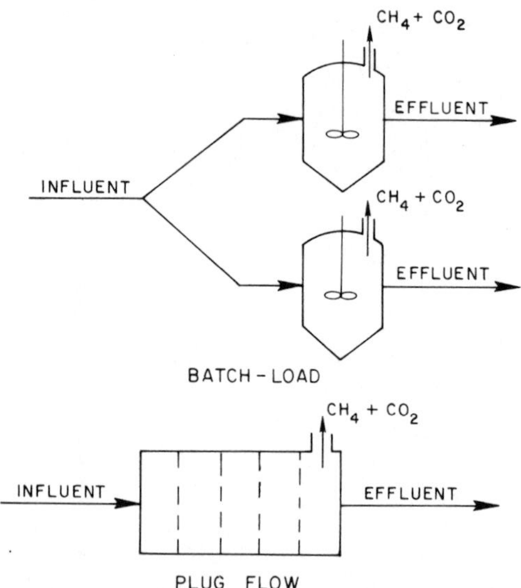

FIG. 5. Types of anaerobic digestion processes.

from manufacturers of digesters for municipal sludge digestion was obtained for comparison. The costs using municipal sewage sludge equipment are higher. Those engaged in animal production can construct a digester at less cost.

Conservative estimates were used in all assumptions, and the systems were designed for a climate where the temperature averaged below freezing for three months. Annual costs for each system represent the yearly amortization of the initial capital investment and the operation and maintenance costs. Capital costs were amortized at a rate of 9% by assuming a 20 year life for structures and a 10 year life for equipment. Maintenance costs were assumed to be 2% of the initial investment per year. Taxes and insurance were based on a rate of 3·5% of the capital cost per year. A summary of the cost data is shown in Table IV. The capital investment ranged from \$200/cow for the simpler systems to a high of \$550/cow for the 'off-the-shelf' municipal sludge digester equipment. Annual costs ranged from \$45 to a high of \$150/cow.

The relative cost of energy production was obtained by dividing the net available energy by the annual costs (Fig. 6). When these costs are

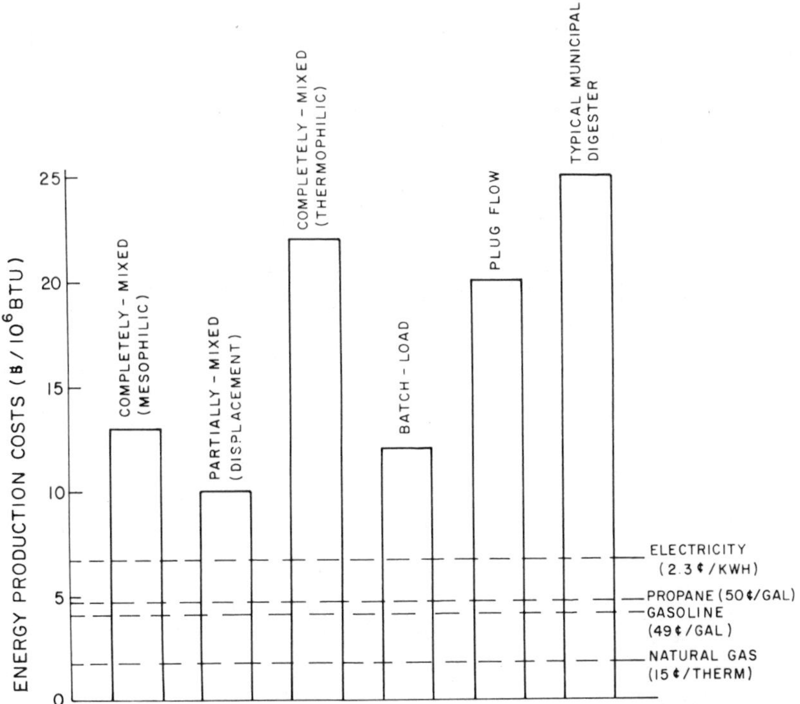

FIG. 6. Cost of net energy production available from anaerobic digestion related to present costs of other energy sources (from ref. 16). (kWh $\times$ 3·6 $\times$ 10^6 = J; gal $\times$ 3·78 = litres; therm $\times$ 1·054 $\times$ 10^8 = J.)

compared with the cost of various types of energy in the USA, it is clear that the cost of energy recovered from animal wastes is not immediately competitive with other energy supplies.

This initial analysis indicated that it would require energy costs above \$8–10/10^6 Btu before the noted systems appear economically competitive. Subsequent analysis of digester design suggests that it may be possible to produce methane at a cost of approximately \$4/10^6 Btu (\$4/10^9 J). Further calculations imply that an operation with as many as 1000 cows may be able

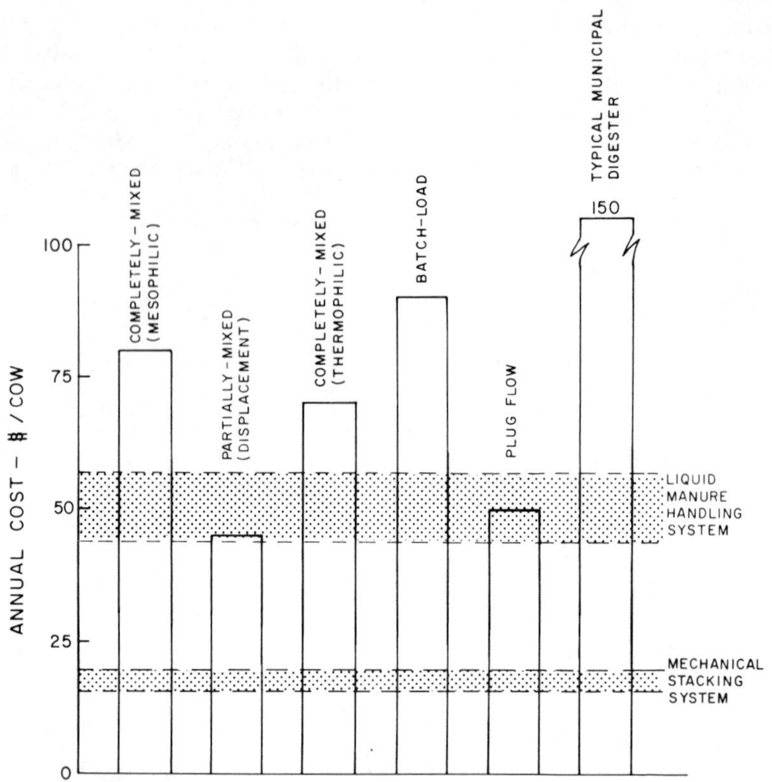

FIG. 7. Total annual cost comparisons of current animal waste management to alternative anaerobic digestion designs (from ref. 16).

to take advantage of the economies of scale and produce energy for as low as $\$2/10^9$ J.

The full-scale plug flow reactor in operation on a 350 head beef feedlot[14] has been operating without difficulty, and the manpower committed to waste management has decreased significantly. The capital cost of the energy (methane) produced in this system has been stated as approximately $\$1.50/10^9$ J.

All animal production operations generate waste, and management of this waste incurs some cost. The costs of anaerobic digestion need not be charged completely to energy production, since some fraction of the costs would be incurred for waste management in any event. The important cost is the incremental costs that would be incurred over and above those costs associated with normal waste management.

TABLE IV

COST COMPARISONS OF FIVE ALTERNATIVE ANAEROBIC FERMENTATION SYSTEMS FOR A 100 COW DAIRY
(from ref. 16)

	Typical municipal digester	*Completely mixed (mesophilic)*	*Partially mixed (displacement)*	*Completely mixed (thermophilic)*	*Batch load*	*Plug flow*
Capital investment:						
Total $	55 000	27 000	20 000	21 000	31 000	22 000
$/cow	550	270	200	210	310	220
Annual costs:						
Total $	15 000	8 000	4 500	7 000	9 000	5 000
$/cow	150	80	45	70	90	50
Net annual gas (10^6 Btu/year)	600	600	450	325	755	250
Energy production costs ($/$10^6$ Btu)	25	13	10	22	12	20

(Btu × 1·054 × 10^3 = J)

A comparison of the costs of conventional animal waste management systems and a digester for a 100 cow dairy was made to identify possible incremental costs. The cost of solid manure storage and mechanical stacking is about $20/cow/year and liquid manure storage and spreading is $40–60/cow/year. The cost of several digestion processes is in the same range as liquid manure storage. In such cases, the incremental costs of the digestion processes are low (Fig. 7).

ON-SITE BIO-FUEL UTILIZATION AND RESIDUE DISPOSAL

Thus far, the general feasibility of generating energy from animal wastes has been discussed without regard to an important factor: the possibilities for using the generated fuels. It is difficult to envisage uses of the products of pyrolysis in animal production operations. Because of this fact, it has been suggested that the oil and char products be combined and that this material be used for power generation. However, because of the high water content, little net energy would be available in treating most animal wastes.

The use of by-products from anaerobic digestion is possible. The wet digested residue is valuable as a fertilizer since most of the original nutrients fed to the animals will be present in the residue. Nitrogen conservation is a desirable characteristic which is part of anaerobic digestion.

Handling of untreated animal wastes and aerobic treatment can result in the loss of 50 % or more of the total nitrogen content of the manure.[12, 18] During anaerobic digestion, much of the organic nitrogen is converted to ammonia and is held in solution. Special attention must be given to the soluble nitrogen during storage and upon soil application to prevent losses to the atmosphere by ammonia volatilization. The nutrient value of the residue and the value of the energy in the wastes can be compared (Table V).

TABLE V
COMPARISON OF THE VALUE OF THE ENERGY DERIVED AS METHANE (AT $2/10^6 BTU) AND THE NITROGEN (AT $0·25/LB N) IN ANAEROBIC DIGESTION OF ANIMAL WASTES

Animal	*By-product value*	
	Nitrogen (*$/animal-day*)	*Energy* (*$/animal-day*)
Beef (1 000 lb)	0·10	0·052
Dairy (1 000 lb)	0·045	0·034
Poultry (5 lb)	0·0017	0·00034
Swine (100 lb)	0·013	0·0068

(10^6 Btu = $1{\cdot}054 \times 10^9$ J; lb × 0·454 = kg)

In all cases the value of the fertilizer exceeds the value of the energy. This comparison, however, is incomplete since the real value of the fertilizer is related to the net gain which is obtained over other management schemes. It is probable that the actual value of the nitrogen is less than one-quarter of that noted.

Perhaps most important is the question of the possibilities for utilizing the energy produced during anaerobic fermentation. Potential uses of methane are for space heating of residences and buildings and heating of hot water. To indicate more clearly the types of fuels used in animal production operations and the degree to which they might be replaced with methane, it is necessary to know the types of fuels used, their quantities and the rates at which they are used. Annual sources and uses are estimated in Fig. 8 and monthly energy demand in Fig. 9. Energy demands on dairies in cold climates are more uniform than on beef feedlots.

Many options exist for utilizing methane gas in agricultural operations. One possibility is to replace fossil fuels in internal combustion engines. However, to utilize the methane it would be necessary to compress the gas to a volume that would fit on a vehicle or adjacent to a stationary engine. For example, for a litre of compressed methane to have the same energy as a litre of gasoline, the methane would have to be compressed to about 34

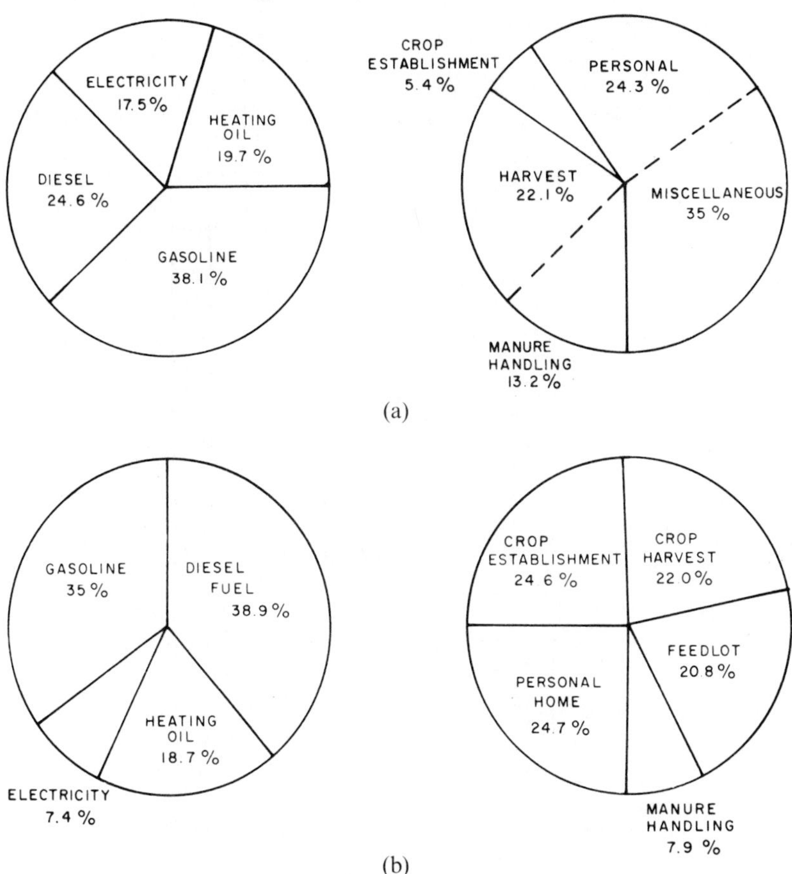

FIG. 8. Distribution of annual energy consumption. (a) 100 cow dairy farm, total direct energy consumption 300×10^6 kcal/year (1200×10^6 Btu/year). (b) 1000 head beef feedlot, total direct energy consumption 671×10^6 kcal/year (2670×10^6 Btu/year).

$\times 10^6$ N/m^2. At lower pressures, larger compressed methane fuel tanks would be needed. The high pressures required to achieve reasonable methane fuel tank size are not encouraging in the consideration of methane as a mobile energy source. Because methane cannot easily replace fossil fuels in internal combustion engines, the gas may replace less than half the energy sources on dairies and less than this on beef feedlots.

Another possible use is the production of electricity using methane. In this possible use, the cost of converting the methane to electricity results in an energy cost that is higher than that available from other sources. As the

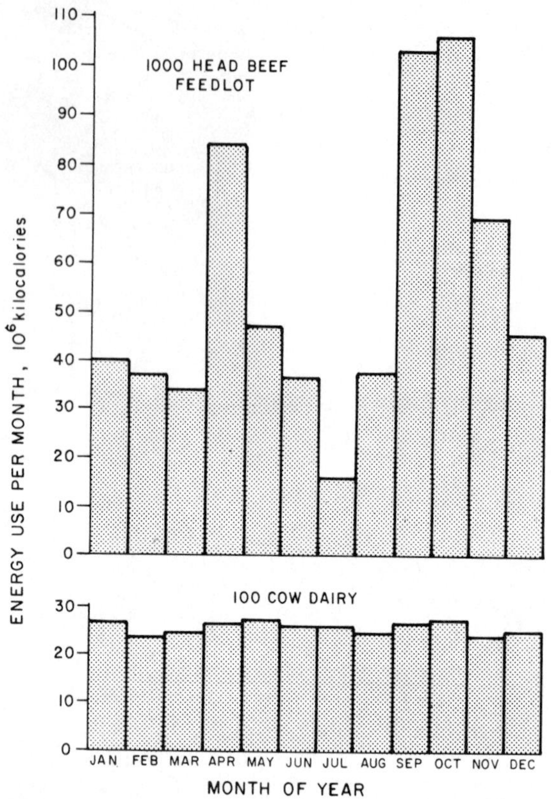

FIG. 9. Monthly energy distribution (fossil fuels and electricity) on 100 cow dairy farm amd 1000 head beef feedlot.

cost of more traditional energy sources increases, the costs of electricity production from methane may become more attractive.

The final feasibility questions remain to be answered. The cost of energy will continue to increase. The world does not have the resources to meet the energy demands 25 years in the future, even if energy demands do not increase and the population grows at a minimum rate. All energy sources will have to be used with discretion, and extravagant uses of energy curtailed. As mankind approaches this situation, it is imperative that all energy sources are used wisely. Animal wastes can serve as a ready supply of energy, and efforts should continue to develop means of reclaiming this energy.

Although general information on the design of anaerobic digesters is available and has been identified earlier, the state-of-the-art of anaerobic digestion of animal wastes is not such that all animal producers can successfully operate an anaerobic digester or utilize the resultant methane gas. A large amount of research and full-scale testing needs to be completed before enough information is available to show clearly the practicality and feasibility of energy recovery from the anaerobic digestion of animal wastes.

REFERENCES

1. Appell, H. R., Wender, I. and Miller, R. D., Hydrogenation of municipal solid waste with carbon monoxide and water. In: *Proc. National Industrial Solid Wastes Management Conference*, 1970, pp. 375–9.
2. Bailie, R. C., Doner, D. M. and Henry, J. D., Potential impact of energy farming for supplementing energy requirements of food production. In: *Proc. Symposium on Energy Requirements of Food Production*. American Institute of Chemical Engineers, New York, 1974.
3. Beekmans, J. M. and Ng, P. C., Pyrolyzed sewage sludge: its production and possible utility. *Environmental Science and Technology*, **5** (1971) 69–71.
4. Burton, R. S., Bailie, R. C. and Alpert, S., Municipal solid waste pyrolysis. American Institute of Chemical Engineers Symposium Series, **70**(141) (1974) 116–23.
5. Drobny, N. L., Hull, H. E. and Testin, R. F., *Recovery and Utilization of Municipal Solid Wastes: A Summary of Available Cost and Performance Characteristics of Unit Processes and Systems*. US Public Health Service Publication No. 1908, US Government Printing Office, Washington, DC, 1971, 118 pp.
6. Finney, C. S. and Garrett, D. E., The flash pyrolysis of solid wastes. *Energy Sources*, **1** (1974) 295–314.
7. Garner, W., Bricker, C. D., Ferguson, T. L., Wiegand, C. S. W. and McElroy, A. D., Pyrolysis as a method of disposal of cattle feedlot wastes. In: *Proc. Agricultural Waste Management Conference*, pp. 101–24. Cornell University, Ithaca, NY, 1972.
8. Jewell, W. J., *Energy from Agricultural Wastes: Methane Generation*. Agricultural Engineering Bulletin 397, College of Agriculture and Life Sciences, Cornell University, Ithaca, NY, 1974, 13 pp.
9. Jewell, W. J. and Morris, G. R., The economic and technical feasibility of methane generation from agricultural wastes. In: *Proc. Symposium on Uses of Agricultural Wastes*, University of Regina, Saskatchewan, 1974.
10. Jewell, W. J., Morris, G. R., Price, D. R., Gunkel, W. W., Williams, D. W. and Loehr, R. C., Methane generation from agricultural wastes: review of concept and future applications. ASAE Paper No. NA74-107, St Joseph, Mich., 1974.
11. Knight, J. A., Tatom, J. W., Colcord, A. R., Elston, L. W. and Har-Oz, P. H., Pyrolytic conversion of agricultural wastes to fuels. ASAE Paper No. 74-5017, Oklahoma, 1974.
12. Lauer, D. A., Limitations of animal waste replacement for inorganic fertilizers. In: *Proc. Cornell University 7th Agricultural Waste Management Conference: Energy, Agriculture and Waste Management*, Syracuse, NY, 1975.

13. Loehr, R. C., *Agricultural Waste Management*. Academic Press, New York and London, 1974, 576 pp.
14. Malstrom, J., personal communication, Rt. 2, Beaune Road, Ludington, Mich., 18 Apr. 1975.
15. McCarty, P. L., Anaerobic waste treatment fundamentals. *Public Works*, Sept.–Dec. 1964, pp. 107–12, 123–26, 91–4, 95–9.
16. Morris, G. R., Jewell, W. J. and Casler, G. L., Alternative animal wastes anaerobic fermentation designs and their costs. In: *Proc. Cornell University 7th Agricultural Waste Management Conference: Energy, Agriculture and Waste Management*, Syracuse, NY, 1975.
17. Nelson, D. M. and Loehr, R. C., Conservation of the energy and mineral resources in waste materials through pyrolysis. In: *Proc. Cornell University 7th Agricultural Waste Management Conference: Energy, Agriculture and Waste Management*, Syracuse, NY, 1975.
18. Prakasam, T. B. S., Srinath, E. G., Anthonisen, A. C., Martin, J. A., Jr, and Loehr, R. C., Approaches for the control of nitrogen with an oxidation ditch. In: *Proc. Agricultural Waste Management Conference: Processing and Management of Agricultural Wastes*, pp. 421–35. Cornell University, Ithaca, NY, 1974.
19. Tatom, J. W., Knight, J. A., Colcord, A. R., Elston, L. W. and Har-Oz, P. H., A mobile pyrolytic system for conversion of agricultural and forestry wastes into clean fuels. In: *Proc. Cornell University 7th Agricultural Waste Management Conference: Energy, Agriculture and Waste Management*, Syracuse, NY, 1975.
20. Tietjen, C., From biodung to biogas: historical review of European experience. In: *Proc. Cornell University 7th Agricultural Waste Management Conference: Energy, Agriculture and Waste Management*, Syracuse, NY, 1975.
21. Williams, D. W., McCarthy, T. R., Price, D. R., Gunkel, W. W., and Jewell, W. J., Analysis of energy utilization on beef feedlots and dairy farms. In: *Proc. Cornell University 7th Agricultural Waste Management Conference: Energy, Agriculture and Waste Management*, Syracuse, NY, 1975.
22. Wong-Chong, G. M., Dry anaerobic digestion. In: *Proc. Cornell University 7th Agricultural Waste Management Conference: Energy, Agriculture and Waste Management*, Syracuse, NY, 1975.
23. White, R. K. and Taiganides, E. P., Pyrolysis of livestock wastes. In: *Livestock Waste Management and Pollution Abatement*, pp. 190–1. ASAE, St Joseph, Mich., 1971.

23

Utilization of Livestock Wastes as Feed and Other Dietary Products

DONALD L. DAY

Professor of Agricultural Engineering, University of Illinois at Urbana-Champaign, Urbana, Illinois, USA

INTRODUCTION AND BACKGROUND

Livestock manure has historically been utilized by various methods. The major methods have been (a) spreading manure on cropland to furnish plant nutrients and to build the soil, (b) the consuming of manure by animals (coprophagy) for dietary nutrients, and (c) using it as a fuel for heating (manure packs and dried dung). Even now the major forms of manure utilization include use for plant nutrients, soil builders, animal nutrients and fuels.

It is known that spreading of livestock wastes on to cropland for its fertilizer and soil-building value is the most economical and energy-saving method of waste management.[1,2] Such analyses, however, do not assign any value to having suitable access to cropland. Competitive land-use demands, as well as concerns for environmental quality and the regulations resulting therefrom, offer justification and opportunity for re-use at the feedlot site, thus bypassing the historical land–plant–animal waste recycle method.

This chapter discusses utilizing livestock wastes as nutrients in animal diets, commonly referred to as 'refeeding'. Usually, though, the manure has been processed into various new products so that the utilization method is not a refeeding of the raw manure.

FEED NUTRIENTS IN WASTES

Several literature reviews have been published on the nutritive value of animal wastes[3–5] and on options and potentials for using processed wastes in livestock feed.[6] Livestock wastes contain many of the same classes of chemical compounds found in feeds. Some form of processing of the wastes for refeeding is desirable to make the nutrients more available, to control odours and insects, and to control disease problems. Also, refeeding

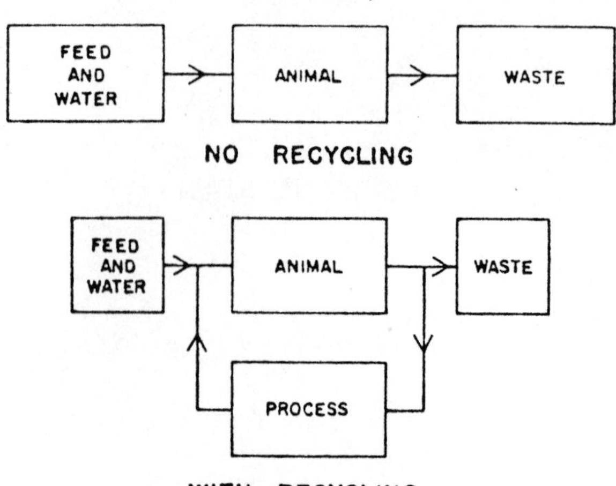

FIG. 1. Potential for reduction of feed and waste with recycling (from ref. 12).

processed wastes offers the possibility of reducing the amounts of new feed required, resulting in the production of fewer excess wastes (Fig. 1).

Three basic categories of information must be known in evaluating a product for its refeeding benefit.[7] These are: (a) establishing the nutritive value (efficacy), (b) determining safety to the health of the animals and (c) establishing that food from animals consuming such a product is safe for man to eat. Also, the product must be acceptable to be eaten or it will not be consumed by the animals. If the above factors are satisfactory, a ration must be formulated around the product to give a resultant adequate diet for the target animals. Finally, equipment and procedures must be established to (a) harvest the raw waste material, (b) process it, (c) incorporate it into the

TABLE I
DISTRIBUTION OF NITROGEN IN FAECES AND URINE OF LIVESTOCK
(from ref. 4)

Species	*% of total N*	
	Faeces	*Urine*
Beef cattle	50	50
Dairy cattle	60	40
Sheep	50	50
Swine	33	67
Poultry	25	75

animal diet and (d) refeed it to the animals. Costs and returns, in terms of both energy and money, are also of interest, as well as its impact on pollution control. Regulations concerning refeeding must be observed and public acceptance of the refeeding scheme must be promoted. Obviously, a team of interdisciplinary specialists must work together to accomplish a successful programme for refeeding a waste material.

Manure is a by-product of livestock production, so properties of manure are greatly influenced by many factors, including ration fed, species of livestock, and methods of manure handling, storage and treatment. Also, manure may include bedding, wasted feed, extra water and soil. Thus the final state of the manure when ready for utilization depends on many aspects of its past history, and it is not surprising that properties vary considerably. However, present livestock management practices are controlled to support a high degree of animal performance and a uniform and consistent waste product. This facilitates plans for utilization by reducing manure variability.

Typical bio-engineering properties of feedlot wastes have already been

TABLE II

AMINO ACID CONTENT OF LIVESTOCK FAECES

(from ref. 4)

Amino acids	*Amino acid g/16 g N*			
	Beef cattle	*Sheep*	*Swine*	*Poultry*[a]
Arginine	1·1	3·4	3·1	4·2
Histidine	0·7	1·3	1·8	1·8
Isoleucine	1·3	3·7	4·8	4·5
Leucine	3·8	4·4	7·2	7·3
Methionine	2·9	5·1	5·1	4·4
Phenylalanine	0·6	1·5	2·7	0·8
Threonine	0·0	3·4	4·0	4·0
Trytophan	1·8	4·8	3·7	4·5
Valine	—	0·9	—	—
Aspartic acid	2·3	4·9	4·8	5·7
Serine	4·3	—	6·3	9·7
Glutamic acid	1·5	—	2·7	4·7
Proline	3·8	—	15·6	14·0
Glycine	1·8	—	4·2	4·9
Alanine	2·7	—	7·0	7·5
Cystine	4·0	—	5·3	9·7
Tyrosine	0·2	—	3·0	2·3
Total amino acids (g)	32·5	33·4	81·9	100·0

[a] g amino acid/16 g true protein nitrogen.

presented in a previous chapter.[8] The amount of nitrogen in manure is of particular interest for refeeding as it is a major component of protein. Distributions of nitrogen in faeces and urine for typical livestock are given in Table I. A large proportion of nitrogen is found in the urine, which makes it desirable to retain the urine for optimizing production of protein in many refeeding schemes. Also, a high percentage of the urea in animal wastes is in the urine, and urea is a common feed additive for cattle rations. Generally, only a small proportion of the total excreted amino acids are in the urine. Over 90% are of faecal origin (Table II).

For refeeding considerations, it should be noted that nitrogen must be in the amino acid form for utilization by non-ruminant animals. Ruminant animals, however, can utilize non-protein nitrogen, so conversion of nitrogen to amino acids is not necessary if the product is to be fed to cattle and sheep.

REFEEDING METHODS

Numerous methods of processing livestock wastes for refeeding have been studied. The utilization methods range from simply mixing raw manure with new feed to complex waste processing schemes that yield several usable products including protein, vitamins, minerals and water. The processed wastes are fed to the same and to different species of livestock (including poultry). The more common methods of processing for refeeding will be discussed in this chapter. However, it needs to be emphasized that these methods have not been commonly practised on extremely large livestock enterprises.

Drying

Probably the oldest method of processing wastes for refeeding is the drying of poultry manure, with heated or natural air, and incorporating it into feed for cattle. There are two main reasons for this: fresh poultry excrement has a lower moisture content than manure from other livestock, so that drying of poultry manure is more feasible; and poultry manure has a high amount of nitrogen relative to other livestock manures and cattle can utilize this nitrogen better than non-ruminant animals. However, dehydrated poultry wastes have been successfully fed to poultry as well as to cattle and sheep.

After a detailed analysis of the various types of equipment suitable for drying poultry manure, rotary drum and batch agitated driers have been found technically most suitable.[9]

Nutrient levels in dried manure are given in Table III and bulk densities in Table IV. Performance data and results of drying poultry, cattle and swine wastes are given in Table V. Example installation and operating costs are given in Table VI and expected animal capacities in Table VII.

TABLE III

NUTRIENT LEVELS BEFORE AND AFTER DRYING, PERCENTAGE OVEN-DRY BASIS
(from ref. 10)

Type		*N*	*P*	*K*	*Ash*	*Crude fibre*	*Ether extract*	*Protein*	*N-free extract*
Poultry	Wet	4·86	2·48	2·27	19·43	17·16	2·99	30·39	30·03
	Dry	3·61	2·39	1·88	19·24	16·40	2·81	22·56	38·99
Dairy	Wet	2·65	0·74	0·65	8·30	28·25	2·55	16·59	44·31
	Dry	2·64	1·15	0·81	11·53	28·47	2·51	16·50	40·99
Swine	Wet	4·24	2·08	1·72	13·31	12·67	12·75	26·46	34·81
	Dry	3·57	2·27	1·40	15·56	14·03	9·02	22·33	39·06

TABLE IV

BULK DENSITIES OF DRIED ANIMAL WASTES
(from ref. 10)

Animal	*kg/m³*
Dairy and beef cattle	192
Poultry	273
Swine	320
Undried animal excreta	1 000

TABLE V

EXAMPLE PERFORMANCE DATA OF A DRIER
(from ref. 11)

Excreta	*Fresh excreta feeding rate (kg/hr)*	*Moisture initial (%)*	*Moisture final (%)*	*Fuel consumption (litres/hr)*	*Electrical consumption (kW)*	*Efficiency (%)*
Poultry	155	76·3	11·1	9·1	4·2	71·8
Bovine (2% straw)	110	82·4	12·0	9·9	4·2	51·6
Swine	100	72·2	12·5	9·1	4·2	44·1

For 100 000 laying hens it would take 10 driers of the capacity used in the above example (Table VI), assuming that some pre-drying can be done with ventilation air. This would allow the manure to arrive at the driers at slightly less than 76% moisture content so that each drier can process the wastes from 10 000 layers in a 40 hr week. Only about 40% of the dried manure can be fed back to these hens; the remainder would be utilized elsewhere.

TABLE VI

EXAMPLE INSTALLATION AND OPERATING COSTS OF A DRIER
(from ref. 12)

Conditions of operation:

Costs are based on a 70 % moisture content input and dry material output of 12–14 % moisture content. Rated capacity is a moisture removal rate of 900 kg of water/hr.

Operating costs:

Parameter	*US$/tonne based on dry product*
Fuel	17·86
Electricity	1·98
Labour	11·02
Depreciation	2·91
Maintenance	2·20
Total operating costs	35·97

Estimated capital costs:

Item	*US$*
Drier with afterburner	69 000
Live bottom wet manure pit (11·6 tonne capacity)	8 000
Installation of machine and pit	4 500
Concrete work for the base	1 500
Storage hopper for the dry product	4 500
Elevator conveyor	4 000
Total	91 500

Dried layer waste (DLW) from caged layers is produced by drying the raw product at temperatures of up to 760 °C. The heat destroys the pathogens and leaves a granulated product with about 10–15 % moisture content and a density of about 273 kg/m^3. Each kg of fresh manure yields about 0·25 kg of DLW.[1] Some conclusions reached from a study of this process are:

(a) Feeding DLW and broiler wastes is technically feasible, especially for large operations (80 000 layers or more).
(b) Feeding a ration to caged layers composed of up to 12·5 % DLW does not adversely affect the performance of laying hens. However,

TABLE VII

EXPECTED NUMBER OF ANIMALS THAT CAN BE SERVED WITH A MANURE DRIER (BASED ON TABLE VI) OPERATING 40 HOURS A WEEK (from ref. 12)

Animal	*Initial moisture (%)*	*Weight of animal (kg)*	*Final moisture (%)*	*Animals served by drier*
Hens	76·3	2	11·1	7 800
Dairy and beef	82·4	635	12·0	15
		454		22
		340		29
Swine	72·2	80	12·5	102
		45		184

body weight and egg production decline when 20 % or more DLW is included.

(c) Complete recycling of DLW in the ration reduces performance. At the 12·5 % refeeding level, only about 40 % of the manure can be refed to the same layers.

(d) Direct land application of poultry manure is the least-cost method studied, assuming land is available. Moreover, land disposal costs can be reduced for large operations when about 40 % of the waste is utilized as feed.

(e) Feeding DLW to beef and dairy cattle can reduce the protein requirement of new feed. Rations containing up to 30 % DLW can be fed without significant problems of performance and carcass quality. However, palatability appears to be a major problem in feeding DLW to cattle.

(f) Broiler waste, both ground and ensiled, has been fed to beef and dairy cattle. It has had little effect on carcass quality, even though feed efficiency tends to be lower when ground broiler waste is added to the ration. Ensiled broiler waste can be fed at a ratio of 2:1 without significantly affecting performance.

(g) Very little data are available on feeding poultry waste to animals on a commercial-size scale.

Wastelage

Wastelage is the product produced by collecting fresh manure from cattle feedlots, mixing it with hay, and ensiling the mixture.[13,14] The manure and hay mixture is loaded into the top of an airtight silo and the ensiled mixture is removed from the bottom. The ensiled mixture is utilized by mixing it with other feed to make a proper diet and then refeeding (Fig. 2).

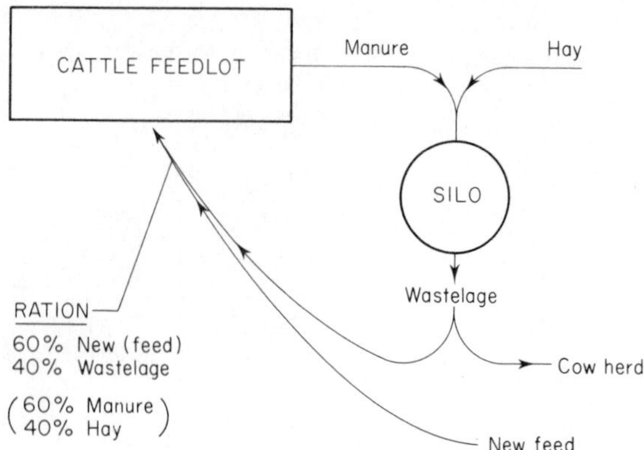

FIG. 2. Flow diagram of the wastelage process.

Mixtures other than hay and manure are also possible. Extensive tests have shown that common feed ingredients containing sufficient carbohydrates to support an acid microbial fermentation can be blended with manure and ensiled to make an effective animal feed. A balanced complete feed mixture (basal) can be blended 1·5:1 with wet feedlot manure, ensiled, and fed to yearling cattle to produce the same rate of animal performance as obtained when the basal ration is fed alone. *Salmonella* and coliform counts are greatly decreased due to the ensiling process.[15] The moisture content of the fresh manure determines to a large extent what feed must be mixed with the manure, as the resultant moisture content should be no more than about 50 %. Figure 3 depicts a process of ensiling new feed (corn, etc.), corn silage and manure.

Only 25–40 % of the manure produced by finishing cattle can be fed back

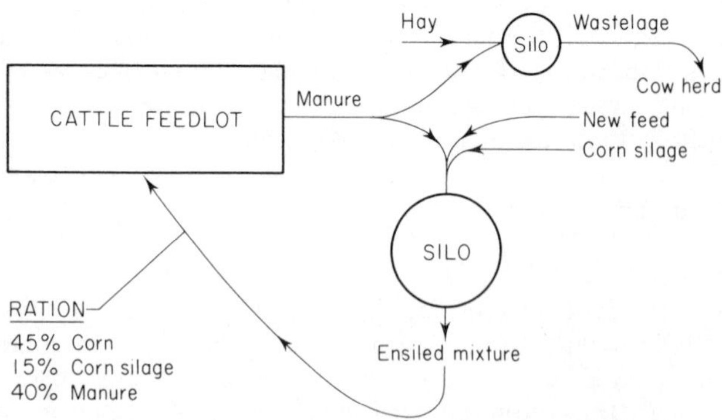

FIG. 3. Flow diagram of wastelage and another ensiling process.

Design data:

Average weight of animals = 375 kg
Ration is same as in Fig. 3
Storage time in airtight silo = 14 days
Manure produced = 6% of body weight per day at 88% moisture = 22·5 kg/head/day
40% of the manure is refed to the feedlot animals = 9 kg/head/day

Ration	*Weight/head/day*		
	Wet basis (kg)	*Dry basis*	
		(kg)	*(%)*
45% corn at 12% moisture	10·13	8·91	81·0
15% corn silage at 70% moisture	3·38	1·01	9·5
40% manure at 88% moisture	9·00	1·04	9·5
Totals	22·51	10·96 = 2·9% of body weight	100·0

Silo capacity (kg) per 1000 animals:

Corn = 10 130
Corn silage = 3 380
Manure = 9 000

Total 22 510 kg/day × 14 day storage time = 315 tonnes

Approximate size of airtight silo for feedlot = 6 m × 16 m.
The circular silo should be top loading with bottom unloading.

FIG. 4. Example material flow for a 1000 head beef cattle feedlot module using wastelage system shown in Fig. 3.

to the same group of cattle. The surplus is fed to another herd, usually dry cows or a cow–calf herd as in Fig. 3.

Example material flow (feed and manure) calculations are shown in Fig. 4 for a 1000 head module of beef cattle in the finishing stage of growth. Estimates for larger feedlots can be made from multiples of the 1000 head module.

Oxidation Ditch Mixed Liquor

Oxidation ditch mixed liquor (ODML) is a method of refeeding aerobically treated liquid wastes as a nutrient-rich drinking water.[16,17] Manure falls through slotted floors into an oxidation ditch in a confinement building (Fig. 5). The aerobic action of the oxidation ditch (an odourless process)

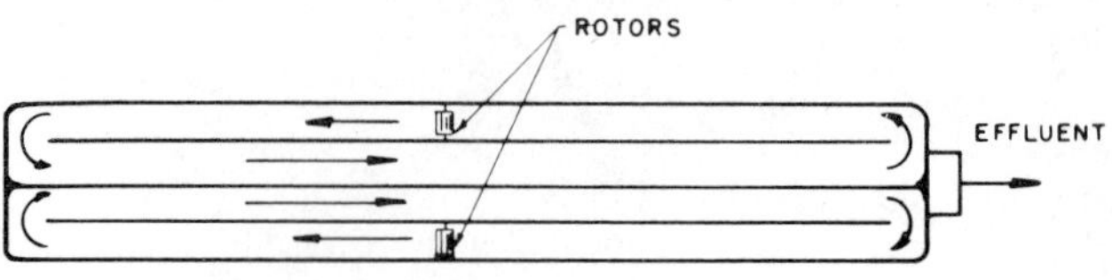

PLAN VIEW OF AN OXIDATION DITCH

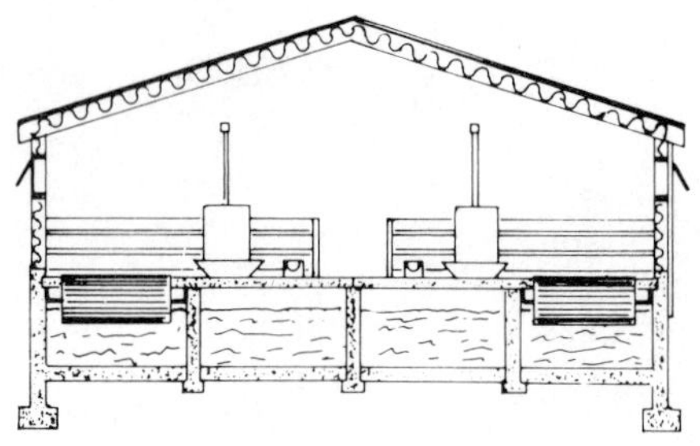

ELEVATION VIEW

FIG. 5. Swine confinement building with an oxidation ditch beneath the totally slotted floor (from ref. 17).

converts organic matter to single-cell protein. The ODML is refed as a nutrient-rich drinking water, and no other water is provided (Fig. 6). The main feed, with protein content reduced by about 15 %, is fed in the usual manner. Table VIII gives typical properties of swine ODML, and Table IX gives typical amino acid contents of various ODMLs compared with corn and soybean meal. Design criteria and energy requirements for livestock oxidation ditches are given in Table X.

TABLE VIII
TYPICAL PROPERTIES OF AEROBICALLY TREATED SWINE WASTES
(from ref. 17)

Parameter	*Mean*	*Range*
pH	7·7	6·0–8·0
Dissolved oxygen (mg/litre)	1·3	0·3–4·3
Temperature (°C)	27	17·5–37·0
Chemical oxygen demand (mg/litre)	29 423	18 425–55 300
Dry matter (%)	3·4	2·1–4·0
Nitrogen (% dry matter)	7·9	5·1–10·0
Ash (% dry matter)	41·7	36·1–48·7

TABLE IX

AMINO ACID CONTENTS (PERCENTAGE DRY MATTER) OF AEROBICALLY TREATED LIVESTOCK WASTES COMPARED WITH SWINE FAECES, PIG REQUIREMENTS, CORN AND SOYBEAN MEAL
(from ref. 16)

	Beef ODML[a]	*Poultry ODML*[b]	*Swine ODML*[c]	*Swine ODML*[d]	*Swine faeces*[e]	*Pig requirement*[f]	*Corn*	*Soybean meal*
Arginine	1·85	1·43	3·49	1·73	0·44	0·28	0·46	3·09
Cystine	0·56	0·43	0·51	1·30	—		0·12	0·42
Glycine	1·49	1·69	3·70	2·15	—		—	—
Histidine	1·03	0·80	1·39	0·45	0·14	0·24	0·22	1·00
Isoleucine	1·19	1·40	2·96	1·66	0·52	0·27	0·31	2·21
Leucine	2·28	2·48	4·53	2·91	0·92	0·74	1·04	3·69
Lysine	1·68	1·76	3·46	1·64	0·60	0·79	0·26	2·69
Methionine	0·62	0·69	1·38	1·41	—	0·53	0·18	0·63
Phenylalanine	1·34	1·34	3·58	1·62	0·81	0·58	0·42	2·39
Threonine	1·35	1·54	3·13	1·86	0·53	0·49	0·34	1·93
Tryptophan	—	—	—	—	—	0·13	0·058	0·69
Tyrosine	0·96	1·15	1·96	1·36	—	—	0·36	1·73
Valine	1·72	2·11	3·30	2·26	0·58	0·50	0·45	2·36
Aspartic acid	2·75	3·15	5·35	3·82	—	—	0·61	5·82
Glutamic acid	3·79	3·61	9·56	5·37	—	—	1·61	7·74
Crude protein (N × 6·25)	—	—	—	45·6	—	16	8·69	45·7

[a,b] Grab samples passed through a 200 mesh screen, 1971.
[c] Average of two grab samples passed through a 200 mesh screen, 1971.
[d] Grab sample, 1967.
[e] Fresh swine faeces.
[f] Weanling pig.

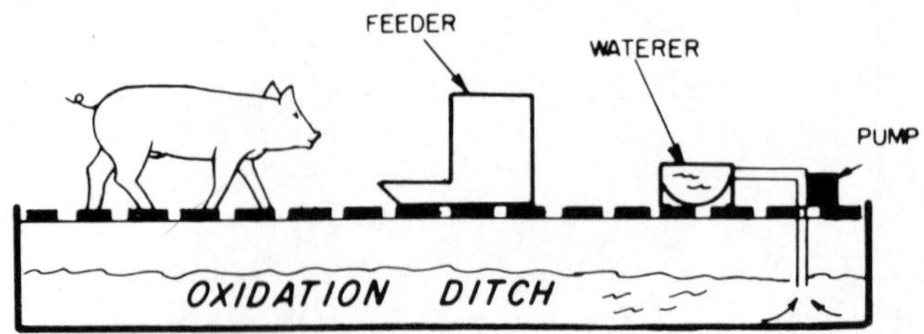

FIG. 6. ODML is pumped from the oxidation ditch directly into a watering trough, and no other water is provided (from ref. 17).

This method can also be used for cattle and poultry. Since cattle can utilize non-protein nitrogen, however, organic wastes do not need to be converted into protein, and the energy required for this method does not seem justified for refeeding to cattle.

With proper circulation and aeration, the oxidation ditch gives good odour control and refeeding ODML keeps surplus waste materials to a

TABLE X

DESIGN RECOMMENDATIONS FOR LIVESTOCK OXIDATION DITCHES BENEATH SLOTTED FLOORS

(from ref. 18)

Animal	*Animal weight (kg)*	*Requirements per animal*			
		Daily BOD_5[a] (kg)	*Daily oxygenation capacity[b] (kg)*	*Ditch volume[c] (m^3)*	*Daily power requirement[d] (kWh)*
Swine:					
Sow with litter	170·1	0·36	0·716	0·671	0·83
Growing pig	29·5	0·064	0·127	0·119	0·15
Finishing hog	68	0·145	0·281	0·272	0·33
Dairy cattle:					
Dairy cow	589·7	1·002	2·00	1·88	2·33
Beef cattle:					
Beef feeder	408·2	0·612	1·22	1·13	1·42
Sheep:					
Sheep feeder	34	0·023	0·05	0·045	0·06
Poultry:					
Laying hen	2·04	0·009	0·018	0·017	0·021

[a] Use specific production data when known.
[b] Twice the daily BOD_5.
[c] Based on 0·53 kg of daily BOD_5/m^3.
[d] Based on 0·862 kg of oxygen/kWh.

TABLE XI
BUILDINGS AND ROTORS FOR A FARROW-TO-FINISH SWINE FARM PRODUCING 13 000 HOGS PER YEAR
(from ref. 19)

Building	*Size (m)*	*Animal capacity*	*Number of rotors*[a]
Gestation 1	9·75 × 50·60	250 sows	2
2	10·97 × 48·77	250 sows	2
3	10·97 × 73·15	410 sows	2
Farrowing 1	6·71 × 42·67	50 sows, 450 piglets	1
2	6·71 × 42·67	50 sows, 450 piglets	1
3	7·92 × 45·11	50 sows, 450 piglets	1
Nursery 1	9·75 × 42·67	1 100 pigs	2
2	9·75 × 42·67	1 100 pigs	2
Growing 1	10·97 × 60·96	800 pigs	4
Finishing 1	10·97 × 121·92	1 500 hogs	8
2	10·97 × 152·40	1 800 hogs	12
			Total 37

[a] Rotors are 1·5 m diameter, 0·81 m wide, have 5 h.p. motors, and run at 100 rev/min. Liquid depth is maintained at 32 cm.

minimum. Ordinarily, water has to be added to the oxidation ditch to maintain the liquid depth of up to 60 cm when all hogs in a building are receiving ODML. Occasionally, however, the ash component of the ODML builds up too high, over 50 %, and then material must be removed and water added to dilute the ash build-up. Two health problems have been noticed: one was a build-up of intestinal worm eggs in the ODML, and the other was too high nitrate levels following a period of excess aeration. Proper management can control both these potential problems.

Building sizes, capacities and number of aeration rotors are given in Table XI for a swine farm producing 13 000 hogs per year.[19] This farm uses only oxidation ditches and evaporation ponds for the complete waste management system. However, refeeding has been experimented with in only a few pens of a finishing building.

COMMERCIAL SYSTEMS

There are three major commercial methods of processing cattle wastes for refeeding in the USA. They are the Cereco, Corral and Grazon systems.*

* Naming companies that have these processes is for information only and does not imply endorsement or recommendation. Furthermore, this is not an exhaustive list of commercial systems in the world.

The Cereco and Corral systems involve complex expensive equipment that limits their use to large feeding operations. Grazon can be used for small operations as well as for large ones.

In all three methods, manure is collected frequently and processed before major losses of nutrients can occur. The frequent collecting of manure helps reduce odours in the feedlots or in confinement buildings. These methods have been developed primarily for beef cattle, but they can also be adapted for other livestock. Details of equipment required may be obtained from the manufacturers.

Cereco System

Figure 7 shows the basic flow diagram of a typical Cereco plant, and Fig. 8 shows a partial plant designed for small operations.

The system can be described as follows. Manure is collected from the feedlot every 4–10 days by a self-loading truck that skims fresh manure from the pens. The truck is equipped with a grinder which grinds the manure as it is unloaded into slurry pits. After 24 hr in the pits, the slurry is pumped into a liquid–solid separator and then pressed. The resulting product is ensiled.

The liquid portion from the press is fermented to encourage the production of single-cell protein and then dried, pasteurized, blended with other feed ingredients and pelleted. The resulting fraction resembles alfalfa

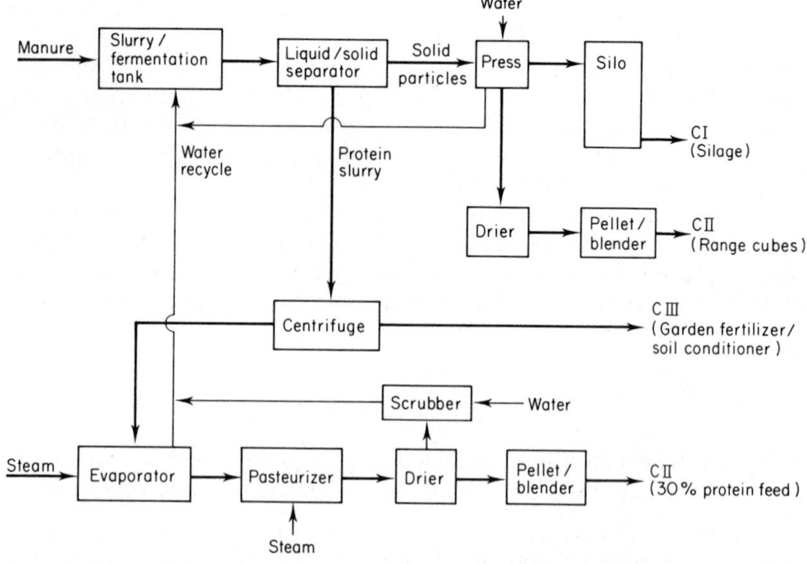

FIG. 7. Flow diagram of a Cereco plant (from ref. 20).

pellets. The slurry from the initial separation in the separator is centrifuged, and a fertilizer product is produced.

Since Cereco plants are designed to meet the specific conditions of an individual feedlot, it is impossible to describe the economics of the process except in very general terms. However, the following generalizations can be made:[20] total capital requirements vary from US$70 to US$90 per ton of

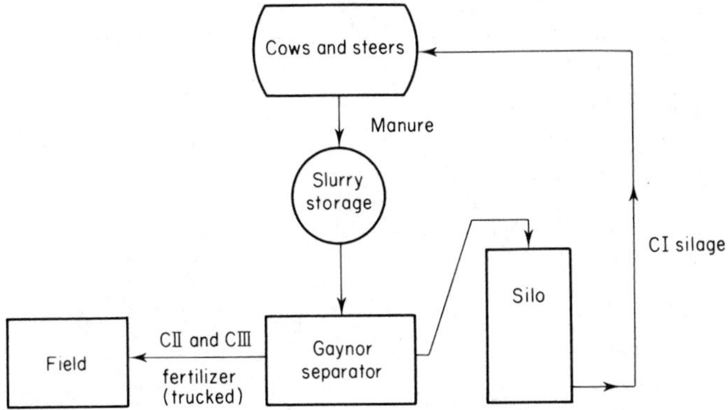

FIG. 8. Partial Cereco plant producing silage and liquid fertilizer (from ref. 20).

dry matter input or per head of cattle feedlot capacity; and total fuel and electrical costs comprise at most 20% and as little as 7% of the value of the feed products obtained. Feeding some of the products to cattle, poultry and fish proved satisfactory.[21]

Corral System

The Corral system can be used with open feedlots or confinement buildings.[22] In a confinement building, fresh urine and manure are collected daily by a scraper which operates underneath slotted floors. The wastes flow by gravity to a holding pit where they are agitated. From this pit the wastes are pumped into a mechanical unit that separates the large solids from the liquid. This unit consists of a vibrator screen and a press. The liquid fraction goes to a holding pond for storage until it is applied on to crops for fertilizer by flood irrigation or by a sprinkler system.

Solids from the press are moved from the separation unit to be composted, ensiled or disinfected. This process sterilizes the product, which then can be used for refeeding in a cow herd or sent to a feed mill to be mixed into the ration for the feedlot cattle. A diagram of the system with confinement cattle is shown in Fig. 9, the solids separation unit is shown in Fig. 10, and a typical materials flow diagram for 10 000 cattle is shown in Fig. 11.

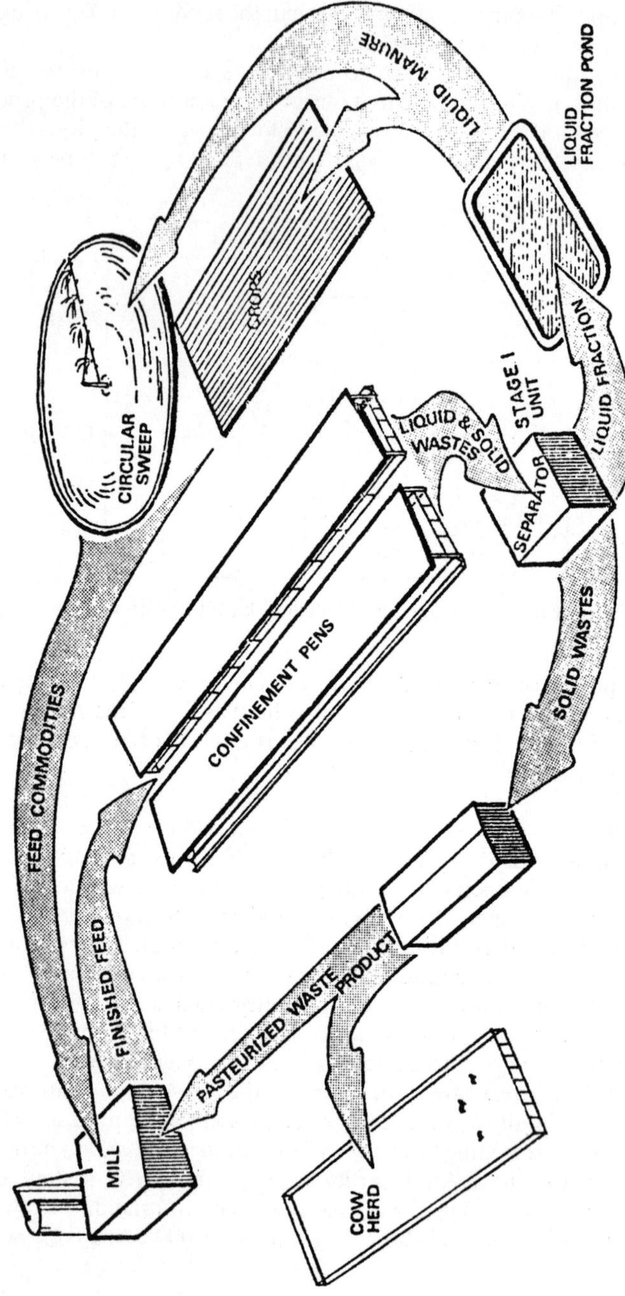

FIG. 9. The Corral system for confinement beef cattle (from ref. 22).

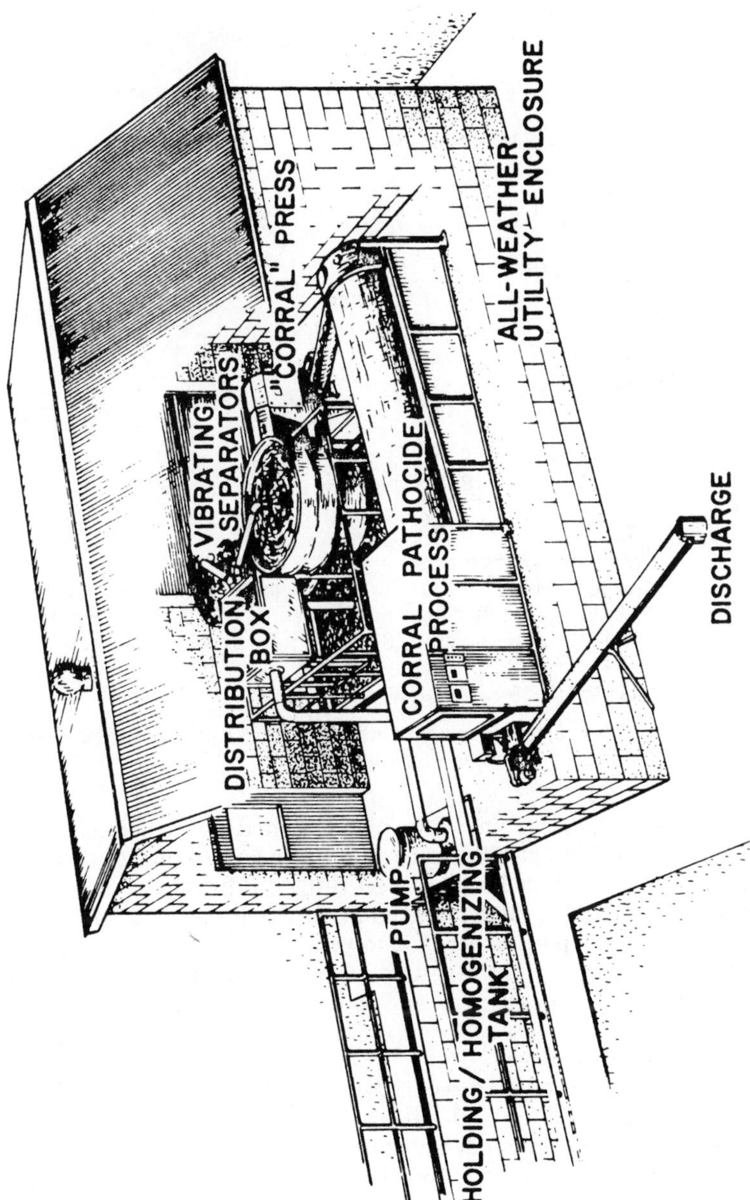

FIG. 10. The Corral solids separation unit (from ref. 22).

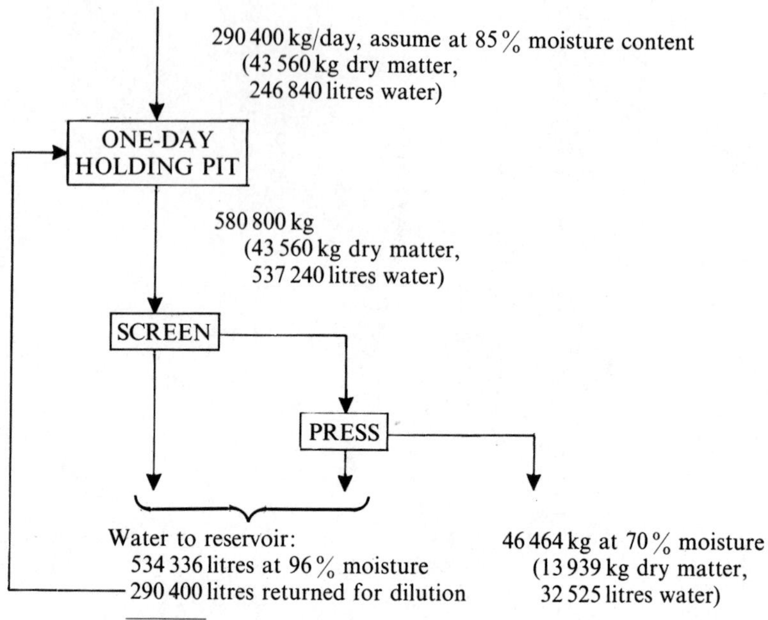

FIG. 11. Typical flow diagram for 10 000 head of beef cattle using the Corral system (from ref. 22).

Costs of the mechanical unit, including associated facilities, range from US$50 000 to US$150 000. The higher price includes two presses that will handle manure from 20 000 head of beef cattle on slotted floors.

Cattle feed trials indicated that consumption, weight gain and conversion rates from a test group being fed 25 % recycled solids were comparable to rates obtained with a control group being fed to a standard high-concentrate ration during the growing and part of the finishing phase.[22]

Grazon Process (Formulage)

The Grazon process involves frequent collection of cattle manure, adding a Grazolin chemical solution and mixing in a truck or wagon to obtain formulage.[23] The chemical is designed to improve the palatability of the manure, to kill the pathogenic organisms and to control odours. This chemically treated manure, formulage, is then mixed with new feed and refed on a regular basis as depicted in Fig. 12.

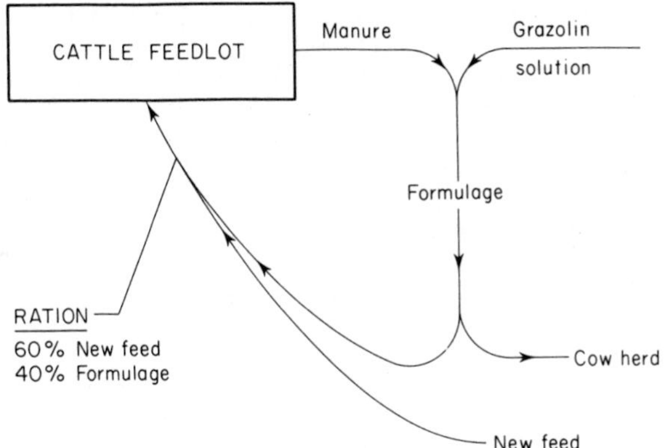

FIG. 12. Flow diagram of the Grazon process (from ref. 23).

The Grazolin solution may be purchased from the Grazon process distributor for the particular area. The approximate cost of preparing formulage (wet weight) is US$5.00/ton for chemicals. This does not include the cost of mixing.

Formulage can be used in proportions of up to 50% for maintenance diets and up to 25% for finishing diets.[23] Refeeding trials have shown that most of the pathogens are killed by the Grazon process. The trials also showed that diets containing up to 50% treated manure were completely acceptable to cattle.

REFERENCES

1. Economic Research Service, *Recycling Poultry Waste as Feed: Will It Pay?* Agricultural Economic Report 254, US Dept of Agriculture, Washington, DC, 1974.
2. Kim, H. C. and Day, D. L., Energetics of alternative waste management systems. In: *Proc. International Symposium on Livestock Wastes*, pp. 398–401, ASAE, St Joseph, Mich., 1975.
3. Smith, L. W., Calvert, C. C., Frobish, L. T., Dinius, D. A. and Miller, R. W. *Animal Waste Reuse: Nutritive Value and Potential Problems from Feed Additives—a Review*. Agricultural Research Service 44-224, US Dept of Agriculture, Washington, DC, 1971.
4. Smith, L. W., Recycling animal wastes as protein sources. *Proc. Alternative Sources of Protein for Animal Production*, pp. 146–73. National Academy of Sciences, Washington, DC, 1973.
5. Whetstone, G. A., Parker, H. W. and Wells, D. M., *Study of Current and Proposed Practices in Animal Waste Management*. US Environment Protection Agency 430/9-74-003, US Government Printing Office, Washington, DC, 1974.

6. Yeck, R. G., Smith, L. W. and Calvert, C. C., Recovery of nutrients from animal wastes: an overview of existing options and potentials for use in feed. In: *Proc. International Symposium on Livestock Wastes*, pp. 192–4. ASAE, St Joseph, Mich., 1975.
7. Taylor, J. C., Regulatory aspects of recycled livestock and poultry wastes. In: *Proc. International Symposium on Livestock Wastes*, pp. 291–2. ASAE, St Joseph, Mich., 1975.
8. Taiganides, E. P., Bio-engineering properties of feedlot wastes (this volume, pp. 131–53).
9. Akers, J. B., Harrison, B. T. and Mather, J. M., Drying of poultry manure: an economic and feasibility study. In: *Proc. International Symposium on Livestock Wastes*, pp. 473–7. ASAE, St Joseph, Mich., 1975.
10. Surbrook, T. C., Sheppard, C. C., Boyd, J. S., Zindel, H. C. and Flegal, C. J., Drying poultry wastes. In: *Proc. International Symposium on Livestock Wastes*, pp. 192–4. ASAE, St Joseph, Mich., 1971.
11. Sheppard, C. C. (Ed.), *Poultry Pollution Problems and Solutions*. Research Farm Science Report 117, Michigan State University, East Lansing, Mich., 1970.
12. North Central Regional Research Publication 222, *Livestock Waste Management with Pollution Control*. Midwest Plan Service, Iowa State University, Ames, Iowa, 1975.
13. Anthony, W. B., Cattle manure: re-use through wastelage feeding. *Proc. Conference on Agricultural Waste Management*, pp. 105–13. Cornell University, Ithaca, NY, 1969.
14. Anthony, W. B., Cattle manure as feed. *Proc. International Symposium on Livestock Wastes*, pp. 293–6. ASAE, St Joseph, Mich., 1971.
15. McCasky, T. A. and Anthony, W. B., Health aspects of feeding animal waste conserved in silage. *Proc. International Symposium on Livestock Wastes*, pp. 230–3. ASAE, St Joseph, Mich., 1975.
16. Day, D. L. and Harmon, B. G., Nutritive value of aerobically treated livestock and municipal wastes. *Proc. The Use of Wastewater in the Production of Food and Fiber*, pp. 240–5 (Oklahoma State Dept of Health, Oklahoma City, Okla.). US Environmental Protection Agency 660/2-74-041, US Government Printing Office, Washington, DC, 1974.
17. Day, D. L. and Harmon, B. G., A recycled feed source from aerobically processed swine wastes. *Trans. ASAE*, **17**(1) (1974) 82–4, 87.
18. Jones, D. D., Day, D. L. and Dale, A. C., Aerobic treatment of livestock wastes. *Agr. Exp. Sta. Bulletin* 737, University of Illinois, Urbana, Ill., 1971.
19. Smart, P., McGain, F., Day, D. L. and Harmon, B. G., Oxidation ditch waste management system for a large confinement swine farm. *Proc. International Symposium on Livestock Wastes*, pp. 190–1. ASAE, St Joseph, Mich., 1975.
20. *Cereco System*. Ceres Ecology Corp., Denver, Colo., n.d.
21. Ward, G. M., Johnson, D. E. and Kienholz, E. W. Nutritional properties of feedlot manure fractionized by Cereco process. *Proc. International Symposium on Livestock Wastes*, pp. 208–10. ASAE, St Joseph, Mich., 1975.
22. Bunger, R. E., Animal waste handling and 'can the tail wag the dog?' Paper for the Beef Confinement Workshop, Feed Ingredients Assoc., 1974. Available from Corral Industries, Phoenix, Ariz.
23. Grazon, *Cattle Diets*. Grazon, Champaign, Ill., 1975.

24

Utilization of Animal Wastes for Crop Production

MILAN SKARDA

Crop Research Institute, Ruzyné-Prague, ČSSR

INTRODUCTION

The use of liquid manure for crop fertilization is the most common and effective way of utilizing this waste material.

The transition from traditional production of dung and urine to one single product, liquid manure, tends to improve working conditions in animal houses and should eliminate the wasteful cycle of moving straw from field to stable to dung-heap and back to field. However, such changes will also affect traditional patterns of operation of crop production. Changes in the composition of farmyard manure will require corresponding reorganization of the thinking and habits of farmers who are used to using green manuring schemes for fertilization.

The utilization of liquid manure as fertilizer cannot be considered solely as an agronomic matter. Hygienists and water supply engineers are also concerned to an unprecedented degree, along with their respective agencies. Their concern lies mainly in their anxiousness over underground-water contamination, disease transmission and environmental pollution. Utilization of manure on cropland requires the harmonious combination of principles from soil science, public health, groundwater hydrology and veterinary medicine.

Crop fertilization with liquid manure must be based on the following principles:

1. Protection and conservation of the environment.
2. Preventing applications of liquid manure in excess of what both the soil and crops can effectively utilize.
3. Avoidance of excessive losses of the fertilizer value of liquid manure.
4. Limiting animal numbers per unit of production to the quantities of manure which can be effectively utilized on the cropland belonging to the animal unit.
5. Increasing the productivity of soils through a systematic utilization of liquid manure as well as a soil amendment.

LIQUID MANURE FERTILIZER CONTENT

Table I shows the typical composition of cattle manure collected from the farmyard and containing straw, or collected as liquid manure. Quantities are the same in both cases; moisture content varies.

TABLE I

COMPOSITION OF CATTLE FARMYARD AND LIQUID MANURE AND DAILY PRODUCTION PER ANIMAL EQUIVALENT (AE)[a]

	Faeces	*Urine*	*Added*		*Total*
			Straw	*Water*	(% *of total*)
			(*kg/AE/day*)		
Farmyard manure	26	4	3	5	38
	(68)	(11)	(8)	(13)	(100)
Urine	—	11	—	5	16
	—	(69)	—	(31)	(100)
Total	26	15	3	10	54
	(48)	(28)	(6)	(18)	(100)
Liquid manure	26	15	—	14	55
	(48)	(27)	—	(25)	(100)

[a] 1 AE is an animal of 450 kg live weight.

The quantity of water added to manure has the most pronounced influence on the composition, quality and chemical properties of liquid manure. By increasing the amount of water added, the liquidity of the manure increases and so does its tendency to separate into three distinct layers during storage; nutrient concentration decreases, together with the concentration of the organic substances, and the storage capacity of tanks must also be enlarged to accommodate the larger volumes.

Organic Matter

Table II shows that organic matter and nutrient content of liquid manure are positively correlated.

About 70–80% of the total solids is organic matter. This organic matter is more easily decomposed by microbial action (as indicated from respiration tests) than farmyard manure. Its carbon content can be utilized for humification, when partly mineralized, and in this way it favourably influences the balance of humic substances in the soil.

Liquid manure is distinguished by its high biological activity. This characteristic is due to the nutrients present and the fact that they are in

TABLE II
AVERAGE COMPOSITION OF LIQUID MANURE[a]

	Young beef			*Milking cows*	*Average for cattle*	*Mean for swine*
	Calves	*Heifers*	*Steers*			
Units are % of total wet waste:						
Dry matter	9·6	10·9	9·9	8·1	9·2	8·3
Organic substances	7·4	8·6	7·6	5·9	6·9	6·7
C_t (total carbon)	3·8	4·1	4·8	2·8	3·5	2·9
N_t (total nitrogen)	0·49	0·44	0·62	0·35	0·43	0·61
N_{NH_4}	0·20	0·19	0·34	0·18	0·22	0·36
P_2O_5	0·28	0·29	0·39	0·17	0·24	0·33
K_2O	0·31	0·50	0·65	0·51	0·50	0·23
CaO	0·41	0·35	0·29	0·21	0·28	0·26
MgO	0·15	0·09	0·13	0·07	0·09	0·08
Na_2O	0·05	0·08	0·07	0·07	0·07	0·06
Units are weight ratios:						
C:N ratio	7·8	9·3	7·7	8·0	8·1	4·8
$N:P_2O_5:K_2O$	1:0·57 :0·63	1:0·66 :1·14	1:0·63 :1·05	1:0·49 :1·46	1:0·56 :1·15	1:0·54 :0·38
Specific weight	1·01	1·02	1·02	1·02	1·02	1·02
pH (no units)	7·4	7·4	7·7	7·6	7·5	7·0

[a] Data obtained from several farms in Czechoslovakia.

suspension. Mineralization rate depends to a significant degree on the $C:N_t$ ratio.

During summer, liquid manure enhances soil microflora activity more pronouncedly than during winter. This induces a more intensive decomposition of organic matter during summer months than during winter months. Mineral nitrogen is nitrified more readily during summer than during the winter period.

All nutrients in liquid manure are in forms easily available to plants.

Nitrogen

Of the total amount of nitrogen, 50–60 % is in ammonium form, instantly utilizable by plants and by the soil microflora. Ammonia nitrogen increases with progressive dilutions of liquid manure with water. Nitrogen uptake in soil is higher for liquid manure than for farmyard manure, or even for composted manure. This indicates a relative surplus of N in relation to C.

Other Nutrients

Phosphorus is mostly bound in the organic component of manure. It is readily available to plants. Potassium is also in a form readily available to plants. The content of partly organic-bound magnesium can be critically important. It is fairly available for the plants and is able to cover not only the immediate requirements of the plants, when regularly applied, but manure magnesium can replace the application of mineral magnesium fertilizers. Magnesium is one of the most important soil nutrients. Minor fertilizer elements present in liquid manure are boron, copper, zinc, manganese, molybdenum, etc. The pH of liquid manure varies in the range of 6·5–8·5, which is a good range of most soils.

When faeces and urine are stored together in liquid systems, losses of organic matter and nutrients occur. When compared with the traditional composting of farmyard manure where, due to improper handling, 40–60 % of organic substances and 30–40 % of nitrogen may be lost, storage losses with liquid manure systems are three times lower than solid waste handling systems.

The easy availability to crops of the nutrients in liquid manure makes this system highly effective. However, heavy losses may be incurred especially under improper application methods. The nitrogen which was saved during storage may easily be lost subsequently during land disposal operations. Liquid manure applied on land and ploughed-under five days later resulted in 30 % greater nitrogen losses than losses expected had manure been ploughed-under immediately after spreading.

METHODS OF FERTILIZING WITH LIQUID MANURE

Liquid manure can be utilized for fertilizing purposes with or without mineral fertilizers and/or with straw.

TABLE III

TONS OF LIQUID MANURE TO BE APPLIED FOR VARIOUS LEVELS OF FODDER CROP ROTATION

(N = 0·4 % of wet manure)

60 % fodder crops		*75 % fodder crops*		*100 % fodder crops*	
Crop	*Rate (tons/ha)*	*Crop*	*Rate (tons/ha)*	*Crop*	*Rate (tons/ha)*
Ley[a]	100	Clover–grass	75	Ley[b]	120
Potatoes	60	Green–cereal mixture	30	Corn for ensiling	75
Average rate	66		45		87
Equivalent kg N/ha	260		180		350

[a] Grass mixture or ryegrass.
[b] Cocksfoot.

The level of supplemental mineral fertilizing may depend on the following factors: the liquid manure dosage, the nutrient content of the liquid manure, crop needs for fertilizers, the types of crops cultivated, the timing of the application, the soil type and the nutrient supply state of the soil, and the specific nutrient content of the liquid manure. But it is necesary to underline the fact that it is possible to reach considerable savings in mineral fertilizers, provided regular fertilizing with liquid manure is maintained; for example, when fertilizing with cattle manure, mineral potassium fertilizers can be omitted because of the high potassium content of cattle manure.

Crops which seem to be the most suitable for an effective utilization of liquid manure are crops which have prolonged growth periods, crops which need copious quantities of fertilizers, fodder crops and grain crops (grain corn and corn for ensiling, fodder beet, clover–grass mixtures, potatoes, oats for ensiling, etc.). These crops are to be preferred in the crop rotations if liquid manure is used. Crops with small fertilizer needs may be nourished with liquid manure during their early stage of growth.

Grass crops consume the highest amount of nutrients from the soil. Furthermore, forage crops form a canopy over the land surface, thus reducing rainfall runoff and decreasing the potential for water pollution

TABLE IV

LIQUID MANURE APPLICATION RATES ON HIGH-FERTILITY SOILS AND GRAIN CROPS

100% grain crops		*78% cereal crops*		*63% grain crops*	
Crop	*Rate (tons/ha)*	*Crop*	*Rate (tons/ha)*	*Crop*	*Rate (tons/ha)*
Beans	60	Alfalfa	30	Alfalfa	30
Winter wheat	—	Alfalfa	—	Alfalfa	—
Winter wheat[a]	30	Winter wheat	—	Winter wheat	—
Spring barley[b]	—	Grain corn[a]	75	Spring barley	—
		Winter wheat (spring)	—		75
		Spring barley	—	Winter wheat	—
		Grain corn[b]	75	Sugar beet[b]	75
		Winter wheat (spring)	—		
		Spring barley	30		
Average rate	23		23		23
Equivalent N (kg N/ha)	90		90		90

[a] Straw ploughed-under.
[b] Green manure.

from land spreading of animal wastes. A limiting factor in the application of manure on forages is the quality of the forage as an animal feed more than yield per hectare.

Growing fodder crops in the crop rotation permits application rates of liquid manure equivalent to 350 kg N/ha (see Table III). Thus it is possible to dispose of more waste per unit of land and thus attain a higher concentration of animals on a smaller land area. Growing cereals in the crop rotation limits the quantities of liquid manure applied to what is equivalent to 90 kg N/ha (see Table IV).

TIMES OF LIQUID MANURE APPLICATION

During the year the times to fertilize with the liquid manure may be divided into five periods: August–September, October–November, December–February, March–May (in early spring before seeding) and June–July. On cropland, liquid manure may be applied as top-dressing during the growth season on perennial feeding crops, on the stubble during winter periods, and may be incorporated into the soil between the rows of the crops after crop emergence.

On grassland, liquid manure may be applied after first mowing, provided a time-interval of 21 days from the last fertilizer application until the harvest is maintained. It is not always possible to apply manure when it is most convenient and expect that period also to be optimal for crop utilization.

To achieve the optimal fertilizing effectiveness, it is recommended that liquid manure be applied at the following times, while respecting the production conditions:

(a) On light and medium sandy loam soils, apply in early spring prior to sowing; in dry regions apply during late autumn and winter.

(b) On medium loamy soils apply during late summer and in autumn. In dry regions and lowlands, in winter. Apply in early spring prior to sowing in high-moisture areas.

(c) On heavy soils apply during late summer and autumn. In lowlands one can also apply manure in winter. Heavy soils should not be fertilized in spring.

(d) Liquid manure should be simultaneous with ploughed-under straw on light soils, particularly the first time it is applied.

(e) During winter applications of liquid manure on all soil types, it is possible to plough-under green manure at the same time.

(f) In drier districts, namely in corn and sugar-beet belts, and on level ground, it is possible to fertilize with the liquid manure during December–February if croplands are not frozen below 10 cm. This would also apply to permanent meadows.

EFFECTIVENESS OF LIQUID MANURE

Table V shows coefficients of effectiveness of liquid manure fertilizer components for different soils and application times.

Liquid manure nitrogen effectiveness is 30% of that of commercial fertilizers when applied to light soils and 90% when spread on heavy soils. Some increase is noted with application time changing from autumn to spring. Potassium effectiveness varies from 60% to 100%. Table V shows that liquid manure is as effective as commercial fertilizer irrespective of the soil type or time of application.

TABLE V

EFFECTIVENESS COEFFICIENTS (COMMERCIAL FERTILIZERS = 100) OF FERTILIZER ELEMENTS OF LIQUID MANURE APPLIED[a] ON DIFFERENT SOILS AND AT VARIOUS TIMES OF THE YEAR

Soil type	*August–September*			*October–November*			*December–February*			*Spring*		
	N	P_2O_5	K_2O	N	P_2O_5	K_2O	N	P_2O_5	K_2O	N	P_2O_5	K_2O
Light	40	100	60	60	100	70	70	100	80	90	100	100
Medium:												
sand-loamy	50	100	70	70	100	80	80	100	90	90	100	100
Loamy	60	100	80	80	100	90	90	100	100	90	100	100
Heavy	80	100	100	90	100	100	90	100	100	90	100	100

[a] Incorporated into the soil within 24 hr of application.

APPLICATION RATES

Factors affecting application rates include the properties of the liquid manure, the type of crop rotation, the topography of the land to be fertilized and the soil fertility. Generally, application rates are determined on the basis of the quantity of liquid manure needed to provide the required nitrogen. As was shown in Tables III and IV, liquid nitrogen could account for 100% of all nitrogen needs for 100% annual feeding crops, but only 50% of winter cereals and 70% of spring cereals. Furthermore, the amount of nitrogen which is to be supplied by liquid manure should be calculated on the basis of the effectiveness coefficients given in Table V.

Table VI shows liquid manure application rates when considering the nitrogen content of the waste and for typical crops grown on continental climates. When nitrogen content is 0·4% (wet basis), maximum application rate is 120 tons/ha of liquid manure on ley cocksfoot.

TABLE VI

RECOMMENDED APPLICATION RATES OF LIQUID MANURE (TONS PER HECTARE)

Crop	*Nitrogen content in the liquid manure*		
	0·4%	0·6%	1·0%
Cereals: winter[a] spring[b]	30	20	12
Leguminous mixtures			
for grain: winter	30	20	17
spring	25–35	15–20	10–20
Oil rapes	30	20	12
Poppy	30	20	12
Potatoes: early	70	50	27
others	60	40	24
Sugar beet: technical	75	50	30
feeding	85	60	34
Feeding beet	85	60	34
Cabbage feeding	85	60	34
Maize: for ensiling	75	50	30
feed maize	75	50	30
Feed carrot	60	40	24
Clover–grass	75	50	30
Clover, alfalfa	30	20	12
Oats for ensiling and for haylage	45–60	30–40	18–24
Green rye	30	20	12
Green mixtures: winter	30	20	12
spring	40–50	25–35	15–20
Beans for drying	40	25	15
Ley: grass mixture or			
ryegrass	100	65	40
cocksfoot	120	80	48
Permanent grass:			
grass	60	40	24
clover–grass	40	25	15
Pastures:			
grass pastures and clover–grass pastures	50	35	20

[a] The liquid manure doses cover 60–70% needed N of N to be supplied by commercial fertilizers.

[b] Higher doses for winter feeding.

Lower liquid manure doses should be applied on soils with a high absorption capacity. On light soils the doses may be increased by 10–20 tons/ha of liquid manure. Autumn applications may meet 50% of total N requirements; the remaining part of the total N necessary could be provided in the form of commercial fertilizers, or again in the form of liquid manure in the spring prior to sowing, or as top-dressing during the growing period. On heavy soils and on medium loamy soils it is possible to provide the total N needed in the form of liquid manure at the time straw is ploughed-under in autumn.

The direct and residual effect of liquid manure at very high application rates of 615 tons/ha have not as yet been determined.

Generally about 1·5 tons/ha/year of organic matter should be applied on cropland. On heavy soils with high-yielding perennial feeding crops, organic matter application may be decreased to 1·0 ton/ha/year. On heavy soils without perennial fodder crops and on light soils, application rates should be in the range of 1·5–2·0 tons/ha/year.

Liquid Manure Applications Under Special Conditions

Application rates of liquid manure are also limited by environmental pollution and health considerations. Land areas are usually classified according to the pollution potential from applied wastes considering both the intended use of the water resources of the area and the vulnerability of those resources to contamination. For example, in areas where runoff water is collected into reservoirs for human water supply, surface applications of liquid manure without immediate soil incorporation would be banned. Also, on land areas with highly permeable soils, application of liquid manure without mixing it with straw might be forbidden so as to prevent groundwater contamination. For specific areas and specific local conditions, application rates might be limited to specific rates.

Lands with high groundwater levels which have not been drained previously can be fertilized with liquid manure application not exceeding 30 tons/ha per single application per year.

Lowlands subject to flooding and water inundation may be fertilized with liquid manure in the late spring after the danger of flooding has passed and soils have thawed.

On sloping land, liquid manure may be applied at rates determined by crop needs and soil types. However, manure should be incorporated into the soil soon after surface spreading. On steep lands, over 8° slope, it might be necessary to place strips of meadow grass between row crops to prevent excessive runoff, particularly if the land drains directly into a stream. If the application rate exceeds 30 tons/ha/year, the total manure should be applied in several doses during the year, with each dose being less than 30 tons/ha. On lands with 8° slope or higher, no liquid manure should be applied during winter or when the ground is frozen.

TABLE VII

EXAMPLE OF A SCHEDULE FOR MANURE APPLICATION FOR A TYPICAL 1000 DAIRY COW 300 HA FARM

Farm	Chvaly			Kyje			Kyje										
Block	I			III			V						Farm: Počernice				
Size (ha)	121			84			77										
Terrain config.	R			R			R						Year: 1976–7				
Soil type	T			S_H			T										
Forecrop	Pš			Cu (40 ha) + Ks (44 ha)			Jč						Total				
Crop	Cu (48 ha) + Ks (73 ha)			Ks			Cu (50 ha) + Ks (27 ha)			Consumption			Production			Stored	
Manure	KjS	Sl	ZH	KjS	Sl	ZH	KjS	Sl	ZH	KjS	Sl	ZH	KjS	Sl	ZH	KjS	Sl
Dose	Cu = 75	5	—	70	—	—	Cu = 72	4	12								
(tons/ha)	Ks = 70						Ks = 70						(tons)				
Month:																	
August	3 600	605						308		3 600	913		1 705	913		3 165	
September							3 595			3 595			1 650			1 220	
October										—			1 805			2 925	
November				4 500					924	4 500		924	1 650		924	75	
December										—			1 705			1 780	
January										—			1 705			3 485	
February										—			1 540			5 025	
March				1 380			1 890			3 270			1 705			3 460	
April	5 110									5 110			1 650			—	
May										—			1 705			1 705	
June										—			1 650			3 355	
July										—			1 705			5 060	
Total (tons)	8 710	605		5 880	—	—	5 485	308	924	20 075	913	924	20 075	913	924		

Manure: KjS = cattle liquid manure, Sl = straw, ZH = green manure.
Terrain: R = plain.
Crops: Pš = winter wheat, Cu = sugar beet, Jč = spring barley, Ks = maize for ensiling.
Soil types: T = heavy, S_H = medium loamy soil.

Scheduling Manure Application

Table VII shows a typical scheduling of manure application for a typical farm and given conditions and for each month of the year. Such records provide the farm manager with the kind of information that is extremely useful in the day-to-day and year-to-year management of the farm.

Table VII indicates that the 20 075 tons of manure produced per year may be utilized in a scientifically programmed crop rotation. Note in the last column of Table VII that the maximum quantity which needs to be stored is 5060 tons (approximately 5161 m^3). This maximum quantity is reached in February in the middle of the winter and in July in the middle of the growing season.

SUMMARY

General findings based on four years of research may be summarized as follows:

1. Application of liquid manure to land at a proper rate with ploughed-under straw and/or in combination with mineral fertilizers tends to increase dry matter yields of crops, improve the agrochemical properties of the soil, and enhance the soil nutrient availability.
2. Liquid manure may easily replace farmyard manure and composted materials, particularly if used in combination with straw.
3. When compared with equivalent nutrient amounts contained in commercial fertilizers, liquid manure is highly competitive, as is shown in Table VIII. For potatoes, liquid manure effectiveness exceeds that of commercial fertilizers.
4. Timing of the application is of critical significance as far as pollution potential and fertilizer effectiveness are concerned.
5. Graduated doses of liquid manure increased nutrient availability by a minimum of 9 % and a maximum of 96 %, as is shown in Table IX, which also shows that forest loam soils would benefit more than chernozem soils.
6. The fertilizing effectiveness of liquid manure is higher for fodder crops than for cereal crops.
7. Adding supplementary nitrogen to fields treated with 50 tons/ha/year of liquid manure is not necessary. Liquid manure is an economical fertilizer when compared with farmyard manure. The economic return is estimated at two to six times better than the cost of applying liquid manure.
8. An application rate exceeding 200 tons/ha inhibits seed germination and crop emergence.

TABLE VIII

COEFFICIENT OF EFFECTIVENESS (CE) OF LIQUID MANURE AS COMPARED WITH NUTRIENT AVAILABILITY OF COMMERCIAL FERTILIZERS (CE = 100)

Swine liquid manure					*Cattle liquid manure*				
Production area	*Soil type and kind of soil*	*Doses*			*Production area*	*Soil type and kind of soil*	*Doses*		
		Light	*Medium*	*Heavy*			*Light*	*Medium*	*Heavy*
A. Fodder crops rotation (100 %):									
Sugar beet area	HM_{ph}	86	85	90	Sugar beet area	$ČM_{gh}$	94	100	89
Potato area	HP_{ph}	92	81	87	Maize area	$ČM_{ph}$	89	100	96
Potato area	HP_{jh}	99	107	111	Hill country area	HP_{h}	77	71	78
B. 50–100 % cereals in crop rotation:									
Sugar beet area	HM_{ph}	98	96	87	Sugar beet area	$ČM_{jh}$	91	90	99
Potato area	HP_{ph}	101	94	101	Maize area	$ČM_{ph}$	94	93	98
Potato area	HP_{jh}	65	81	90	Hill country area	HP_{h}	88	72	101

ČM = chernozem, HM = grey-brown podzolic soil, HP = brown forest soil.
ph = sand-loamy soil, jh = clay-loamy soil, h = loamy soil.

TABLE IX

RELATIVE FERTILIZER EFFECTIVENESS OF CATTLE MANURE

Graduated liquid manure doses	*Clay-loamy chernozem (sugar beet area)*	*Sand-loamy chernozem (maize area)*	*Loamy brown forest soil (hill country)*
Not fertilized	100	100	100
I	114	109	133
II	121	126	164
III	111	124	196

9. The simultaneous ploughing-under of straw with liquid manure allows high manure application rates on light soils and reduces nitrogen losses by 30–50 %.
10. Applying composted swine manure in a mixture with city solid wastes produces yields slightly higher than liquid manure, as is shown in Table X. The added cost of composting before land application has not been evaluated.

TABLE X

COMPARISON OF CROP YIELDS FROM THE USE OF FARMYARD MANURE AND COMPOST FROM CITY WASTES

Waste	*Rate (tons/ha/yr)*	*Control*	*Annual PK-fertilizing*		*Stock PK-fertilizing for 5 years*	
			$N_1P_1K_1$	$N_2P_2K_2$	$N_1P_1K_1$	$N_2P_2K_2$
Manure	9	105	111	115	116	117
Compost	9	103	112	119	116	118

11. There are no significant differences in the availability of nitrogen between nitrogen found in the solid matter of liquid manure and that found in the liquid fraction of liquid manure. On light soils the solids fraction of liquid manure is equivalent to the farmyard manure.
12. Poultry liquid manure is an efficient organic fertilizer for heavy and light soils. It has a slower nutrient release train than swine waste. Nutrient release rate is faster on light than on heavy soils.
13. Liquid manures from modern animal units are a valuable organic fertilizer for plant production; they return nutrients and organic substances back to the land, increasing soil fertility and productivity and improving land ecology.

25

Land Disposal of Feedlot Wastes by Irrigation in Czechoslovakia

R. KURC

Irrigation Research Institute, Bratislava, ČSSR

The simplest method of transporting animal wastes to the field is by tank wagons. Another method is direct application of the liquid waste by irrigation.

When liquid wastes are transported to the field via pipes, settling of solids in the pipe reduces the effective flow diameter. The resultant decrease in capacity is highly undesirable, especially when the pipes are buried beneath the soil surface. Such problems have been largely prevented by separating the solids from the waste and transporting only the liquid portion of the waste by pipe.

Costs for disposal by tank or irrigation systems vary. If tanker trucks are used to transport cattle wastes from 2 to 6 km, costs range from 1·3 % to 4·2 % of the production costs. If liquid waste is diluted with water in a ratio of 1:5 and then applied by irrigation techniques, disposal costs amount to approximately 1 % of production costs. Transport of swine wastes by tanker trucks amounts to 1·2–3·7 % of production costs, while irrigation disposal costs are approximately 1 % of production costs. When dairy cow manure is disposed of via tanker truck, the disposal costs are 1·1–3·4 % of the value of the milk produced. Disposal costs by irrigation are approximately 0·6 % of the value of the milk produced.

One problem often encountered in irrigation disposal systems is the availability of fields during the growing season. Application of waste to the field in large quantities or at certain periods in the growth cycle can damage the crops. Similarly, rain may make the soil wet and thus prevent access to the field. In such instances the waste must be stored until such time as it can be applied. During the growing season the available cropland area is reduced because some of the crops cannot tolerate application of wastes. The liquid waste cannot be applied to many crops, including fodder plants, several weeks prior to harvest. Tank trucks cannot always enter the fields to spread liquid waste when row crops are growing. Finally, application of the waste by means of movable irrigation piping is undesirable for the operator of the irrigation system, since he must move through the field after waste has been applied.

Similar problems arise when animals contract an infectious disease. In such instances the housing unit is quarantined and all waste must be held within the feedlot until proper disinfection is accomplished. During winter months snow covers the ground. If the waste is applied over the snow, the spring snow melt and concurrent water runoff may carry the waste contaminates into water supplies. Thus, storage capacity for approximately 120 days should be provided. Cost of such additional storage capacity is eventually translated into increased production costs.

Irrigation of feedlot wastewater is feasible when solids are separated by mechanical and/or other physical treatment techniques. The first two steps in such treatment involve the mechanical separation of suspended particles from the liquid waste. This involves separation of both suspended and settleable solids. Tests with centrifugal filters indicated that such filters effectively separate settleable solids.

The colloidal material in the liquid portion of the centrifugal waste can be removed by coagulation with lime. The precipitant settles rapidly and has a favourable consistency which allows it to be easily handled. After coagulation, the liquid fraction of the waste contains only dissolved solids which present no problem in pipe transport.

Mechanical and physical treatment of the animal waste allows more effective utilization. For cattle wastes, separated solids amount to approximately 25–30% of the volume of the original waste and have a moisture content of 65–70%. The solid fraction produced from treatment of swine waste is approximately 20% of the original volume.

Separation of the solid and liquid phases of the animal waste enables them to be applied separately to the soil. The solids can be spread upon the land while liquids can be added to irrigation water as liquid fertilizers.

Chemical and physical analysis of the liquid fraction from the centrifugal separators indicated that the maximum potassium content was 1500 mg/litre, while dry matter content was 2·5–3%. A comparison of the chemical analyses from the original waste and the liquid fraction separated from the waste indicated that only about 50% of the potassium in the original waste was present in the liquid fraction. Apparently the potassium ions are bonded to the calcium and the suspended solids which are separated during the coagulation process.

One major advantage of the separation process is the reduction in potassium content which is achieved. This allows higher application rates for the separated liquid than could be used for the original waste. Additional increases in application rates can be achieved by crop management techniques. Application rates as high as 5000 m^3/ha of a 1:5 mixture of waste and water have been used on land planted with potassiumphylous crops such as grass and clover. This is an application rate of 1000 m^3/ha of raw liquid animal waste. The higher application rates are achieved because these crops tend to absorb large amounts of potassium.

Assuming that the waste is applied at a rate of 1000 m^3/ha, 1 ha of grass or clover could accommodate the wastes from 70 head of adult cattle or 500 head of swine. Irrigation at a rate of 1000 m^3/ha of sludge water is roughly equivalent to 1430 tons of liquid waste.

After the potassiumphylous crop is harvested from the area being irrigated with animal waste, the waste is applied to the area previously irrigated with water. The potassiumphylous crop is ploughed-under and followed with another potassiumphylous crop such as potatoes.

High application rates of animal wastes can cause groundwater pollution. Liquid waste applied at a rate of 5000 m^3/ha resulted in 93–96 % of the nitrogen, 80 % of the phosphorus and roughly 90 % of the potassium being held within the top 40 cm. Approximately 97–98 % of the BOD of the liquid waste was removed. These results imply that application of the separated liquid, which has lower percentages of N, P and K, should not be as big a threat to groundwater pollution. Permanently installed irrigation pipes cost more, but labour problems are reduced. It may be undesirable to require workers to operate a system which involves moving pipes which have been coated with a layer of waste.

Irrigation equipment presently on the market which offers promise are the central pivot systems which are self-mobile and may cover over 50 ha/day. Such systems require semi-permanent installation and high initial investment. Disposal of liquid wastes by irrigation costs approximately one-third to one-half the costs of trucking.

PART IV

Economics of Animal Waste Management

26

Costs of Liquid Handling and Treatment of Large Swine Feedlot Wastes

P. ŠIMERDA

Swine Breeding Research Institute, Kostelec nad Orlicí, ČSSR

COSTS FOR LAND DISPOSAL

The most effective way of disposing wastes from modern large swine feedlots is by disposal on cropland. The most common methods of land disposal are liquid handling systems. The costs of such systems were studied for one year (1973–4) in swine feedlots ranging from 1200 to 20 000 pigs capacity.

Table I shows costs incurred in liquid manure handling in 8 large swine feedlots ranging in capacity from 2480 to 20 530 pigs of average capacity. Two types of wagons were used: lorry-mounted and tractor-drawn tank wagons with tank capacities of 3·5 and 10 m^3. Cost ranged from 6·91 Kčs ($0·69) to 13·27 Kčs ($1.33) per cubic metre of liquid waste hauled and spread on cropland. Costs varied with tank capacity (the 10^3 capacity being

TABLE I

OPERATING COSTS FOR HAULING AND SPREADING OF SWINE FEEDLOT WASTES

Feedlot location	*Capacity (av. no. of pigs)*	*Hauling distance (km/wagon trip)*	*Cost (Kčs/m^3)*[a]	*Wagon*	
				Capacity (m^3)	*Tank type*
Štěpánovice	2 480		16·55	3·50	Lorry
Valovice	3 808	9·56	11·05	3·50	Lorry
Štěpánovice	4 033	8·75	8·37	10	Lorry
Dobřany	11 707	11·05	13·27	3·50	Lorry
Dobřany	12 288	12·03	6·91	10	Lorry
Milotice	12 934	19·13	8·71	10	Lorry
Milotice	12 934		12·93	10	Tractor
Smiřice	20 530		10·90 19·23[b]		

[a] 1 Kčs = US$0.10 approx.

[b] Pumping costs from storage ponds, plus solids removal costs.

more economical than that of the 3·5 m³ tank) and with the layout of the disposal field with respect to the feedlot buildings and waste storage facilities. At the Smiřice feedlot, 33 % of the slurry was hauled directly on to the land and this operation cost about 10·9 Kčs/m³. However, 67 % of the slurry was centrifuged, separating the solids from the liquids. The liquids were pumped into a field irrigation system, while the solids were hauled to the field. These hauling costs plus the storage pond costs of 350 000 Kčs/year ($35 000/year), but excluding the irrigation system costs, amounted to 20 Kčs/m³ ($2/m³).

In terms of animal live weight, the costs of liquid manure waste handling ranged from 0·13 to 0·24 Kčs/kg of live weight gain ($0·013 to $0·024/kg l.w.). Approximately 12–20 kg of liquid manure were generated per kg of daily weight gain by the animals.

Labour requirements for liquid waste handling are given in Table II. Labour requirements amounted to 0·20–0·40 man-hr/100 kg of live weight gain. The quantity of liquid waste hauled and disposed ranged from 4 to 10 m³/man-hr. All these operations are significantly affected by weather, land conditions and nutrient needs of the crops grown. Thus, manpower needs differ with the time of the year.

TABLE II

LABOUR REQUIREMENTS OF LIQUID MANURE HAULING AND LAND DISPOSAL

Feedlot location	*Wagon*		*Hauling distance (km/wagon trip)*	*Labour requirements*		
	Type	*Capacity (m^3)*		*Man-hr/ 100 kg l.w. gain*	*Pig capacity/ man*	*m^3/man-hr*
Štěpánovice	Lorry	10	8·91	0·27	3 418	6·94
Dobřany	Lorry	10	12·03	0·22	4 322	9·92
Dobřany	Lorry	3·5	11·05	0·40	2 280	4·44
Milotice	Lorry	10	19·13	0·24	4 499	6·04
Milotice	Tractor	10		0·35	3 041	4·09

A 10 m³ capacity lorry tank can accommodate the hauling requirements of 3400–4500 fattening pigs. The investment on such a lorry tank would be equivalent to 100 Kčs per pig capacity ($10/pig). For tractor-drawn tank wagons of the same capacity, the investment is 90 Kčs/pig ($9/pig). Initial investment for storage facilities (2–3 months storage, 0·8 m³/pig capacity) amounts to 500 Kčs/pig capacity ($50/pig).

TREATMENT COSTS

Treatment of feedlot wastes is more expensive than direct land disposal. Both initial investment and operating costs are high for biological treatment of swine feedlot wastes, as is shown in Table III. A comparison of costs in Table III with costs listed in Tables I and II indicates that waste treatment investment costs per pig can be as much as 4 times higher, operational costs per kg of weight gain are 3 to 8 times higher, while labour requirements are increased by a factor of 2.

TABLE III

COSTS FOR TREATMENT OF SWINE FEEDLOT WASTES

Feedlot location	*Capacity (no. of pigs)*	*Treatment system*	*Investment costs (Kčs/pig)*[a]	*Operating costs (Kčs/m³)*	*Operating costs (Kčs/kg l.w. gain)*	*Labour (pigs/man/year)*[b]
Rycholka	1 211	Aeration	704	27·40	0·42	1 331
Doudleby	950	Aeration	2 464	93·65	1·43	2 185
Třeboň	17 000	Anaerobic	785	72·10	1·08	2 550

[a] 1 Kčs = US$0.10 approx.
[b] To operate treatment plant.

VALUE OF LAND-DISPOSED FEEDLOT WASTE

Application of cattle manure and straw at rates of 21–63 tons/ha resulted in a 3·6–4·5 Kčs return in additional yield per Kčs of disposal cost (3·6–4·5 benefit to cost ratio). Similar results were obtained with pig slurries for which the benefit to cost ratio ranged from 2·8 to 4·6 in brown soils where potatoes and rye were grown.

Hauling pig slurry for 150 days per year over a distance of 4 km would cost 23·60 Kčs/ton ($2·36/ton). However, the additional benefit accrued in increased yields amounts to 65–100 Kčs/ton ($6.50–10/ton). However, in large feedlots these economic benefits cannot be realized, because insufficient storage and weather do not allow the application of manure in accordance with the optimal needs of crops for nutrients.

27

Cost of Animal Waste Management

LYNN R. SHUYLER

Program Leader, Animal Wastes, US Environmental Protection Agency, Ada, Oklahoma, USA

INTRODUCTION

Legislation, rules and regulations are being passed in most countries establishing effluent guidelines for discharge by animal feedlots. In the United States of America, guidelines for effluent discharge require that runoff from all storms smaller than a 24 hr, 25 year storm be retained and not discharged into a stream.

The legalistic approach for controlling water pollution places increasing importance on studies to analyse the economic impacts of alternative effluent guidelines and suggested management practices. These costs must be compared with expected societal benefits in order to appraise the desirability of pollution control measures.[1]

This paper is limited to confined feedlots, eliminating all range and pasture production of animals. Discussion of confined production is subdivided into open and covered feedlots. Open feedlots are those which expose both animals and waste within the pens to precipitation. Covered feedlots are those constructed to prevent precipitation from contacting the animals in their pens. Therefore, only open cattle and swine feedlots need to be discussed in terms of effluent guidelines to prevent runoff from reaching natural water bodies. Poultry units usually do not have discharges into streams. Most of the large open feedlots have already constructed runoff control facilities and can meet the effluent guidelines, as is shown in Table I.

This paper combines information from several studies and presents the data and comparisons on a cost-per-head basis; however, the information does not represent all conditions that may be encountered. There is a large variation in costs relative to size of production unit, species being produced, geographical location and topography of the site.

Control of pollution from open feedlots consists of the following steps: (a) to contain storm water runoff from the pen surface; (b) to divert any upslope runoff around the pens, therefore not polluting this water by allowing contact with animal manure; and (c) to store runoff from pens until it can be applied to soil for final disposal.

Open cattle feedlot effluent pollution control investments range from $1.46 to $48.00 per head produced and annual costs range from $0.24 to $7.60 per head produced. Covered confinement production investments for pigs range from $9.02 to $38.40 per head produced for pollution control, and annual costs range from $0.77 to $3.89 per head produced. The ranges are far too wide to allow their use on a specific production unit. However, if

TABLE I
ESTIMATED CURRENT EFFLUENT CONTROL STATUS OF THE UNITED STATES FEEDLOT INDUSTRY AS OF 1974
(from ref. 2)

	% of total production	*% meeting effluent guidelines*
Beef (26 million head):	100	
Open feedlot		
less than 1000 head/feedlot	35	20–30
more than 1000 head/feedlot	65	>70
Swine (55 million head):	100	
Dirty/open lot	60	5–10
Concrete/open lot	25	75–80
Totally confined	15	100
Dairy (11·5 million head):	100	
Stanchion barns	60	25–35
Loose housing	15	60–70
Cow yards	25	60–70

costs are calculated for a specific production unit, they should fall within these ranges. If calculated costs for a feedlot exceed the upper limits of the above ranges, careful consideration should be given to other less costly locations for the feedlot, or other methods of control should be investigated.

Cost analysis of waste management systems is not a science, since each location is different, construction cost varies greatly, and the cost of materials is constantly changing. Cost analysis is simply an organized, educated, guessing game. If the game is played with a strict set of rules and the players are experienced in the field of waste management, the 'guesses can be excellent estimates'.

All the costs presented here are for conditions existing in the United States in 1973. The changes in construction costs and material costs by years are shown in Fig. 1 for the United States. Using this figure, it is simple to use cost data from years past and update it to the present, allowing comparison of old systems with proposed new ones. In the following examples, an attempt has been made to present the systems so that they can be compared with one another. The actual cost for a given system will vary greatly from the information given in this section due to different construction costs, material costs and even money costs. However, the approach used to develop these cost estimates should be the same as the one

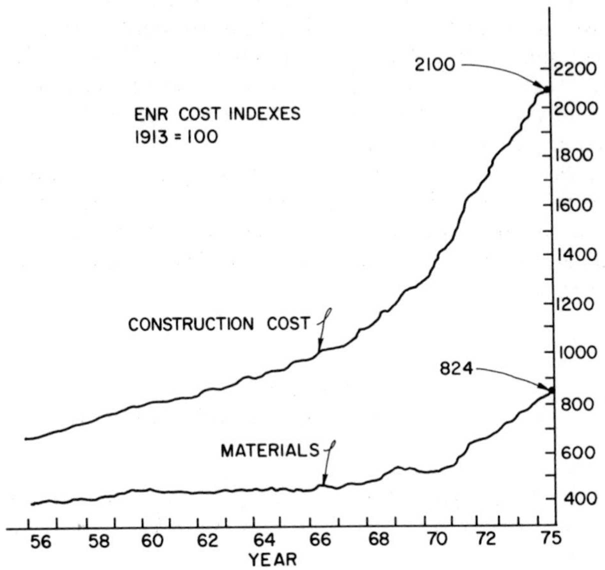

FIG. 1. Construction and materials costs indices (from ref. 3).

presented in this paper. Use of these data for any other location should be done by an experienced, qualified engineer. Hopefully, however, these cost data may serve as general guidelines in national policy-making related to environmental quality around animal feedlots.

COST CALCULATION METHODS

Open Feedlots

Two basic assumptions are made: (a) livestock producers, at present, have the required equipment and land for the removal and disposal of solid manure so that no additional costs for solid waste removal are required; and (b) the only investment costs required are for construction of lagoons to store runoff and for equipment to remove and dispose lagoon contents.

These costs are computed for beef cattle, hog and dairy feedlots on the basis of the following:[2]

1. The surface area is estimated for each type and size of feedlot.
2. Runoff volumes are estimated using 25 year, 24 hr rainfall, with a 100% runoff factor.
3. Lagoon construction costs are estimated from general cost relationships for various capacity ponds.
4. Investment costs for removal and disposal of liquids are estimated assuming different levels of technology for various size feedlots. For small cattle feedlots, costs are based on the purchase of a pump and liquid spreader, while costs for larger feedlots include an irrigation system, as shown in Table II.
5. Unless otherwise stated in the tables, operation and maintenance costs are assumed to equal 5% and 10% of the initial investment for lagoons and equipment, respectively. Depreciation costs are calculated by assuming lagoon life of 20 years and equipment life of 5 years.

TABLE II

COMPONENTS OF EFFLUENT CONTROL SYSTEMS FOR OPEN FEEDLOTS

(from ref. 2)

Species and size of open feedlot	*Holding pond or lagoon*	*Pump*	*Travelling gun*	*Tank spreader*	*Centre pivot*	*Hand carry sprinkler*
Beef:						
100	×	×				
500	×		×			
1 000	×		×			
5 000	×				×	
10 000	×				×	
20 000	×				×	
Swine:						
100	×	×				
300	×			×		
900	×		×			
Dairy:						
25	×	×				
50	×	×				
100	×			×		
200	×					×
500	×		×			
1 000	×		×			

Covered Feedlots

The covered systems selected for analysis are completely confined finishing swine units and poultry units in which manure is handled with no bedding added.

A covered finishing feedlot may be divided into three sub-systems: (a) the conversion (feed and water) system, (b) the environmental control system, and (c) the waste handling system. For the purpose of this paper, everything above the floor (walls, roof, fan and pen partitions) is considered part of the environmental control system and everything on the floor and below is considered part of the waste handling system.[4]

Components comprising the waste handling system are functionally classified by the process they might perform: (a) floor, (b) collection, (c) transfer, (d) storage, (e) separation, (f) treatment, (g) transport to disposal, and (h) disposal. By classifying each component it is possible to evaluate each component for investment, operating cost, labour and management requirements. It should be pointed out that each system studied does not necessarily have a separate component for each of the classifications.

CALCULATED COSTS

Material costs vary between production units due to variations in local prices and date of construction. In order to make valid comparisons, all costs were updated to 1973 costs, and the same material costs were applied to each system studied.

Open Beef Feedlot

In open lot systems, cattle are held in high-density lots with about 18 m^2 of space per animal. Shelter may or may not be provided depending on the location and size of the feedlot; for example, farmer feeders in areas of cold wet weather provide some shelter, while cattle in large feedlots located in drier climates are exposed to all precipitation.

Feedlots are normally constructed on land with sloping topography and lots on flat land are mounded to provide some degree of slope to facilitate the removal of liquid wastes from the lots via gravity flow.

The technology used in disposing of liquid and dry manure varies greatly depending on the climatic location of the feedlot and its size. Generally, all feedlots follow similar practices for disposal of solid wastes while considerable variation exists in liquid runoff handling. Solid wastes are usually removed from lots and applied to cropland as fertilizer. Management of liquids varies from collection systems using holding basins with land application of the wastes at a later date, to uncontrolled discharge of the liquids into natural watercourses.

The costs for meeting effluent guidelines consist of investments in lagoons

TABLE III

INVESTMENT COSTS REQUIRED FOR EFFLUENT CONTROL AND ANNUAL COSTS AND TOTAL PRODUCTION UNIT INVESTMENT COSTS BY SIZE OF BEEF FEEDLOT IN THE UNITED STATES (from ref. 2)

	Capacity (number of head)[a]					
	100	500	1 000	5 000	10 000	20 000
Investment costs:						
Holding pond	$1 550	$1 550	$2 100	$6 400	$13 500	$26 000
Dispensing system						
Pump	500					
Travelling gun		3 000	6 000			
Centre pivot				24 000	32 000	37 500
Total investment	$2 050	$4 550	$8 100	$30 400	$45 500	$63 500
Investment cost/head marketed	$20.50	$9.10	$3.75	$2.80	$2.10	$1.46
Annual costs:						
Operating and maintenance						
Holding pond	$77	$77	$105	$320	$675	$1300
Equipment	50	300	600	2 400	3 200	3 750
Depreciation						
Holding pond	77	77	105	320	675	1 300
Equipment	100	600	1 200	4 800	6 400	7 500
Total annual cost	$304	$1 054	$2 010	$7 840	$10 950	$13 850
Annual cost/head marketed	$3.04	$2.10	$0.95	$0.75	$0.50	$0.32
Production unit investments costs:						
Base replacement + Investment	$6 000	$43 092	$76 780	$427 000	$610 000	$960 000
Investment for effluent control	2 050	4 550	8 100	30 400	45 500	63 500
Total investment	$8 050	$47 642	$84 880	$457 400	$655 500	$1 023 500
% Increase for effluent control	34·2%	10·5%	10·5%	7·1%	7·5%	6·6%

[a] Assume turnover rate of 1 for 100 and 500 head lots. For all other lots a turnover rate of 2·16 was used.

and liquid dispensing systems. These costs, as shown in Table III, vary from \$2050 for a feeder marketing 100 head annually to \$63 000 for feeders marketing 43 000 head. Minimum costs of \$1550 for holding ponds are assumed for feedlots marketing 500 head or less. Holding pond costs beyond this point increase, although not at a proportionate rate, as size of feedlots increases, but holding pond costs per head decrease as feedlot capacity increases, going from \$15.50 to \$0.60 per head.

Economies of scale also exist in annual costs for effluent control. Annual cost per head marketed varies from \$3.04 for the smallest feeder to \$0.32 per head of the largest feeder.

Investment in effluent control systems as a percentage of the original investment varies from 34·0 % for feeders marketing 100 head to 6·6 % for feeders marketing 43 000 head. This ratio decreased for all size segments as feedlot capacity increased.

Open Dairy Feedlot

Management systems used in handling dairy cows involve two basic technologies: (a) full confinement during winter months with cows on pasture during grazing periods, and (b) open lot systems with cows held in confinement. The former practice prevails in the north and north-east United States while the latter prevails in the south, south-west and western United States.

In using management systems involving full confinement with summer pasture, the technology required for manure disposal requires methods for both removing and spreading solid wastes from barns and controlling effluent runoff. Effluent runoff includes runoff from holding areas adjacent to barns plus water used in cleaning milking equipment and parlours.

The technology required for manure disposal for open lots is similar to the confinement–pasture system with the exception of increased requirements for controlling effluent discharges. In open lots, all excretions are exposed to precipitation since little protection is provided for animals.

Costs for effluent control for dairymen include investment costs for lagoons or holding ponds and various dispensing systems. Total costs vary from \$1200 for a 25 cow herd to \$6000 for a 1000 cow herd (Table IV). Minimum costs for lagoons or holding ponds of \$1000 were assumed with costs increasing as herd sizes increased.

Total investments per cow in effluent control decreased as size of herds increased. Investment costs per cow vary from \$48 for 25 cow herds to \$6.00 per cow for 1000 cow herds (Table IV). Annual costs per head vary from \$7.60 for the small dairies to \$1.20 for the large dairies. These data can also be seen graphically in Fig. 2.

Investment in effluent control systems represents an insignificant portion of total investment in dairy operations. A \$1200 investment for a 25 cow

TABLE IV

INVESTMENT COSTS REQUIRED FOR EFFLUENT CONTROL AND ANNUAL OPERATING COSTS AND PRODUCTION UNIT INVESTMENT COSTS BY SIZE OF DAIRY HERDS IN THE UNITED STATES
(from ref. 2)

	Number of dairy cows[a]					
	25	50	100	200	500	1 000
Investment costs:						
Lagoon or holding ponds	\$1 000	\$1 200	\$1 600	\$2 000	\$2 100	\$3 000
Dispensing systems						
Pump	200	200				
Tank spreader			2 000			
Travelling gun					3 000	3 000
Hand carry sprinklers				2 200		
Total investment costs	\$1 200	\$1 400	\$3 600	\$4 200	\$5 100	\$6 000
Investment cost/head	\$48.00	\$28.00	\$36.00	\$21.00	\$10.20	\$6.00

Annual costs:						
Operating and maintenance						
Lagoons or holding ponds	$50	$60	$80	$100	$105	$150
Equipment	20	20	200	220	300	300
Depreciation						
Lagoons or holding ponds	50	60	80	100	105	150
Equipment	40	40	400	440	600	600
Total annual costs	$160	$180	$760	$860	$1 110	$1 200
Annual cost/head	$6.40	$3.60	$7.60	$4.30	$2.22	$1.20
Production unit investment costs:						
Buildings (replacement)	$9 000	$13 000	$18 000	$32 000	$78 000	$144 000
Equipment (replacement)	8 500	10 000	28 000	50 000	46 000	66 000
Effluent controls	1 200	1 400	3 600	4 200	5 100	6 000
Cattle	12 500	25 000	50 000	100 000	250 000	500 000
	$31 200	$49 400	$99 600	$186 200	$379 100	$716 000
% Increase for effluent control	4·0%	2·9%	3·8%	2·3%	1·4%	0·8%

[a] Management systems used varied by number of milk cows:
Herds of 25 and 50 cows were assumed to be housed in stanchion barns. Cows are confined in winter and held on pasture during grazing periods.
Herds of 100 and 200 cows were assumed to be housed in loose barns during winter and held on pasture during grazing periods.
Herds of 500 and 1 000 cows were assumed to be confined throughout the year in open lots. Lot density = 1 cow/37·2 m^2.

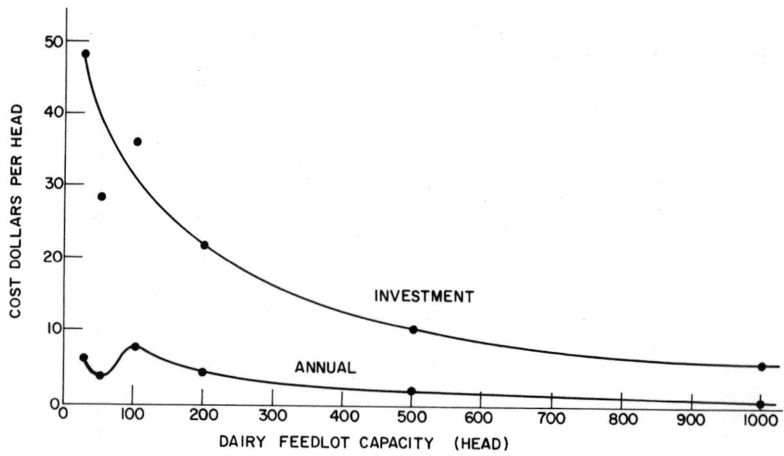

FIG. 2. Costs for open dairy feedlots (from ref. 2).

herd represents a 4% increase in investment. A $6000 investment for a 1000 cow herd represents less than a 1% increase in investment.

Open Pig Feedlot

Three management systems prevail in swine production: (a) open lot pasture, (b) concrete with full or partial roofing, and (c) full confinement. It is assumed that the type of management system used is correlated with the number of hogs raised annually. Open lots are assumed to be used extensively by small producers selling 100–300 head annually. It is also assumed that solid concrete floors with roofs are being used by most producers selling 900 head and full confinement systems are being used by most larger producers.

In open lot systems, hogs are held on dirt or pasture and limited shelter is provided, thus exposing the hogs to precipitation. Density rates vary from 25 to over 250 hogs/ha. All excretions are normally left on the land and subjected to effluent runoff.

Producers feeding hogs on concrete with pens partially or fully protected use different levels of technology for manure disposal than do open lot producers. Solid wastes may either be scraped from the pens or 'flushed out' with water. Both methods necessitate removal and spreading of wastes on fields or the use of holding basins for later disposal on cropland to prevent effluent runoff. Control of liquid wastes is accomplished through holding basins or lagoons.

To meet effluent guidelines, the costs consist of investments in lagoons and liquid dispensing equipment. These costs, as shown in Table V, vary from $2050 for the operator producing 100 hogs annually to $12 000 for the

TABLE V

INVESTMENT COSTS REQUIRED FOR EFFLUENT CONTROLS AND ANNUAL COSTS AND TOTAL PRODUCTION UNIT INVESTMENT COSTS BY NUMBER OF HOGS MARKETED ANNUALLY[a]
(from ref. 2)

	Number of hogs marketed annually			
	100	300	900	10 000
Investment costs:				
Lagoons or holding ponds	$1 550	$1 550	$2 100	$6 000
Dispensing system				
Pump	500			
Tank spreader		2 000		
Travelling gun			3 000	6 000
Total investment	$2 050	$3 550	$5 100	$12 000
Investment cost/head marketed	$20.50	$11.80	$5.70	$1.20
Annual costs:				
Operating and maintenance				
Lagoons or holding ponds	$77	$77	$105	$300
Equipment	50	200	300	600
Depreciation				
Lagoons or holding ponds	77	77	105	300
Equipment	100	400	600	1200
Total annual costs	$304	$754	$1 110	$2 400
Annual cost/head marketed	$3.05	$2.50	$1.23	$0.24
Production unit investment costs:				
Replacement investment for facilities and equipment	$1 500	$7 500	$25 000	$250 000
Investment for effluent controls	2 050	3 550	5 100	12 000
Investment costs	$3 550	$11 050	$30 100	$262 000
% Increase for effluent control	136·7%	47·3%	20·4%	4·8%

[a] All enterprises were assumed to be farrow-to-finish hog operations.

operator producing 10 000 hogs annually. Minimum lagoon cost of $1550 is assumed for the two smaller hog operations producing 300 or less hogs annually. Lagoon cost of $2100 is assumed for the operation producing 900 hogs, and $6000 is assumed for 10 000 hogs.

Economies of scale exist in annual costs for effluent control. Annual costs per head produced vary from $3.05 for the smallest hog operation to $0.24 for the operation producing 10 000 hogs annually.

Investment in effluent control systems, as a percentage of original investment, varies from 136·7% for the operator producing 100 hogs annually to 4·8% for the operator producing 10 000 hogs. For the four operations being considered, this percentage decreased as the size of the operation increased. The high of 136·7% is explained by the fact that the operator will be required to have an investment larger than his original investment.

Covered Swine Feedlot

Producing hogs in full confinement is a relatively new technology and is used primarily by the larger producers. These systems, while providing complete protection from precipitation, must still provide storage of the wastes in pits, oxidation ditches or lagoons. It can be assumed that nearly all these systems meet effluent standards, when only storm water runoff is concerned.

This section concentrates on examining a number of unique swine waste handling systems in order to present these alternatives and make quantitative comparisons.

Components of these systems have attracted attention because of the merits they might offer the producer. A description of each component and a detailed cost analysis of each are presented in Table VI.

The components listed in Table VI have been grouped by functions as follows:

1. *Floors.* This is important because different collection, transport and separation systems may require different floor types.
2. *Collection.* In most cases the floor or a very shallow pit under the slats will serve to collect wastes; however, there are cases when a small collection pit may be necessary for scraping systems.
3. *Transfer.* This refers to the method of movement of the wastes away from the animals and may be hydraulic or mechanical in nature.
4. *Separation.* Many systems will not use separation of the wastes from the liquid; however, when it is used it could be by screen or by settling.
5. *Storage.* Storage of the wastes may be for short periods of time, such as holding pits for process systems, or it may involve many days retention in a treatment lagoon.
6. *Treatment.* Treatment systems may be very simple, such as anaerobic lagoons, or complex, such as mechanical treatment for discharge or processing for other uses.
7. *Disposal.* This will be considered to be land disposal of any waste or effluent not discharged or used in a process of some type.
8. *Processing.* Items mentioned in this section are capable of utilizing part or all of the wastes to produce some product such as fuel or feed.

Several systems will be developed from the list of components in order fully to understand the use of economic data.

System A

This is a low capital cost system with considerable labour involved in water management. The building has wood slats in the floor and the pit under the floor has concrete side walls and an earthen bottom. This type of pit construction could allow groundwater pollution and is therefore very site-specific. There is no treatment of the wastes and they are periodically hauled to adjoining agricultural land using several 5670 litre vacuum tank wagons. The detailed cost analysis for this system is shown in Table VII. The investment cost for this system is \$9.12 per head, while the annual cost per head marketed is \$0.87 per head.

System B

This system uses a combination flooring system: part of it is solid concrete while the remainder is wood slats. Under the wood slats is a shallow pit, 0·61 m deep by 2·44 m wide, running the length of the buildings. These pits overflow into a central sump equipped with float-controlled pumps which pump the overflow to the primary lagoon. In operation, the pits are pumped out as flush water is admitted to the other end and then left with about 150 mm of liquid in the pit to cover fresh manure.

Liquid is hauled from the lagoons to adjacent farmland by truck when weather permits. The cost analysis is shown in Table VIII. The investment cost for this system is \$8.99 per head capacity, while the annual cost per head marketed is \$0.73.

System C

This system has a partial slatted floor with 5 m of solid concrete and 7·5 m of slatted floor over a 1·5 m pit in each pen. The solid portion of the floor is never scraped and the liquid in the pit is delivered to the field with an irrigation system.

An air diffusion system in the pits consists of a vane-type compressor, 4 cm galvanized and plastic pipe, and a series of valves. The diffusion system is operated for 15 min per hour. The solenoid gate switches from one pit section to the other every hour. In effect, the waste in each pit is being aerated for 15 min every 2 hr in one-eighth of the time.

The cost analysis for this system is shown in Table IX. Since odour control is obtained by mechanical means, the investment cost is considerably higher than the first two systems. It is \$16.94 per head capacity and the annual costs are \$1.36 per head marketed.

TABLE VI
HOG WASTE MANAGEMENT COMPONENTS FOR SYSTEMS USED FOR 35 000 HEAD CAPACITY UNITS

Process	*Item*	*Unit*	*Cost/unit*	*Total investment*	*Annual costs*		
					Total[a]	*\$/head cap.*	*\$/head*
Floor	Concrete floor	2 640 m^3	\$65.35/m^3	\$172 500	\$27 600[a]	\$0·790	\$0·316
	Concrete slats	26 000 m^2	\$13.45/m^2	\$349 700	\$56 000[b]	\$1·600	\$0·640
	Wood slats	26 000 m^2	\$3.77/m^2	\$98 200	\$28 000[d]	\$0·800	\$0·320
	Support wood slats	26 000 m^2	\$2.15/m^2	\$56 000	\$8 960[a]	\$0·256	\$0·102
Collection	Concrete pits	1 145 m^3	\$63.35/m^3	\$74 800	\$11 970[a]	\$0·342	\$0·137
Transfer	Flush system (slat floor)			\$20 600	\$4 020[c]	\$0·115	\$0·046
	Oper. costs	40 hr/wk	\$4.00/hr		\$8 320	\$0·238	\$0·095
	Flush system (solid floor)			\$37 620	\$7 340[c]	\$0·210	\$0·084
	Oper. costs	40 hr/wk	\$9.40/hr		\$18 000	\$0·514	\$0·206
	Drag system (slat floor)			\$40 000	\$7 800[d]	\$0·223	\$0·089
	Oper. costs	8 hr/day	\$4.00/hr		\$2 920	\$0·082	\$0·033
Storage	Holding pit (concrete)	2 000 m^3	\$65.35/m^3	\$130 700	\$20 900[a]	\$0·600	\$0·240
	Holding pond	39 700 m^3	\$0.65/m^3	\$25 800	\$4 130[a]	\$0·118	\$0·047
	Holding pond	90 500 m^3	\$0.65/m^3	\$58 830	\$9 400[a]	\$0·269	\$0·107
	Prim. lagoon	247 800 m^3	\$0.65/m^3	\$161 000	\$25 760[a]	\$0·736	\$0·294
	Sec. lagoon	105 500 m^3	\$0.65/m^3	\$68 600	\$10 976[a]	\$0·314	\$0·126
	Aeration pit (concrete)	3 220 m^3	\$65.35/m^3	\$210 430	\$33 670[a]	\$0·962	\$0·385
	Oxidation ditch	7 500 m^3	\$65.35/m^3	\$490 100	\$78 400[a]	\$2·240	\$0·896

Separation	Screen			$20 000	$3 200[a]	$0·091	$0·036
	Oper. costs	24 hr/day	$0.20/hr		$1 752	$0·050	$0·020
Treatment	Oxidation wheel	175 m	$850/m	$148 750	$29 000[d]	$0·829	$0·332
	Oper. costs	365 days	$0.94/m/day		$60 040	$1·720	$0·688
	Diffused air system			$160 000	$31 200[d]	$0·891	$0·356
	Oper. costs		$3 500/mo.		$42 000	$1·200	$0·480
	Spray runoff (15 ha)			$40 000	$11 400[a]	$0·326	$0·130
	Oper. costs		$1 000/mo.		$12 000	$0·343	$0·137
	Floating aerators	550 h.p.	$300/h.p.	$165 000	$32 175[d]	$0·919	$0·368
	Oper. costs		$243/day		$88 330	$2·524	$1·000
	Barriered land filter	4·0 ha	$20 000/ha	$80 000	$15 600[c]	$0·446	$0·178
	Oper. costs		$1 000/mo.		$12 000	$0·343	$0·137
Transport	Irrig. system			$10 000	$1 950[b]	$0·056	$0·022
	Oper. costs		$300/mo.		$3 600	$0·103	$0·041
	Vacuum wagon	4	$2 100 each	$8 400	$1 470[b]	$0·042	$0·017
	Tractor and labour	3 150 hr/yr	$4.00/hr		$12 600	$0·360	$0·144

* Annual cost as percentage of total investment: [a] 16·0 %, [b] 17·5 %, [c] 19·5 %, [d] 28·5 %.

TABLE VII

WASTE MANAGEMENT SYSTEM COSTS FOR SYSTEM A: DEEP PIT WITH WOOD SLATS AND EARTH BOTTOM, 35 000 HEAD CAPACITY

Process	*Item*	*Unit*	*Cost/unit*	*Total investment*	*Annual costs*		
					*Total**	*\$/head cap.*	*\$/head*
Floor	Wood slats	26 050 m²	\$3.77/m²	\$98 200	\$28 000[d]	\$0.800	\$0·320
	Support	26 050 m²	\$2.15/m²	\$56 000	\$8 960[a]	\$0·256	\$0·102
Storage	Concrete walls	2 000 m³	\$65.35/m³	\$130 700	\$20 900[a]	\$0·600	\$0·240
	Excavation	39 700 m³	\$0.65/m³	\$25 800	\$4 130[a]	\$0·118	\$0·047
Transport	Vacuum wagon (4)	5 670 litres		\$8 400	\$1 470[b]	\$0·042	\$0·017
	Tractor and labour	3 150 hr	\$4.00/hr		\$12 600	\$0·360	\$0·144
Totals		3 150 hr		\$319 100	\$76 060	\$2·18	\$0·87

* Annual cost as a percentage of total investment: [a] 16·0 %, [b] 17·5 %, [c] 19·5 %, [d] 28·5 %.

TABLE VIII

WASTE MANAGEMENT SYSTEM COSTS FOR SYSTEM B: PARTIAL WOOD SLATS, SHALLOW PITS AND TWO-STAGE LAGOONS, 35 000 HEAD CAPACITY

Process	Item	Unit	Cost/unit	Total investment	Annual costs		
					*Total**	*$/head cap.*	*$/head*
Floor	Concrete floor	1 955 m^3	$65.35/$m^3$	$127 800	$20 450[a]	$0·584	$0·234
	Wood slats	6 530 m^2	$3.77/$m^2$	$24 600	$7 000[d]	$0·200	$0·080
Collection	Concrete pit	1 145 m^3	$65.35/$m^3$	$74 800	$11 970[a]	$0·342	$0·137
Transfer	Flush system			$20 600	$4 020[c]	$0·115	$0·046
	Oper. costs	40 hr/wk	$4.00/hr		$8 320	$0·238	$0·095
Storage	Primary lagoon (excavation)	90 500 m^3	$0.65/$m^3$	$58 830	$9 400[a]	$0·269	$0·107
Transport	Tank truck			$8 000	$1 560[c]	$0·045	$0·018
	Oper. costs	330 hr	$4.00/hr		$1 320	$0·038	$0·015
Totals		2 410 hr		$314 630	$64 040	$1·83	$0·73

* Annual costs as percentage of total investment: [a] 16·0 %, [b] 17·5 %, [c] 19·5 %, [d] 28·5 %.

TABLE IX

WASTE MANAGEMENT SYSTEM COSTS FOR SYSTEM C: PARTIAL SLATS AND AIR DIFFUSION OXIDATION, 35 000 HEAD CAPACITY

Process	*Item*	*Unit*	*Cost/unit*	*Total investment*	*Annual costs*		
					*Total**	*\$/head cap.*	*\$/head*
Floor	Concrete floor	1 590 m^3	\$65.35/m^3	\$103 900	\$16 625[a]	\$0·475	\$0·190
	Concrete slat	8 050 m^2	\$13.45/m^2	\$108 680	\$17 400[a]	\$0·497	\$0·199
Storage	Concrete pit	3 220 m^3	\$65·35/m^3	\$210 430	\$33 670[a]	\$0·962	\$0·385
Treatment	Aeration system			\$160 000	\$31 200[b]	\$0·891	\$0·356
	Oper. costs	12 mo.	\$1 230/mo.		\$14 760	\$0·422	\$0·169
Transport	Irrigation system			\$10 000	\$1 950[b]	\$0·056	\$0·022
	Oper. costs	12 mo.	\$300/mo.		\$3 600	\$0·103	\$0·041
Totals				\$593 010	\$119 205	\$3·41	\$1·36

* Annual costs as a percentage of total investment: [a] 16·0 %, [b] 17·5 %, [c] 19·5 %, [d] 28·5 %.

The air diffusion system is used to reduce odours and reduce the volume of waste that must be hauled to the field.

System D

This system has a slatted floor with an oxidation wheel for partial treatment and odour control. In addition, the system has a concrete storage tank for the overflow. This is to provide for waste storage and still allow the level in the oxidation ditch to remain the same.

The initial cost of the aeration wheels is assumed to be $850 per metre of wheel. The yearly cost of operation is figured at 19·5% of initial costs. The electrical costs are figured at $0.02/kWh. The waste is to be transported by 5670 litre vacuum wagon to cropland for disposal and would require 3150 hr per year.

This system has an investment cost totalling $29.32 per head capacity and an annual cost equalling $2.77 per head marketed (Table X). These high costs are the result of the large amounts of concrete used for construction and the use of high-cost, high-energy treatment of the wastes.

The ability of this system to work properly is highly dependent upon the management inputs. The labour beyond maintenance and disposal is minimal.

System E

This system has a partial slatted floor, with 5 m of solid concrete and 2·5 m of slatted floor over a shallow flushing pit. The wastes are flushed over a screen to remove solids. The liquid then flows into an oxidation ditch for extended treatment and is then discharged into a lagoon for discharge either into a stream or for re-use as flushing or drinking water for the hogs.

The cost analysis for this system is shown in Table XI. The investment cost for this treatment system is $33.18 per head capacity, while the annual costs per head marketed are $3.00. While these costs seem high, the system does allow discharge of all water used for flushing. The costs could be greatly reduced by recycling most of the flushing water prior to treatment, thereby greatly reducing the size of the treatment portion of the system.

System F

This system utilizes a solid floor with flushing gutters through each pen along one side of the pen. The waste is flushed from the pen into the gutter and the wastes are then transported to the screen. The solids are removed at the screen and the liquids are then transferred to the Barriered Landscape Water Renovation System for treatment and re-use or discharge.

The cost analysis for this system is shown in Table XII. The investment cost for this system is $8.86 per head capacity. The annual operating cost for this system is $0.976 per head marketed.

TABLE X

WASTE MANAGEMENT SYSTEM COSTS FOR SYSTEM D: CONCRETE SLATS AND PIT WITH OXIDATION DITCH AND OUTSIDE STORAGE, 35 000 HEAD CAPACITY

Process	*Item*	*Unit*	*Cost/unit*	*Total investment*	*Annual costs*		
					*Total**	*$/head cap.*	*$/head*
Floor	Concrete slat	26 000 m^2	$13.45/$m^2$	$349 700	$56 000[a]	$1·60	$0·640
Storage	Concrete (ditch)	4 100 m^3	$65.35/$m^3$	$267 900	$42 900[a]	$1·22	$0·488
	Concrete (storage pit)	3 850 m^3	$65.35/$m^3$	$251 600	$40 250[a]	$1·15	$0·460
Treatment	Aerator	175 m of wheels	$850/m	$148 750	$29 000[c]	$0·829	$0·332
	Electricity	365 days	$0.94/m/day		$60 040	$1·72	$0·688
Transport	Vacuum wagon (4)	5 670 litres		$8 400	$1 470[b]	$0·042	$0·017
	Tractor and labour	3 150 hr	$4.00/hr		$12 600	$0·360	$0·144
Totals				$1 026 350	$242 260	$6·92	$2·77

* Annual cost as percentage of total investment: [a] 16·0%, [b] 17·5%, [c] 19·5%, [d] 28·5%.

TABLE XI

WASTE MANAGEMENT SYSTEM COSTS FOR SYSTEM E: PARTIAL SLATS, FLUSHING, SCREEN, OXIDATION DITCH, LAGOON WITH DISCHARGE, 35 000 HEAD CAPACITY

Process	*Item*	*Unit*	*Cost/unit*	*Total investment*	*Annual costs*		
					*Total**	*\$/head cap.*	*\$/head*
Floor	Concrete floor	1 590 m^3	\$65.35/m^3	\$103 900	\$16 625[a]	\$0·475	\$0·190
	Concrete slat	8 080 m^2	\$13.45/m^2	\$108 680	\$17 400[a]	\$0·497	\$0·199
Transfer	Flush system (slat floor)			\$20 600	\$4 020[c]	\$0·115	\$0·046
	Oper. costs	40 hr/wk	\$4.00/hr		\$8 320	\$0·238	\$0·095
Separation	Screen			\$20 000	\$3 200[a]	\$0·091	\$0·036
	Oper. costs	24 hr/day	\$0.20/hr		\$1 752	\$0·050	\$0·020
Treatment	Oxidation ditch	10 570 m^3	\$65.35/m^3	\$690 750	\$110 520[b]	\$3·16	\$1·26
	Oxidation wheel	175 m	\$850/m	\$148 750	\$29 000[d]	\$0·829	\$0·332
	Oper. costs	365 days	\$0.94/m/day		\$60 040	\$1·72	\$0·688
	Lagoon	105 500 m^3	\$0.65/m^3	\$68 600	\$10 976[b]	\$0·314	\$0·126
Disposal	Discharge						
Totals				\$1 161 280	\$261 853	\$7·489	\$2·996

* Annual cost as percentage of total investment: [a] 16·0%, [b] 17·5%, [c] 19·5%, [d] 28·5%.

TABLE XII

WASTE MANAGEMENT SYSTEM COSTS FOR SYSTEM F: SOLID FLOOR, FLUSHING, SCREEN, BARRIERED LANDSCAPE WATER RENOVATION (BLWR), 35 000 HEAD CAPACITY

Process	*Item*	*Unit*	*Cost/unit*	*Total investment*	*Annual costs*		
					*Total**	*\$/head cap.*	*\$/head*
Floor	Concrete floor	2 640 m^3	\$65.35/m^3	\$172 500	\$27 600[a]	\$0·788	\$0·315
Transfer	Flush system			\$37 620	\$7 340[c]	\$0·210	\$0·084
	Oper. costs	40 hr/wk	\$9.40/hr		\$18 000	\$0·514	\$0·206
Separation	Screen			\$20 000	\$3 200[a]	\$0·091	\$0·036
	Oper. costs	24 hr/day	\$0.20/hr		\$1 752	\$0·050	\$0·020
Treatment	BLWR	4·0 ha	\$20 000/ha	\$80 000	\$15 600[c]	\$0·446	\$0·178
	Oper. costs		\$1 000/mo.		\$12 000	\$0·343	\$0·137
Disposal	Discharge						
Totals				\$310 120	\$85 492	\$2·442	\$0·976

* Annual cost as percentage of total investment: [a] 16·0 %, [b] 17·5 %, [c] 19·5 %, [d] 28·5 %.

Confined Poultry Systems

Poultry waste management systems have been developed with three main concepts:

(a) The systems handling liquid waste have been used and many producers find them very easy to manage; however, the problem of odour control has caused many producers to seek other methods.
(b) Dry manure systems seem to solve the odour problem of the liquid system partially, if not completely, but they usually require more labour and in some cases more equipment than the liquid systems.
(c) The third system is a dry manure system with some method of processing the manure for product recovery.

A list of components used for poultry systems and their costs is shown in Table XIII.

System A

This system is a liquid manure system with a shallow pit under the cages which is flushed periodically. The waste is then transferred to an oxidation ditch and finally into an anaerobic lagoon prior to re-use or discharge. The solids are removed from the treatment components and spread on cropland with a vacuum wagon. The cost analysis for this system is shown in Table XIV. The investment and annual costs per bird are \$3.86 and \$0·956, respectively.

System B

This system is a dry manure system with a deep pit. The pit is equipped with a rake system which mixes the wastes and exposes it to air movement provided by fans located in the pit. The manure is periodically removed from the pit and spread on agricultural land. The manure, when removed from the pits, has had enough moisture removed from it for it to be handled as a dry solid, and it has no odour. The cost analysis for this system is shown in Table XV. The investment and annual costs per bird are \$1.28 and \$0.38, respectively.

Drying System for Poultry

There has been considerable interest in processing poultry wastes for use as a feed for other animals. This has been accomplished by drying the wastes and mixing it into rations for feeding. There are several types of driers for this use; however, one system will be evaluated for costs, which are shown in Table XVI for a 100 000 bird unit.

Manure Fractionation for Dairy Wastes

This process takes semi-solid dairy wastes blended into a slurry for short-term fermentation, then the wastes proceed to a liquid/solid separation

TABLE XIII

POULTRY WASTE MANAGEMENT COMPONENTS FOR SYSTEMS USED FOR 100 000 BIRD CAPACITY UNITS

Process	*Item*	*Unit*	*Cost/unit*	*Total investment*	*Annual costs*	
					*Total**	*$/bird capacity*
Collection	Concrete pit	780 m^3	$65.35/$m^3$	$50 973	$8 160[a]	$0·082
Transfer	Solid floor flushing system			$20 000	$3 900[c]	$0·039
	Oper. costs	20 hr/wk	$8.00/hr		$8 320	$0·083
	Mechanical drag, 'rake', fans			$75 000	$14 625[c]	$0·146
	Oper. costs	12 mo.	$1 250/mo.		$15 000	$0·150
Storage	Oxidation ditch	3 420 m^3	$65.35/$m^3$	$223 500	$35 760[a]	$0·358
Treatment	Oxidation wheel	50 m	$850/m	$42 500	$12 110[d]	$0·121
	Oper. costs	365 days	$0.94/m/day		$17 155	$0·172
	Anaerobic lagoon	72 700 m^3	$0.65/$m^3$	$47 300	$8 280[b]	$0·083
Transport	Vacuum wagon	1	$2 100	$2 100	$410[c]	$0·001
	Tractor and labour	8 hr/wk	$4.00/hr		$1 670	$0·017
	Spreader		$1 800	$1 800	$350[c]	$0·001
	Tractor and labour	4 hr/wk	$4.00/hr		$830	$0·001

* Annual cost as percentage of total investment: [a] 16·0 %, [b] 17·5 %, [c] 19·5 %, [d] 28·5 %.

TABLE XIV

WASTE MANAGEMENT SYSTEM COST FOR SYSTEM A: SHALLOW PIT, FLUSHING, OXIDATION DITCH, LAGOON AND DISCHARGE, 100 000 BIRD POULTRY UNIT

Process	*Item*	*Unit*	*Cost/unit*	*Total investment*	*Annual costs*	
					*Total**	*$/bird capacity*
Collection	Concrete pit	780 m^3	$65.35/$m^3$	$50 973	$8 160[a]	$0·082
Transfer	Flushing system			$20 000	$3 900[c]	$0·039
	Oper. costs	20 hr/wk	$8.00/hr		$8 320	$0·083
Storage	Oxidation ditch	3 420 m^3	$65.35/$m^3$	$223 500	$35 760[a]	$0·358
Treatment	Oxidation wheel	50 m	$850/m	$42 500	$12 110[d]	$0·121
	Oper. costs	365 days	$0.94/m/day		$17 155	$0·172
	Anaerobic lagoon	72 700 m^3	$0.65/$m^3$	$47 300	$8 280[b]	$0·083
Transport	Vacuum wagon	1	$2 100	$2 100	$410[c]	$0·001
	Tractor and labour	8 hr/wk	$4.00/hr		$1 670	$0·017
Disposal	Discharge					
Totals				$386 373	$95 765	$0·956

* Annual cost as percentage of total investment: [a] 16·0%, [b] 17·5%, [c] 19·5%, [d] 28·5%.

TABLE XV

WASTE MANAGEMENT SYSTEM COST FOR SYSTEM B: DEEP PIT, PIT DRYING RAKE AND FANS, LAND SPREADING FOR DISPOSAL, 100 000 BIRD POULTRY UNIT

Process	*Item*	*Unit*	*Cost/unit*	*Total investment*	*Annual costs*	
					*Total**	*$/bird capacity*
Collection	Concrete pit	780 m^3	$65.35/m^3	$50 973	$8 160[a]	$0·082
Transfer	Mechanical drag, 'rake' and fans			$75 000	$14 625[c]	$0·146
	Oper. costs	12 mo.	$1 250/mo.		$15 000	$0·150
Transport	Spreader	1	$1 800	$1 800	$350[c]	$0·001
	Tractor and labour	4 hr/wk	$4.00/hr		$830	$0·001
Disposal	Cropland					
Totals				$127 780	$38 970	$0·380

* Annual cost as percentage of total investment: [a] 16·0 %, [b] 17·5 %, [c] 19·5 %, [d] 28·5 %.

TABLE XVI

COST OF OPERATING A MANURE DRYING SYSTEM FOR A 100 000 BIRD UNIT

Process	*Unit*	*Cost/unit*	*Total investment*	*Annual cost*	
				*Total**	*$/bird capacity*
Drier			$70 500	$11 280[a]	$0·113
Storage pit and production tanks			$16 500	$2 640[a]	$0·026
Fuel	1 200 tonnes/yr	$17.86/tonne		$21 432	$0·214
Electricity	1 200 tonnes/yr	$1.80/tonne		$2 160	$0·022
Labour	2 800 hr/yr	$5.00/hr		$14 000	$0·140
Totals			$87 000	$51 512	$0·515

* Annual cost as percentage of total investment: [a] 16·0%, [b] 17·5%, [c] 19·5%, [d] 28·5%.

system. The solids are pressed to remove moisture and transferred to a silo for storage until they are used as feed. The liquid from the press is dried and the residue is pelleted as a feed. The liquid from the initial solid separation unit is centrifuged, with the solid being used as a soil conditioner and the liquid being dried and pelleted for feed. According to the manufacturer, the costs are presented in Table XVII for a 10 000 head dairy system. Systems

TABLE XVII

COST OF OPERATING A MANURE FRACTIONATION PROCESS FOR A 10 000 HEAD DAIRY UNIT

(from ref. 5)

Item	*Total investment*	*Annual cost*	
		*Total**	*$/bird capacity*
Capital cost: land, utilities, building, 4 slurry tanks and 1 liquid tank	$120 000	$19 200[a]	$1.92
Equipment cost, *i.e.* pump, separators, centrifuges, driers, etc.	$400 000	$64 000[a]	$6.40
Operating costs:			
Fuel		$120 000	$12.00
Labour		$95 000	$9.50
Totals	$520 000	$298 200	$29.82

* Annual cost as percentage of total investment: [a] 16·0%, [b] 17·5%, [c] 19·5%, [d] 28·5%.

for other types of animal wastes are being designed; however, costs were not available at this time.

The total cost of $29.80 per head seems very high; however, this system, as do many of the others described in this section, provides a product which has some real cash value. If any system is to be operated for any purpose other than disposal, it must produce enough cash return to lower the cost per head to at least equal to the cost of operating a disposal system or the process will be considered too costly for use.

GENERAL CONSIDERATIONS

The preceding sections have discussed many types of waste treatment systems and their costs. These costs have, in some cases, seemed excessive; however, they must be considered along with other animal production costs. Figure 3 gives some perspective to many of the costs involved in

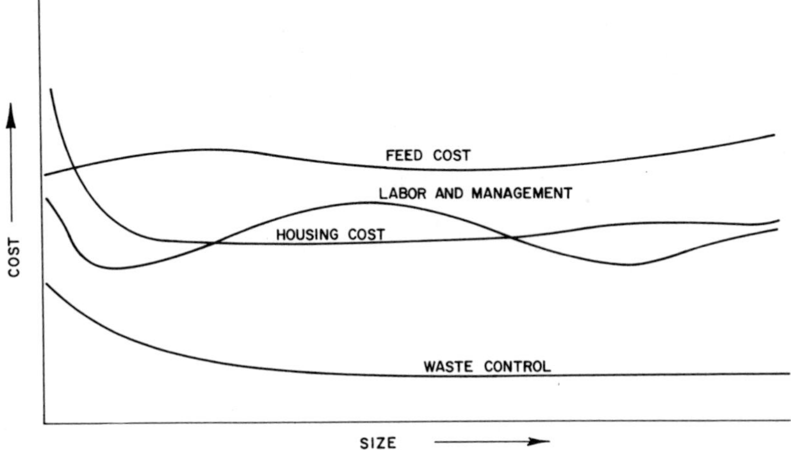

FIG. 3. Costs for covered feedlots.

animal production using covered feedlots. It can be seen from this figure that there are economies of scale. It can also be seen that feed cost is the largest single cost, and, as has been shown in the past, it can vary greatly. Feed costs in the United States have nearly doubled in the last few years. This is not to say that waste control costs are not important because they are. It is presented to help put all costs related to animal production in perspective.

REFERENCES

1. Forster, D. L., Connor, L. J. and Johnson, J. B., Economic impacts of alternative water pollution control rules on beef feedlots of less than 1000 head capacity. *Michigan Agricultural Experiment Station Journal*, Article No. 7163, USDA, East Lansing, Mich., 1975, 17 pp.
2. David, M. L., Seltzer, R. E. and Eickhoff, W. D., *Economic Analysis of Proposed Effluent Guidelines, Feedlots Industry*. Report No. EPA-230/1-73-008, US Environmental Protection Agency, Washington, DC, 1973, 300 pp.
3. Anon., Wage rates for key construction trades. *Engineering News Record*, 1975.
4. Horsfield, B., Gottbrath, J. and Kadlec, J., Swine waste management systems. Presented at 1973 Winter Meeting of ASAE, Chicago, Ill., 11–14 Dec., 1973. ASAE Paper No. 73-4517, St Joseph, Mich., 1973, 33 pp.
5. Harper, J. M. and Seckler, D., Engineering and economic overview of alternative livestock waste utilization techniques. Colorado State University, Fort Collins, Colo., 1975. Presented at International Symposium on Livestock Wastes held at Urbana-Champaign, Ill., 21–24 Apr. 1975, 17 pp.

PART V

Feedlot Waste Management in Selected Countries

28

Land Disposal of Animal Wastes in the GDR

H. KORIATH

Fertilization Research Institute, Potsdam, German Democratic Republic

With the progressive establishment of factory-like livestock units and the remodelling of old animal facilities, the amount of slurry produced has increased over the past decade. The term 'slurry' defines a waste product of livestock production which is a mixture of faeces and urine and may contain spilled or other production wastewaters.

In 1975 almost 40 million tons of slurry were produced in the German Democratic Republic. In this connection it should be borne in mind that on the transition to large industrialized livestock units, where the animals are kept on a no-litter system, slurry production will be highly concentrated in certain places. Taking the biochemical oxygen demand as the conversion factor, slurry production per large animal equivalent (1 large animal equivalent = 500 kg animal live weight) of cattle would correspond to the wastewater load from about 10–12 population equivalents, while one large animal equivalent of pig would amount to about 18–20 population equivalents. This implies that a dairy unit housing 2000 cows causes the same wastewater load as a town of about 25 000 inhabitants. A fattening unit for 250 000 pigs even compares with a town of about 100 000 inhabitants.

Considering the population density and the strained water regime as well as the great demands on the productive power and the recreational value of the territory of the German Democratic Republic, the Constitution of the GDR as well as the Environment Act contain highly stringent regulations for keeping its waters clean and for protecting the landscape. Therefore, all the various methods applied for treatment and disposal of animal wastes are geared to the requirements of environmental protection. Another basic demand is that waste products must not simply be 'cleared away', but that rather they should be supplied for economically efficient utilization, *i.e.* for the production of useful material. In consideration of these aspects, the disposal of animal wastes in the GDR is based on the following guidelines:

1. The considerable amounts of plant nutrients, organic matter and water contained in slurry must be maximally and most efficiently used for reproducing soil fertility and boosting crop production.

2. Technologies for the treatment and disposal of slurry and wastewaters must comply with the requirements of environmental protection.
3. Methods for the treatment and disposal of waste products must provide for the innocuous integration of factory-type livestock units into their respective landscapes.

In order to choose an optimal and economical technology for the land disposal of animal wastes, the proper siting of livestock production units is of decisive importance. Neglect or undervaluation of this aspect would entail increased expenditures for slurry disposal in crop production according to the demands of environmental protection which, under unfavourable conditions, might jeopardize the overall economic efficiency of the livestock unit concerned. Waste disposal is therefore of prime importance when planning and designing factory-type livestock units.

Major parameters underlying the design of disposal systems are the amounts of faeces, urine and wastewaters produced. Whilst the production of faeces and urine varies within rather narrow limits, and is controlled first of all by the animal species concerned and by the method of rearing as well as by the feeding regime, the production of wastewaters and the water inflow into the slurry vary greatly depending on the conditions prevailing in the given livestock units. From the results obtained in GDR livestock units we obtained the average values quoted in Table I.

TABLE I

AVERAGE AMOUNTS OF FAECES AND URINE, WASTEWATERS AND SLURRY PER LARGE ANIMAL EQUIVALENT (500 kg ANIMAL LIVE WEIGHT) AND YEAR IN LIVESTOCK UNITS

	Faeces and urine (tons)	*Wastewaters (tons)*	*Slurry (tons)*
Milk production	18·3	27·8	46·1
Youngstock rearing (cattle)	16·4	11·7	28·1
Calf rearing	14·6	4·7	19·3
Cattle fattening	14·6	12·2	26·8
Pig breeding	9·1	31·9	41·0
Pig fattening	9·1	11·0	20·1

Slurry composition depends on the animal species concerned, on the purpose of the animal's use and on the feeding regime, but most of all it is controlled by the amount of wastewaters included. Therefore, the dry matter and nutrient contents often vary over a wide range. From slurry analysis in a large number of factory-type livestock units we obtained the average values listed in Table II.

Because of the great demands on cleaning and disinfection, water inflow into slurry is very high. In cattle units, especially those for milk production, wastewaters double the amount of slurry produced, and in pig units every ton of faeces–urine mixture turns into 3–5 tons of slurry. The increase in volume due to water inflow means a big technological and economic load for the methods of slurry storage, application and utilization. Therefore,

TABLE II

AVERAGE DRY MATTER AND NUTRIENT CONTENTS OF SLURRY FROM FACTORY-TYPE LIVESTOCK UNITS

Unit	*Dry matter*		*N* (%)	*P* (%)	*K* (%)
	Faeces and urine (%)	*Slurry* (%)			
Milk production	10	4	0·22	0·04	0·22
Youngstock rearing	12	7	0·35	0·07	0·27
Cattle rearing	9	7	0·48	0·12	0·23
Cattle fattening	11	6	0·35	0·08	0·30
Pig breeding	9	2	0·22	0·03	0·10
Pig fattening	9	4	0·44	0·07	0·20
Layer hens	20	16	1·23	0·41	0·40

the development of cleaning and disinfection methods requiring only small amounts of water is a very important approach.

The adequate sizing of storage facilities is an essential prerequisite for the proper land disposal of animal wastes. The size of the storage facilities depends on the amount of slurry produced daily and on the required storage period. The storage period is dependent upon the time interval during which it is impossible to apply slurry. The intervals during which the waste products cannot be utilized is governed by: (a) cropping aspects, such as cropping ratio and crop rotation; (b) the prevailing weather and soil conditions; and (c) the methods of slurry application.

Moreover, local conditions as well as injunctions imposed by water management or hygiene authorities, respectively, may well influence the times when slurry is applied. Depending on the prevailing weather and soil conditions, on the type of crop production and on the method of slurry application, the appropriate values regarding slurry storage capacity that are listed in Table III are effective under the production conditions of the German Democratic Republic. Irrespective of these aspects, any slurry produced has to be stored for an obligatory period of four days for veterinary-sanitary inspection.

In the German Democratic Republic, slurry applications—*i.e.* slurry transport and placement on farmland—is accomplished by tank trucks, sprinkler irrigation systems or a combined system including pipelines and filling stations for tank trucks in the field. At present, tank trucks account for the largest share in slurry application; however, the share of slurry sprinkling is increasing as the concentration of livestock production increases. In these cases, combined systems for sprinkler irrigation with clean water and slurry are used exclusively. The choice of application method is controlled not only by the location, the structure and the

TABLE III

APPROXIMATE VALUES FOR REQUIRED SLURRY STORAGE CAPACITY (IN DAYS)

Type of crop production	*Lowland*		*Foothill mountains*		*Medium-altitude mountains*	
	T	*V*	*T*	*V*	*T*	*V*
Cereals/root and tuber crops	45	45	60	60	90	90
Root and tuber crops/forage	30	30	60	45	90	70
Forage	30	30	50	40	80	60

T = application with tank.
V = application by sprinkler irrigation.

requirements of the livestock units as well as of crop production, but largely also by aspects of water management, environmental protection and transport. Before permission to erect a livestock unit is granted, the public supervising authorities must be informed of the projected technology for waste placement and disposal. Apart from the requirements of municipal and land hygiene, special importance is attached to those parameters that follow from the threshold loads for water management and soil fertility, including trouble-free crop production.

On the basis of the results of long-term model and field trials with lysimeters installed in soil columns 50, 60 and 100 cm in depth, respectively, the following loads are, for the time being, considered as the water management threshold values for agricultural soils in the German Democratic Republic:

Cattle slurry	up to 6 large animal equivalents/ha
Pig slurry	up to 4·5 large animal equivalents/ha
Slurry derived from layer units	up to 2·5 large animal equivalents/ha

(1 large animal equivalent = 500 kg animal live weight)

The following main criteria were used for assessing fertilization with animal wastes under the aspect of water management: for organic pollution, the BOD_5 (biochemical oxygen demand, mg/litre = amount of dissolved oxygen required for oxidative biological decomposition of organic matter within 5 days at 20 °C), for mineral load, the N and P contents; and for hygienic rating, the amount of coliform bacteria. The threshold values taken from water management are effective on the understanding that in the case of slurry load exceeding 4 large animal equivalents/ha, the overall quantity has to be applied in several batches during the course of the year. For reasons of crop production, such as excessive fertilization of the crop species concerned, salt concentration, inadequate plant cover, efficient utilization of the slurry nutrients, etc., the above threshold values are reached in exceptional cases only.

When planning the land for the disposal of animal wastes, we therefore take into account the threshold loads depending on the local and production conditions of plant growing. One example of these approximate values applied in the German Democratic Republic is shown in Table IV. Compared with tank truck application, sprinkler irrigation in most cases allows the application of larger amounts of slurry, since in that case highly fertilizer-consuming crops would be grown and the nitrogen demand of the crops would be higher. Therefore, at equal amounts of slurry from livestock units, a smaller application area is, as a rule, needed if the slurry is applied by sprinkler irrigation.

TABLE IV

APPROXIMATE VALUES FOR THRESHOLD LOADS IN LARGE ANIMAL EQUIVALENTS/HA ON SLURRY APPLICATION BY SPRINKLER IRRIGATION

Livestock unit	*Cereals*	*Cropping ratio (%) Root and tuber crops*		*Field grass*
	60:20:20 (200 *kg N/ha*)	40:20:40 (300 *kg N/ha*)	20:20:60 (400 *kg N/ha*)	0:0:100 (500 *kg N/ha*)
Dairy cows	2·0	3·0	4·0	5·0
Calf rearing	2·5	3·5	4·5	5·5
Youngstock rearing (cattle)	2·0	3·0	4·0	5·0
Cattle fattening	2·0	3·0	4·0	5·0
Pig breeding	2·5	3·5	4·5	5·5
Pig fattening	2·5	3·5	4·5	5·5

Certain threshold values for animal waste load must be observed for protective areas for drinking water as well as for water reservoirs for future groundwater use.

Compared with farmyard manure, slurry contains a higher percentage of easily soluble nutrients. For the most part, more than 50 % of the slurry nitrogen is present in a soluble form (mainly as ammonium carbonate). Most of the remaining nitrogen contained in the slurry organic matter is being mobilized and thus becomes available to the plants in the course of the growing season.

Potassium is contained in the slurry almost exclusively in a plant-available form, while phosphorus is found mostly in organic compounds. In addition, slurry adds magnesium and micro-nutrients to the soil; with regard to the micro-nutrients, the accumulation of these substances in soils and plants has to be watched carefully on continuous application of large quantities of slurry.

On the basis of results obtained from over 100 field experiments, mineral fertilizer equivalents were established in the German Democratic Republic for the purpose of incorporating slurry into the fertilizing system and in this connection duly considering the nutrient quantities applied in the form of slurry, as well as for dimensioning slurry applications. These fertilizer equivalents indicate the effects of slurry nutrients as compared with those of nutrients from mineral fertilizers applied at the optimal time. The mineral fertilizer equivalents of slurry nutrients are, above all, dependent on crop species, soil and time of application. In addition, the mineral fertilizer equivalents of slurry nitrogen also depend on the amount of slurry applied and on the way in which the slurry is distributed on the field. For slurry nitrogen, the mineral fertilizer equivalents are lowest in the case of autumn application to sandy soils (Table V). The dependence of the slurry effect on the time of application declines as the exchange capacity of the soil increases.

TABLE V

MINERAL FERTILIZER EQUIVALENTS OF SLURRY NITROGEN AS INFLUENCED BY CROP SPECIES, SOIL AND TIME OF APPLICATION

(Example: Potatoes)

Time of application	*Soil*		
	Sand	*Sandy loam*	*Loam/clay*
August–September	20	40	50
October–November	40	50	60
December–February	60	60	60
March	70	70	70

The advantage of higher nutrient supplies (related to equal livestock numbers) from slurry as compared with farmyard manure, which is due to smaller losses, is in part offset on light soils in that slurry has to be applied throughout the year, *i.e.* often also at times unsuitable from the cropping point of view, which may give rise to considerable losses of nutrients, and of nitrogen in particular. These nutrient losses, which in particular occur when slurrying soils of low exchange capacity late in summer or in autumn, were

TABLE VI

NITROGEN LEACHING ON SANDY SOIL AS INFLUENCED BY THE TIME OF SLURRY APPLICATION AND STRAW MANURING

Fertilization of potatoes, 1971–2	*Time of slurry application*	*Nitrogen leaching (kg/ha)*		
		1971–2	1972–3	*Total* (1971–3)
1. 320 kg slurry–N/ha	August	37	30	67
2. 320 kg slurry–N/ha + Straw manuring in August	August	23	17	40
3. 320 kg slurry–N/ha	November	38	26	64
4. 320 kg slurry–N/ha + Straw manuring in August	November	28	17	45

found to be strongly reduced by straw manuring or by growing stubble or winter catch crops, respectively (Table VI).

Only in recent years has large-scale research work been undertaken to study the effect of slurrying on major soil characteristics, above all the soil carbon and nitrogen contents. Experimental results obtained from different places in the German Democratic Republic revealed that slurry application does not exert an adverse effect on the soil humus content. The influence of combined straw and slurry application on the carbon and nitrogen levels of the soil is comparable to that of farmyard manure (Table VII).

In the German Democratic Republic, research and development on problems of animal waste disposal proceed from the objective of closely linking the introduction of modern, factory-type production methods into agriculture with the improvement of the working and living conditions of the working people, of further raising soil fertility, and of adapting the recreational value of the countryside to the increasing demand. With the new Environment Act, the Government of the German Democratic

TABLE VII

INFLUENCE OF STRAW AND SLURRY APPLICATION ON CROP YIELD AND SOIL HUMUS CONTENT AT EQUAL INITIAL QUANTITIES OF ANIMAL EXCREMENTS

(8-year average on lightly loamed sandy soil)

	Dry matter yield	*% in topsoil*	
	(dt/ha)	*C*	*N*
NPK + Farmyard manure	38·1	0·55	0·051
NPK + Slurry + Straw	40·1	0·60	0·053

Republic has attributed great authority to the problems of environmental protection. On this basis, specialists from the fields of agriculture, water management, civil and construction engineering, medicine and hygiene as well as from various branches of the technical sciences and biosciences cooperate in a highly complex way to solve the tasks involved in the effective disposal of animal wastes.

29

Swine Feedlot Wastewater Treatment in Romania

C. A. L. NEGULESCU

Senior Research Officer, Institute for Water Resources Engineering, Bucharest, Romania

INTRODUCTION

In Romania, as in many other countries, industrial-type large pig-fattening feedlots have been constructed. Wastes generated in these 15 000–300 000 pigs capacity units are hydraulically removed with the addition of large quantities of dilution water. The disposal of the resulting wastewater, which is 30–40 litres/pig/day, is limited by the amount of land surrounding the feedlot. Such land is not usually adequate.

Furthermore, since transport of the wastewaters is economically limited to short hauls, extensive treatment of the feedlot wastewater is necessary.

WASTEWATER CHARACTERISTICS

Typical ranges in the values of the parameters of wastewaters from a 150 000 pigs per year feedlot are given in Table I. These values are compared in Table II with published values for similar feedlots in other countries.

TABLE I
WASTEWATER CHARACTERISTICS AT A 150 000 PIGS PER YEAR FEEDLOT

Parameter	*Units*	*Observed range*
Flow	m^3/day	3 300–4 500
Total solids	mg/litre	7 000–8 300
Volatile solids	mg/litre	5 200–6 300
Suspended solids	mg/litre	6 000–7 200
pH	—	7·5–8·5
COD	mg/litre	13 200–14 000
BOD_5	mg/litre	4 700–5 000
Total nitrogen	mg/litre	825–855
Total phosphorus	mg/litre	33–41

TABLE II

COMPARISON OF VALUES OF WASTEWATER PARAMETERS OBSERVED IN ROMANIA WITH VALUES REPORTED IN LITERATURE

Indicators	*Units*	*Values found in literature*	*Values observed in Romania*
Quantity of wastes	kg/AU*/day	28–43	
	litres/AU*/day	32–52	300–380
Total solids (TTS)	kg/AU*/day	2·5–8·1	2·24
Volatile solids (TVS)	kg/AU*/day	2·26–3·64	1·75
	% TTS	73–87%	78
BOD_5	kg/AU*/day	0·64–2·8	0·95–1·15
	kg/kg TVS	0·32–0·64	0·54–0·66
COD	kg/AU*/day	1·63–3·72	1·97–2·84
	kg/kg TVS	1·20–1·76	1·13–1·60
Total nitrogen	g/AU*/day	158–540	175
	% TTS	3·8–9·4%	7·8
Total phosphorus	g/AU*/day	40–98	37
	% TTS	1·95–3·6%	1·7

* AU = Animal Unit = 500 kg of live weight animals.

TREATMENT PROCESSES USED IN ROMANIA

Treatment systems adopted to process wastewaters from these large pig feedlots include primary treatment and secondary treatment with disinfection. Primary treatment is accomplished in horizontal settling tanks equipped with scraping mechanisms. Secondary treatment consists of activated sludge basins with mechanical systems of aeration, or oxidation ponds, followed by final settling tanks for activated sludge systems, and chlorination before stream discharge.

The sludge from the primary settling tanks is dewatered on drying beds or it is stabilized by mechanical sludge aerators. A detention time of 30 min in the primary settling tanks results in 60–70% removal of suspended solids and 40% removal of incoming BOD_5.

Drying beds for primary sludge should be designed at 300 m^2/1000 pigs. After 60 days in dewatering beds, moisture content of the primary sludge is reduced from 92–95% down to 70–75%. On dry matter basis, primary sludge is 80% organic matter with the concentration of fertilizer elements being 2·9% for nitrogen, 0·56% for phosphorus and 0·56% for potassium.

For stabilization of the sludge by aeration, the oxygen demand to be satisfied is 0·1 kg of oxygen per kg of volatile solids per day. A detention time of 10 days is recommended. BOD_5 removal in the activated sludge plants reaches 99 % at loading rates of 0·3–0·8 kg BOD_5/day/kg MLSS (mixed liquor suspended solids).

In practice, removal efficiencies achieved with activated sludge treatment plants are 85 %. They can be as high as 92 % with two-stage aeration basins. Nitrogen and phosphorus content of the effluent is high.

With oxidation ponds, treatment efficiencies achieved were 92 % during summer and 80 % during winter. However, the effluent was high in suspended solids, nitrogen and phosphorus.

The treatment plants where the activated sludge processes have been started up under normal conditions of temperature (spring, summer, autumn) operated satisfactorily throughout the winter, the overall efficiency decreasing by only a small percentage. It is essential that such plants are put in operation several months ahead of cold weather, in order to ensure the development of a stable activated sludge floc.

For final disinfection of fully treated wastewater by the activated sludge process, a chlorine dose of 12–50 mg/litre might be required.

30

Feedlot Waste Management in Upper Silesia, Poland

O. KOSAREWICZ and I. FIRLUS

Environmental Pollution Abatement Centre, Katowice, Poland

Animal production in Poland will be increased through the construction of large feedlots with high concentrations of animals. Already many large feedlots have been constructed in Poland, and also in the densely populated and highly industrialized Upper Silesia region in western Poland. These new modern feedlots no longer practice the traditional methods of using bedding on the pen floors to absorb urine. Instead, wastes are hydraulically removed as liquid wastewaters by flushing them with water. Both the volume of wastewater and its pollutional content are such that pig feedlots are now considered a major source of water pollution.

A swine feedlot, for example, in Rzeczyce, Upper Silesia, with a capacity large enough to market 36 000 pigs per year generates daily 1100 m^3 of wastewater whose BOD averages 8600 mg/litre. The daily BOD load to be discharged is calculated to be $9{\cdot}46 \times 10^6$ g BOD/day. At 54 g BOD/head/day, the unitary value per person in Poland, the wastewater from the feedlot has a population equivalent of 175 200. As a matter of fact, the wastewater from this feedlot constitutes 24% of the BOD load being discharged into the nearby Kłodnica river which drains an area of 304 km^2 inhabited by 719 000 people (1973 census). The average flow of the river is 263 520 m^3/day. The Environmental Pollution Abatement Centre in Katowice has designated this catchment area as a 'model region', and comprehensive environmental studies have been initiated by the Centre.

Several methods of treatment of wastewaters from these new pig feedlots are being installed and others are being studied on a laboratory or pilot plant scale. Very few of the feedlots have sufficient land to dispose of their wastes.[1] Swine wastewaters may be disposed of on land at application rates of 50–100 m^3/ha/year.[5] Thus, the Rzeczyce pig feedlot would need 4000–8000 ha of land; such land area is not available within a feasible proximity of the feedlot. Even if land were available, extensive storage volume would be required, because no wastewater could be discharged during the cold months of the year.

A swine wastewater treatment system which has been installed in several farms is of Hungarian design, the so-called Vidus type. In this system, wastewaters are put through vibrating screens and then into a pre-aeration chamber which also serves as an equalizing surge tank with a detention time of 16 hr. Additional removal of solids is accomplished through coagulation. Al_2SO_4 is applied at the rate of 1000 mg/litre to help settle out solids. The treatments following coagulation in their order of sequence are activated sludge treatment for 14 hr, final sedimentation, and finally, disinfection with chlorine before discharge. Sludges from sedimentation tanks are thickened and then placed on drying beds. Excess activated sludge and supernatant from sludge thickener are recycled to the pre-aeration tank. The solids remaining on the vibrating screens represent about 3·5% of the volume of the wastewater. Their moisture content is 85%. They are disposed of by landfilling. Thickened, post-coagulation sludge constitutes about 5% of the volume of the wastewater. It is 94% water. Attempts to dewater this type of sludge by the use of drying beds have not been successful. Sludge handling in this type of treatment system is a problem requiring extensive study.

The BOD removal efficiency of the aforementioned Vidus type treatment plant can be as high as 98%, reducing the BOD from 8600 to 200 mg/litre. COD is reduced from 14 960 to 440 mg/litre for a COD removal efficiency of 97%. BOD to COD ratio is 0·58 in the influent and 0·46 in the effluent.

Research in animal waste management in Poland is being conducted by Agricultural Institutes, the Institute of Meteorology and Water Management and some universities. Several systems have been proposed and are now being investigated. One system[4] consists of chemical coagulation followed by 3 hr detention in a sedimentation tank, 17 hr aeration, 4·5 hr final sedimentation and 5 days in a stabilization pond. In such a system, and under winter operation, COD was reduced from 21 146 to 610 mg/litre for 97% removal efficiency. The pond removed only 1·7% of the COD, while the activated sludge aeration tank was responsible for 14% of the BOD removal.

Discharge of pig wastewaters into water bodies is opposed by others[2] who believe that the use of pig wastes as fertilizer is the best method of disposal. Therefore, pig waste handling and treatment systems should be designed for effective use in agriculture. To control odour and to absorb the liquid portion of the waste, the addition of peat and brown coal has been suggested.

Control of pathogenic organisms in animal wastewaters may be accomplished by methods similar to those used in treating sewage sludge in preparation for disposal on to agricultural land. *Salmonella* and faecal coliforms do not survive in solid waste compost pits where the temperature ranges between 40 and 69°C.[3] It is not known yet whether such high temperatures must be maintained or for how long, to ensure destruction of pathogens.

Feedlot location, operating conditions and waste treatment efficiencies to attain at each location are all controlled by legislation enacted in 1970 and in force as of 1 January 1971. Water management authorities define water quality classes which in turn determine the level of treatment before discharge. Recent legislation, in effect since January 1975, defines procedures to be followed in the location of swine feedlots and their operation. These regulations specify, for example, that the minimum distance of pig farm buildings from residential and commercial dwellings must be 300 m. The minimum distance from roads and railways depends on the category of the wastewater treatment plant. The minimum distance is 1000 m for Category I plant and 50 m for Category V.

Generally, wastes from poultry farms and from small pig-rearing units in small farms do not present a big disposal problem. The wastes are collected as solids, stored and then disposed on land.

Studies concerning wastewater management from large animal feedlots are aimed at analysing the feasibility of large pig units in view of the high wastewater treatment requirements. The practicality of raising animals without bedding is being examined. To ensure destruction of pathogens and at the same time control odours, thermal treatment of wastewaters by composting and by the use of peat and coal are being researched.

REFERENCES

1. Bartoszewski, K. and Sorko, M., Oczyszczanie ścieków z ferm tuczu świń metoda osadu czynnego. Materialy z XVII Konferencji Naukowo-Technicznej pt. 'Postep techniczny w dziedzinie oczyszozania ścieków', Katowice, czerwiec 1974.
2. Nowosielski, O., Odpady jako środki nawozowe w rolnictwie i ogrodnictwie. Materialy z VII Konferencji Naukowo-Technicznej pt. 'Wykorzystanie odpadów w rolnictwie', Warszawa, 6–7 Dec. 1974.
3. Przybojewska, B. Higieniczne aspekty problemu unieszkodliwiania odpadów. Materialy z Seminarium pt. 'Usuwanie i unieszkodliwianie odpadów'. Kraków, 5–7 Dec. 1973.
4. Rybiński, J., Wysokosprawne metody oczyszczania ścieków z ferm przemyslowych. Materialy konferencyjne Sesji Naukowej pt. 'Wysoko efektywne metody oczyszozania ścieków', Kraków, 19–21 June, 1975.
5. Kutera, J., Obecny stan zaawansowania badań w problemie rolniczego wyloozystania gnojowicy. Materialy seminaryjne, Bydgosecz, 1973.

31

Feedlot Waste Management in Hungary

P. FARKAS

Research Institute for Water Resources Development, Budapest, Hungary

and

T. RACZ

National Water Authority, Budapest, Hungary

The beginning of intensive industrialization of manufacturing in Hungary in the latter part of the nineteenth century and of agriculture in the middle of the twentieth century, and the resulting migration of rural people to urban/industrial centres, have changed Hungarian animal production patterns drastically.

The total animal population in Hungary in 1973 was about 2 million cattle, 7 million pigs and 33 million poultry, as is indicated in Table I. It was estimated that animal farms consumed $70 \times 10^6\ m^3$/year of water, which constituted 10% of the total potable water produced in Hungary. The volume of liquid manure was estimated at $7 \times 10^6\ m^3$/year. This volume of

TABLE I
ANIMAL POPULATION IN HUNGARY, 1973
('000 head)

Species	*Total*	*Traditional farming*	*In modern feedlots*	
			Total	*Liquid manure system*
Cattle total	1 965	1 689	276	18
Cows	763	632	131	7
Hogs total	6 980	4 163	2 817	1 220
Sows	616	321	295	149
Horses total	189	189	—	—
Mares	81	81	—	—
Sheep total	2 259	2 259	—	—
Ewes	1 262	1 262	—	—
Poultry	32 744	32 744	6 920	—

TABLE II

SURVEY OF WASTE SYSTEMS ON 26 FEEDLOTS IN HUNGARY

Pig farms site no.	*Discharge rate* (m^3/*day*)	*Number of animals*	*Wastewater (litres/day/animal)*	*Disposal area (ha)*	*Irrigation area (animals/ha)*	*Technology*[a]
1	62	10 000	6·2	—	—	po-d
2	16	3 200	5·0	—	—	po-d
3	120	12 000	10·0	—	—	ps-i
4	15	3 500	4·3	—	—	po-t-i
5	290	20 000	14·5	86 (fp)	236 (fp)	ps-su-fp-co
6	600	50 000	12·0	—	—	su-d-co
7	127	10 000	12·7	290	29	po-t-co-i
8	10	1 000	10·0	—	—	su-d
9	38	1 500	25·0	—	—	su-f-d
10	192	10 000	19·0	—	—	su-f-d
11	50	4 500	11·0	—	—	su-i
12	80	6 600	12·2	216	31	su-i
13	150	4 500	33·0	—	—	f-po-i
14	110	5 700	19·3	4	1 420	f-fo
15	150	3 300	45·0	—	—	f-po-i
16	200	7 300	28·0	333	22	f-i
17	120	6 600	18·0	—	—	f-i
18	120	4 400	27·0	—	—	ho-i
19	100	4 600	22·0	188	29	ho-i-fo
20	100	4 400	23·0	—	—	ho-i
21	90	—	30·0	—	—	s-d
22	150	4 500	33·0	—	—	la-d-i
23	120	3 000	40·0	—	—	la-d
24	200	5 000	40·0	—	—	su-d
25	90	4 500	20·0	18	250	su-po-fo
Avg.	175·4	8 070	20·45	—	—	
26[b]	45	500	85·0	—	—	su-po-i

[a] Technologies: po = ponding, ps = settling ponds in series, su = settling units, fp = fishpond, f = straw or gravel filters, ho = homogenization, a = activated sludge, la = lime and activated sludge, t = tank wagon transportation.

Disposal: i = irrigation, fo = forest and poplar plantation disposal, co = composting and land disposal, d = discharge into surface waters.

[b] Cattle farm.

liquid manure is expected to double by 1980. Compared with the residential and industrial wastewater discharges of 1200×10^6 m^3/year, a volume of 7–14 $\times 10^6$ m^3/year of liquid manure might not seem too big. However, from the point of view of rural environmental protection, liquid manure discharges are a problem. For example, if the PE (population equivalent for direct discharge into streams) of a pig is taken as 5 and that of a cow as 20, then the daily liquid manure produced in Hungary would be equivalent to the wastewater discharges from 6–7 million people, or 70 % of the water pollution potential from the 10 million people living in Hungary.

A 1973 survey of 25 pig farms and one cattle farm[1] revealed that the average pig feedlot has 8000 pigs. This is 10 times larger than the number of pigs on the average feedlot in Western Europe. However, only 6 of the 24 farms surveyed were larger in capacity than 8000 pigs (see Table II). Most farms have concrete floors which are partially slatted. Feedlot facilities consist of building units housing 1000–2000 pigs per unit. The pigs are fed a dry high-protein feed of cornmeal, vitamin and micronutrient additives, plus ground limestone. Pigs get water from automatic waterers. Manure is deposited in the channels below the slatted floors and is flushed out hydraulically. Most feedlots have their own sows and fatten only their own pigs.

The volume of wastewater generated in Hungarian feedlots range from 4 to 45 litres/day/pig, with the average being about 20 litres/day/pig. For the one cattle farm surveyed, the wastewater volume was found to be 85 litres/day/cow. The concentration of pollutants varies from farm to farm (Table III), but the total quantity generated per animal appears to be similar for all farms. For example, the BOD ranges in concentration from 5300 to 20 000 mg/litre in three farms. However, BOD per pig for the same farms ranges between 200 and 260 g/day/pig (Table IV).

About 65 % of feedlot wastewater is discharged on land. About 82 % of the land-disposed wastewater is pretreated by solids separation via

TABLE III

CONCENTRATION OF POLLUTANTS IN WASTEWATERS FROM PIG FEEDLOTS IN HUNGARY

Site no.	*COD*	*BOD_5*	*SS*	*Total P*	*Total N*	*pH*
	(mg/litre)					
7	35 000	20 000	20 000	380	1 200	8·0
21	10 000	5 300	8 000	80	700	7·5–8·5
22	14 000	7 900	10 000	320	1 100	8·6
26[a]	29 000	14 500	15 600	370	1 750	7·5

[a] Data taken from cattle farm.

lagooning, settling units, straw filters, etc., followed by land irrigation. The rest of the feedlot wastewaters are discharged into rivers after treatment.

Homogenization of feedlot wastewaters through hydraulic jet mixers is preferred because they minimize nutrient losses. Dilutions of 1:10 with water are necessary, however, for proper homogenization so as to be able to spray irrigate the wastewater with sprinkler nozzles.

TABLE IV

QUANTITIES OF POLLUTANTS PRODUCED PER ANIMAL IN HUNGARIAN FEEDLOTS

Site no. and reference	*Wastewater discharged (litres/day/animal)*	BOD_5	*COD*	*SS*	*Total P*	*Total N*	*Total K*
		(g/day/animal)					
Pigs:							
7	12·7	250	450	250	4·8	15·3	—
21	30	200	340	300	3	26	—
22	33	260	460	330	4·6	16	—
Ref. 2	4–20	—	500–1 300	—	27–37	46–68	—
Cattle:							
Ref. 2	75	—	—	4 600	250	115	68

Storage lagoons must provide at least 90 days of hydraulic detention because no irrigation can take place during winter months. These lagoons are 2–3 m deep. Difficulties arise in cleaning the bottom of these lagoons after their liquid supernatant content has been pumped. Mechanical solids separators consisting of centrifuges and stationary and vibrating screens have been used. Loading rates for medium-size screens are 10–20 m^3/hr, and for centrifuges 5 m^3/hr. Centrifuges are expensive. Their main use is in thickening biological sludges.

It has been calculated that the optimum ratio of pigs to land is 18 pigs/ha, in terms of proper manure disposal and fertilization of corn cropland, provided no nitrogen is lost. However, if manure is lagooned then the ratio is 30 pigs/ha of corn cropland. For the surveyed farms (Table V) the ratio is 22–31 hogs/ha of land. Table V gives the land area requirements for various manure handling technologies.[3] In one site where a load ratio of 1420 pigs/ha of poplar trees is attained, tree damage is extensive and soil clogging is common. An upper limit of 300 pigs/ha appears to be desirable, based on loadings of 9·5 kg of organic matter per m^2/year for well-cultivated soil. However, pretreatment of liquid manure determines the upper limit for land applications. In Hungary, disposal on poplar forest land appears to be the easiest, followed by grassland spreading by irrigation after biochemical treatment. Problems of groundwater pollution, particularly those arising from nitrates, salt accumulations, excessive potassium and heavy metals concentrations, must be considered carefully, however.

Ten farms out of the 26 farms surveyed and listed in Table II discharge treated wastewater effluent directly into surface waters. In 7 of these 10 farms the wastewater treatment consists of lagooning whose effluent exceeds several times stream discharge effluent standards.

TABLE V

LAND REQUIREMENTS FOR THE DISPOSAL OF PIG FEEDLOT WASTEWATER RECEIVING VARIOUS PRETREATMENTS IN HUNGARY

Disposal method	*Hydraulic load (mm/year)*	*COD* (*kg/m²/year*)	*N*	*P*	*K*	*Land area (pigs/ha)*
Forest irrigation after homogenization (poplar trees)	120	9·5	0·55	0·155	0·31	33
Cow dung fertilization	—	0·30	0·007	—	—	—
Irrigation with homogenized manure	12 + 120	0·95	0·055	0·0155	0·031	330
Aerobic treatment and soil filtration	800–1 000	1·14	0·80	0·34	0·69	5

In 3 of the 10 farms, treatment of the wastewater is activated sludge treatment. Oxidation ditches have been tried but purification is not adequate. They produce effluent of poor quality, about 1000 mg COD/litre with poor settling characteristics.

A biochemical treatment plant patented by a Hungarian company is being marketed abroad. The system's components and COD removal efficiencies are as follows:

(a) Solids separation with vibrating screens (64 % COD removal for an influent of 10 000–20 000 mg COD/litre).
(b) Pre-aeration in stabilization tank.
(c) Alum coagulation and primary settling (85 % COD removal).
(d) Activated sludge aeration basins.
(e) Final clarification and denitrification in anaerobic packed columns (96 % COD removal).
(f) Chlorination (if required by law) of effluent (effluent BOD = 50 mg/litre).

Furthermore, the solids separated by the screens are conveyed out on to a pile and composted. The solids resulting from the addition of alum in the primary sedimentation tank plus excess activated sludge are placed on drying beds or centrifuged. This sludge is difficult to dewater.

The effluent from the system is a clear yellowish liquid which, however, must be disposed of in rivers with high flow volumes. If sprayed on land, application rates of 40 m^3/day/ha may be used, thus reducing the pigs to land ratio in the range of 1000 to 1300 pigs/ha. As yet, no treatment technologies of feedlot wastewaters are capable of producing effluents which can meet Hungarian effluent standards for all bio-engineering parameters (see Table VI). Treatment of wastewaters requires high capital

TABLE VI

EFFLUENT QUALITIES OBTAINABLE WITH VARIOUS TREATMENT TECHNOLOGIES

Technology	*COD*	*BOD_5*	*SS*	*Total N*	*Total P*	*Total salts*
	(*mg/litre*)					
Anaerobic lagoon	2 000–6 000	1 000–3 000	100–200	400–600	20–50	1 000
Oxidation ditch[a]	800–1 500	100–200	100–200	400–600	20–50	1 000
Lime-activated sludge	300–600	50–100	20–30	300–500	6–8	800–1 200
Alum-activated sludge and denitrification	350–600	50–70	20–30	200–300	6–8	800
Effluent standards	75	—	1 000	40[b]	4[c]	1 000

[a] Estimated from preliminary tests.
[b] Free NH_4 value; for lakes, 20 mg/litre NO_3.
[c] PO_4 value.

investments. They range from US$12/pig for tank wagon transport for land diposal to US$40–60/pig for biochemical treatment (see Table VII). The relative costs of operating a biochemical treatment plant are twice that of disposing wastewaters by irrigation, as is indicated in Table VII. Increasing the size of feedlots may reduce production costs per unit of animal capacity, but waste and wastewater management and disposal problems increase. Therefore, it has been determined in Hungary that the best size of pig feedlot is 8000–10 000 pigs per farm, with the uppermost limit being 30 000 pigs per farm.

Research and development studies are aimed at determining the most economical size of pig feedlots, feasible pretreatment systems before land utilization of wastes and wastewaters, and purification by biochemical treatments of various combinations. Attention is being given to the re-utilization of the wastes for growing fish, for refeeding to animals after

TABLE VII

RELATIVE CAPITAL AND OPERATING COSTS OF PIG FEEDLOT WASTEWATER TREATMENT AND DISPOSAL SYSTEMS IN HUNGARY

Technology	*Relative capital cost per pig capacity*	*Relative operating costs*	
		per pig/year	*per m³ of wastewater*
Tank wagon transport	1·0	1·0	1·0
Irrigation	2·2	0·5	0·8
Biochemical treatment of liquid	4·3	1·3	1·6
Sludge treatment of solids	2·5	1·0	1·1

Note: Relative cost of 1·0 for capital costs is US\$12/pig capacity, and for operating is US\$6/pig/year and US\$1·2/m³.

drying and for soil amendments. Systems are being researched which would provide sufficient treatment for stream discharges in areas where the soil has limestone fissures making land disposal of liquid manure inadvisable.

In Hungary, every activity in connection with water management is governed by the Water Management Act of 1964. This Act states: 'Any kind of contamination and hazardous pollution of waters is prohibited'. This means that 'waters have to be protected from any influents which may impair their biological and chemical properties, natural quality and self-cleaning capability. In the course of storage, handling, delivery and ultimate disposal of materials, care must be taken not to contaminate and pollute natural waters.'

A system of fines was promulgated for polluters of natural waters. Fines are increased progressively in case of negligence. The system of fines and concomitant measures have brought about significant success in water quality conservation in Hungary.

However, the judging of cases of pollution from agriculture required special consideration. In cooperation with agricultural and public health officials, the National Water Authority published in 1974 guidelines for the resolution of the conflict between the public demands for good, inexpensive pork and good, unpolluted water quality. Highlights of the main Principles of Agricultural utilization and Disposal of Liquid Manure are as follows:

(a) Liquid manure is no wastewater, but water with high organic solids content. Therefore, the main objective of treatment should not be purification of the liquid manure, but conservation of its fertilizer value.

(b) A prerequisite to liquid manure utilization is pretreatment. The type and degree of pretreatment will depend on local conditions and constraints.
(c) Liquid manure may cause much greater damage to water quality than municipal wastewater, if proper handling is not provided.
(d) The simplest form of utilization of liquid manure is soil conditioning and its use as a fertilizer. Other alternative methods of utilization are fishponding, biogas production and animal feed recycling.
(e) In planning, constructing and operating waste utilization systems, public and veterinary health as well as water quality must be considered fully.
(f) If, however, local and economic constraints do not permit utilization or land disposal of liquid manure, two alternatives may be followed: (i) either manure is purified to a degree prescribed for municipal and industrial wastes before discharge in surface water: or (ii) no animal feedlot with hydraulic waste flushing system is permitted to operate in such localities.
(g) Liquid manure is generated in the course of the operation of agricultural enterprises. Consequently, the construction and operation of waste disposal and utilization systems is the responsibility of the enterprise, even when the state government is granting financial and technical aid to the enterprise.
(h) Disposal of feedlot waste and wastewaters on pasture or cropland must be done properly according to official guidelines.

The issuance of these guidelines is but the first step in the effort of the Government of Hungary to balance requirements of food production and environmental quality.

REFERENCES

1. Farkas, P. Sertéshizlaldal szennyvizek tisztitása Magyarorszagon (The treatment of pig farm wastes in Hungary). VITUKI Report No. III.1/4/102. 1973.
2. Bartha, I. Korszeru hazai állattarté mezógazdasági uzemegysegek szennyvizeinek kezelése és hasznositáse (Manure handling and utilization techniques of up-to-date animal farms). MELYEPTERV Report, Budapest, 1972.
3. Vermes, L., Hazai higtrágya kezelési eljárasok komplex muszeki-gazdasagossági vizsgálata (The examination of Hungarian manure-handling techniques for their complex technical-economical performance). VITUKI Report No. III.3/4/132. 1972.
4. Farkas, P. Chemical-biological combined treatment tests for pigfarm waste, with special regard to phosphate elimination. Paper II/1, IUPAC Int. Congress on Ind. Waste Water, Stockholm 1970.

32

Feedlot Waste Management in the Soviet Union

B. A. RUNOV

Vice-Minister, USSR Ministry of Agriculture and Food, Moscow, USSR

In the Soviet Union, meat production has increased from 8·7 million tons in 1960 to 12·3 million in 1970 and 14·5 million in 1974. More than 260 large-scale livestock complexes have been built in the Russian Federated Republic over the last few years, largely as a result of a government decision in 1971 to develop livestock production on an industrial basis. It is planned to construct more than 1000 such large livestock complexes in the Soviet Union.

The beef cattle population is about 4 million head, which is only 3·6% of the total cattle population in the Soviet Union. To develop the beef industry further, 140 breeding farms, with a total population of approximately 200 000 head of cattle, have been established. They are mainly large units, housing 2500–30 000 animals each, and they are designed for intensive feeding. Some of the larger units, handling 10 000 head each, are: at Voronovo, near Moscow; at Pashsky, in the Leningrad region; at Yumatovsky, in the Bashkir Autonomous Republic; at Mir, in the Brest area; and at Bratsk (feeding 20 000 animals). Typical results in five such feedlots were: daily gain of 1 kg, feed conversion of 7:1 and 3·5 man-hr direct labour costs.

Since large-scale livestock farms and complexes may be viewed as factories producing high-quality fertilizer in addition to the basic dairy or meat product, practical possibilities exist for establishing continuous flow lines for manure disposal. Thus, in the design of such large farm systems, manure removal, processing, disinfection and storage must ensure the maximum preservation of fertilizer value with minimum labour, transportation and disposal costs. These goals are usually fulfilled by gravity-flow and recirculating manure removal systems, with subsequent fermentation under anaerobic conditions in the thermophilic regime. The gravity-flow system, in operation for many years in dairy farms, revealed a number of advantages, including transportation of manure without added water or the need to liquefy the manure. Recirculation with the hydraulic flushing of excrement has been successfully employed on a number of farms. In this system, the liquid fraction of the settled excrement (or a well-agitated mixture of it) serves as the transportation medium. Such a system has been in effect at the Zhodino complex near Minsk for more than 15 years.

The combined application of gravity-flow and recirculation systems for removal, together with anaerobic fermentation can considerably reduce the output of manure from feedlots. At present, work is nearing completion on a pilot anaerobic treatment plant for 300 cows at the Dzerzhinsky collective farm near Moscow. Working in combination with a recirculating removal system, the plant will be able to produce several types of organic fertilizer for use in the farm's hothouses, along with the heat from the combustible gas. If successful, the plant will be installed at other farms.

The utilization of manure as a supplement in feed rations is practised in the Soviet Union by adding manure to feed rations after drying it in dehydrators. (The Tomilino poultry factory in the Moscow region has been operating a guano drying unit since 1972, with a daily output of approximately 50 tons of dry matter.) Studies undertaken by Soviet scientists show that, by using high temperature drying of manure, it is possible to substitute up to 25% nutrients in the rations of young feeder cattle with no effects on the daily gain or feed conversion values. An original biochemical treatment plant was constructed on the Teleneshty state farm, in the Moldavian Republic. It is a yeast plant built to utilise wastes from a 24 000 head hog-feeding farm. The plant produces about 100 tons of dry feed yeast per year. One hopes that studies on conversion of processed manure into livestock feed will be expanded, particularly with the idea of overcoming the psychological barriers against their application.

Serious efforts are under way within the USSR to develop manure utilization methods which would effectively eliminate the possibilities of environmental pollution. How important an issue it is can be seen in the 1972 decision of the Government and the Party to insist upon a rational treatment of land, water, plant and wildlife resources in an effort to prevent any pollution of the environment due to agricultural production. A comprehensive set of measures is now being worked out by the various departments and agencies and is being implemented for the successful realization of this goal.

33

Animal Waste Management in Greece

F. Plytas, J. Matsoukas and M. Panayiotidis

Ministry of Agriculture and Ministry of Social Services, Athens, Greece

INTRODUCTION

The impressive increase in the demand for meat, milk and eggs after the Second World War, as a consequence of the increase in both consumer income and population, led to the development of animal feedlots near towns and villages. Additional slaughterhouses and meat and milk processing plants had to be built, compounding the environmental health and pollution problem from animal agriculture.

ANIMAL PRODUCTION PATTERNS IN GREECE

As is shown in Table I, between 1965 and 1974 total meat production increased by 104% while milk and egg production increased by 48% and 40% respectively.

Except for horses, mules and donkeys, which were reduced by 68%, the number of animals in Greece has also increased, as is shown in Table II. Poultry and swine production units have expanded rapidly over the last decade and have almost changed to intensive and mass production units. In 1975 there were some 145 farms with 10 000–80 000 layers capacity and 130 broiler farms with capacities ranging from 100 000 to 5 million broilers per year. In swine production there are 1200 units with 40–300 sows producing 600–4500 pigs/year, and 115 feedlots of 300–500 sows, fattening 4500–8000 pigs/year.

This rapid growth and change to intensive and mass production methods resulted from the new techniques and methods which are being applied in poultry and swine production today, as well as from the fact that the successful practice of the modern poultry and swine production industry does not need direct connection with agriculture. Beef and dairy cattle feedlots tend to remain small, although even there some large units are being developed.

TABLE I

PRODUCTION OF ANIMAL PRODUCTS IN GREECE, 1965–1974, IN TONS PER YEAR

Type of product	*Year*		*Change* (%)
	1965	1974	
Meat:			
Beef, veal	61 500	107 500	+75
Mutton, lamb, goat	79 000	111 000	+41
Pork	47 000	107 000	+128
Poultry	25 000	107 000	+330
Rabbit	2 600	6 500	+150
Total meat	215 000	439 000	+104
Milk:			
Cow	444 500	721 500	+66
Sheep	375 500	539 500	+44
Goat	289 500	380 000	+31
Total milk	1 109 500	1 641 000	+48
Eggs:			
Poultry eggs	82 000	115 000	+40

TABLE II

CHANGES IN ANIMAL POPULATION IN GREECE, 1965–1973

Type of animal	*Year*		*Change* (%)
	1965	1973	
Horses, mules, donkeys	948 500	646 000	−68
Cattle (cows and oxen)	1 083 500	1 233 000	+14
Sheep, goats	11 714 000	12 828 500	+10
Pigs	558 500	858 000	+54
Poultry	22 425 000	31 801 000	+42
Rabbits	1 233 000	2 046 000	+66

WASTE MANAGEMENT

In the case of the small-size animal units, manure is collected in small pits outside the stables. When the manure is dried, it is transported and spread on cropland.

In large animal feedlots, various methods of handling, treatment and disposal are used. Liquid manure handling systems are common in swine feedlots. Poultry manure from deep litter houses is dry when removed, so there are no major difficulties with its direct disposal on cropland.

Feedlot wastewater treatment systems incorporating primary treatment for the mechanical removal of solids, biological treatment for the removal of BOD and final disinfection by chlorination are being designed into the newly constructed large feedlots.

LAWS AND REGULATIONS

Animal production units must be built away from towns, villages, roads and tourist establishments, depending on the type of animal feedlot, as is shown in Table III.

In the case of poultry and swine feedlots there are also regulations establishing minimum distances between feedlots. Currently this minimum distance is 100 m, but will be changed to 500–1000 m. This regulation is intended to prevent the concentration of animal feedlots in one locality or region.

Construction of animal production units must be made according to good animal husbandry principles as well as being within laws and regulations concerning environmental quality and public health protection. Feedlot wastewaters come under the general regulations governing effluent discharges into public waterways from industrial sources. In order to upgrade wastewater treatment, the government offers loans and subsidies to animal producers.

TABLE III

MINIMUM DISTANCES (IN METRES) BETWEEN ANIMAL STABLES AND HUMAN ACTIVITY CENTRES

Human activity centre	*Type of animal*				
	Cattle	*Pigs*	*Sheep, goats*	*Poultry*	*Rabbits*
Towns above 5 000 inhabitants	500	700	500	500	500
Towns of 2 000–5 000 inhabitants	200	400	100	100	100
Villages above 1 000 inhabitants	50	100	50	50	50
National roads	50	100	50	50	50
Tourist places, etc.	300	500	300	300	300

34

Animal Waste Management in The Netherlands

A. A. JONGEBREUR

Institute of Agricultural Engineering, Wageningen, The Netherlands

INTRODUCTION

In 1973 about 53 000 Dutch farmers had an income out of the so-called intensive livestock production feedlots. Table I shows animal production figures for The Netherlands. Today, management skills and technology make it possible for one man to manage completely 1000 fattening pigs, 100 breeding sows, 15 000 laying hens, 45 000 broilers or 300 veal calves. It is

TABLE I

NUMBER OF FARMS WITH LIVESTOCK PRODUCTION IN THE NETHERLANDS IN 1973

Farms with fattening pigs	28 219
Average number of pigs per farm	121
Farms with breeding sows	36 000
Average number of breeding sows per farm	21·6
Farms with laying hens	28 942
Average number of laying hens per farm	615
Farms with broilers	2 438
Average number of broilers per farm	14 680
Farms with veal calves	4 200
Average number of veal calves per farm	101·6

expected that the number of animal farms will decrease in the future whereas the number of animals per farm will increase.

NUISANCE FROM FEEDLOTS

A 100-year-old Public Nuisance Act requires that every livestock producer must have a permit, which is given only to a feedlot which meets the

TABLE II

MINIMUM DISTANCE BETWEEN FEEDLOTS AND HUMAN HOUSES

Number of laying hens		*Number of broilers*	*Number of breeding sows*	*Number of fattening pigs or veal calves*	*Minimum distances (m)*	
Dry manure	*Liquid manure*				*Non-agricultural region*	*Agricultural region*
2 000	1 000	7 500	100	75	100	50
6 000	3 000	20 000	300	200	125	80
15 000	7 500	50 000	750	500	175	125
30 000	15 000	100 000		1 000	250	175
60 000	30 000	200 000		2 000	350	250

minimum distance standards given in Table II. In the case of existing feedlots, if the minimum distances are exceeded, odour control procedures must be initiated. Research is now being carried out to correlate odours and odour control with the size and type of animal feedlot, the housing system, the ventilation system, the management of the feedlot and the type of manure storage and handling. An objective of this research is to develop a simple and objective odour measurement technique.

VENTILATION AIR ODOURS

Odours from ventilation exhausts have been successfully controlled through the application of biological air washers. There are two major types of washers: counter-current when water and air are in the opposite direction of each other, and cross-current washers where the washing water and the ventilation air cross each other perpendicularly. Washer water is recirculated until it gets saturated with odours or with aerosols. To improve odour removal, the contact time between the exhaust air and the water is increased by using fill materials in the washer. These fill materials should have a surface area of approximately 200 m^2/m^3.

Inoculation of the washing water with activated sludge promotes the growth of bacteria on the fill material which act as trickling filters in stabilizing the dissolved organic components of the exhaust air. With this method of odour control, a reduction of approximately 60–85% of the concentrations of characteristic smelling components in the ventilation air may be achieved. Until now, biological air washers have been operating satisfactorily. However, this method of odour control is expensive, as shown in Table III, and is thus recommended for farms which are threatened with nuisance law suits.

TABLE III

INVESTMENT AND YEARLY COSTS FOR A BIOLOGICAL AIR WASHER OF 6000 M^3 AIR PER HOUR

	Initial investment	*Operating costs*
Swine feedlots	$22.72/pig capacity	$2.28/finished pig
Poultry feedlots	$1.81/laying hen	$0.40/hen/year

FIELD SPREADING ODOURS

Odour from field spreading operations may be prevented by aeration of the waste before disposal. The minimum volumes for aeration basins are 500

TABLE IV

COSTS OF WASTEWATER TREATMENT IN THE NETHERLANDS

	Fattening pigs		*Laying hens*	
	Storage under the slats	*Aeration and flushing*	*Storage of liquid manure in the house*	*Aeration and flushing*
Number of animals	1 000	1 000	15 000	15 000
Investment per animal	\$2.69	\$11.54	\$0.23	\$0.87
Extra investment per animal capacity	—	\$8.85	—	\$0.64
Extra yearly costs for aeration and flushing:				
per animal capacity	—	\$2.60	—	\$0.19
per delivered pig	—	\$1.04	—	—

litres per fattening pig, 1000 litres per breeding sow and 30 litres per laying hen. Power requirements for aeration by floating surface aerators are 6 W per fattening pig, 12 W per breeding sow and 0·5 W per laying hen. There are about 25 pig farms which apply an aeration system for odour control in The Netherlands. Costs associated with such operations are given in Table IV.

Another method of controlling odours is by direct injection of the waste into the soil. With a good working depth of the injector, the right size of injection teeth and the right amount of manure, soil injection functions satisfactorily. Good results are obtained with injection in both fallow land and in crops, *e.g.* maize, sugar beet and potatoes. Injection in permanent grassland is still being investigated. This method of field spreading is presently limited to small feedlots.

Recent trends are towards multi-storey poultry housing units with facilities to dry up manure by air circulation down to 35–40 % moisture. At this moisture, odours are moderated and subsequent dehydration to 15 % moisture becomes more attractive economically and technically.

MANURE BANKS

To encourage the use of feedlot wastes on cropland, the Development Fund for Agriculture has organized Manure Banks which subsidize the transport of liquid manure to croplands over long distances, thus stimulating its use on arable land not belonging to the feedlot owners. These Manure Banks are also organized to advise farmers on the correct application of the liquid manure. Manure Banks are now operating in three provinces in The Netherlands. The subsidy of a maximum of \$1.15/m^3 over a distance of more than 8 km is given to the enterprise which uses the liquid manure. Manure Banks are free to apply this subsidy for different distances, ranging from \$0.50/m^3 for a 8–15 km distance to \$2.50/m^3 for distances over 50 km.

Quantities transported over 8 km in 1974 and thus eligible for subsidy were 112 000 m^3 in three provinces. In 1975 over 175 000 m^3 were transported in one province alone.

For efficient transport, big tank wagons of 30–40 m^3 capacity are used. Since these cannot go on the fields at all times, it is best to build a storage pit on the arable farm. This pit may be covered with plastic film to prevent groundwater pollution. Filling and emptying of these pits is done by means of pumps located in a concrete pit. The investment costs for these pits for storing liquid manure range from \$3 to \$4/m^3. However, subsidies are also available for the construction of these pits.

35

Feedlot Waste Management in Bulgaria

N. BOGOEV

Research Institute on Water Supply, Sewage Purification and Sanitary Engineering, Sofia, Bulgaria

In Bulgaria, as in other countries, considerable concentration of livestock production has taken place in recent years. Large animal feedlots of up to 100 000 pigs and dairy cattle complexes of up to 1500 head are now operational. This high concentration of livestock production has created many problems which relate to environmental pollution.

Currently, the most critical issue is the problem of what to do with the large wastes and wastewaters generated at the large pig feedlots. In most of the swine feedlots, large quantities of water are used to remove wastes by hydraulic flushing. The wastewater generated is assumed to be 40–100 litres/pig/day. The BOD of this wastewater averages about 750 mg/litre. Only in one large feedlot is dry handling of wastes practised.

In Bulgaria, as in other countries, final disposal is based on the principle that animal wastes are rich in nutrients and therefore must be utilized as soil amendments or as plant fertilizers. However, public health officials concerned with disease and groundwater pollution potential have not authorized the irrigation of land with swine feedlot wastewaters without adequate pretreatment. Experimental research programmes have been initiated to prove that land disposal of pig wastes is safe for both human and animal welfare. In the meantime, however, construction of large pig feedlots is proceeding at a much faster pace than the research programme. Therefore, biological treatment of wastewaters must be given before disposal.

Wastewaters from pig farms must be treated to remove 95–99 % of BOD. This high treatment is needed for farms discharging into rivers with low flows. The first biological treatment plant designed by a foreign firm consisted of a collection pit followed by solids separation with a vibrating screen, aeration, final sedimentation and drying beds for the separated solids. Unfortunately, frequent clogging of the vibrating screen and other failures rendered the plant ineffective.

Pilot plants for two-stage biological treatment, chemical treatment and mechanical dewatering of solids with vibrating screens and centrifuges have been used to obtain design values. Such values are then submitted to design institutes to size treatment units and to determine capital investment costs of pig wastewater treatment. Capital investment costs of \$36 per pig capacity and operating costs of \$4 per pig marketed are estimated for waste treatment. For swine feedlots of 100 000 pig capacity, the capital costs for proper wastewater treatment are excessive, making such large feedlots not economically feasible.

36

Feedlot Waste Management in Czechoslovakia

J. Hojovec

Veterinary Medicine College, University of Brno, Brno, ČSSR

The present stage of the development of agriculture in the Czechoslovak Socialist Republic is characterized by high specialization and concentration of production based on cooperation and integration of the work operations. One of the most highly concentrated and specialized agricultural production operations are the large confinement animal feedlots.

In modern large animal feedlots a great number of animals are kept within a confined space, environmentally optimized through extensive mechanization, where frequent turnover by new animal generations is practised, high-value feeds are fed and a minimum number of workers per animal are employed.

In Czechoslovakia the following animal numbers per feedlot are considered as large-scale intensive feedlots:

Cattle: 500 or more dairy cows
800 or more calves
500 or more young dairy cattle
800 or more beef cattle
Pigs: 5000 or more pigs being fattened
750 or more sows in farrowing facilities
Poultry: laying-hen flocks larger than 100 000 birds
poultry rearing houses with an overall capacity higher than 60 000 birds
broiler houses with an overall capacity higher than 100 000 birds

The most common method of waste handling used in such large confinement feedlots is liquid manure. Liquid manure is an important source of organic and mineral fertilizing elements. It can be used for agricultural purposes, especially for the fertilization of soil. Health hazards arise only when liquid manure is used without due care; these hazards must be avoided.

Based on existing information and knowledge, the following recommendations should be followed in the use of liquid manure systems for large cattle, swine or poultry feedlots:

1. In considering a site, selecting it and in designing a feedlot project: (a) general public health and veterinary sanitation principles should be followed so as to keep the surrounding territory disease-free as much as possible and to prevent gross environmental pollution; (b) liquid manure must be stored in places which are located on the periphery of the farm, preferably on a side opposite to the nearest residential area or to other premises whose sanitation is critical and must be protected.
2. Land application of liquid manure guarantees the most efficient use of the soil-fertilizing value of animal wastes with the lowest investments of capital and costs of operation.
3. Storage pits and sewer pipes must be watersealed and sized to provide at least 100 days of storage, plus additional space for emergency storage, and must have provision for the adequate disinfection of all the stored wastes during times of disease epidemics or when disinfection is essential in the protection of the surrounding land and water areas. Disinfection chemicals residues must not create soil pollution either.
4. Untreated liquid manure can be spread on agricultural fields only on condition that it was produced by disease-free livestock.

The problems which require additional research are the problems of the tenacity of the pathogenic germs in liquid manure, the control of odours and the design of waste treatment systems. It is therefore recommended to research intensively processes which could remove or moderate excessive, noxious smells from animal feedlots. Improvements are needed for thermophilic aerobic stabilization systems to become suitable for liquid manure treatment. Researchers who are working on this system are urged to make their results and findings available through a continuous and prompt network of information exchange between countries and interested institutes.

When deodorizing chemicals are used, they must not inhibit the microbiological activity of soil or deteriorate the quality of surface or groundwater.

To limit the release of odours and to avoid nitrogen losses during land application of liquid manure, direct incorporation of manure into soil is recommended. This method should be practised exclusively in highly populated areas surrounding swine feedlots. Liquid manure can be transported by pumping or in closed tanks when solids have been separated, or in the case of dry poultry manure, these wastes can be transported on open trucks. The pollution of roads must be prevented in either case. Liquid manure may be used for soil fertilization as the only fertilizer, combined with straw or green manure, with or without mineral fertilizers. It is advantageous in crop rotations to use the aforementioned combinations alternately.

It is a realistic estimate to say that gross crop yield will increase by at least 5% after the application of liquid manure, in comparison with the use of manure with bedding. If manuring requirements of different crop rotations are fully considered, liquid manure can be used effectively not only in two-year and three-year cycles but also in annual application systems.

However, systems of fertilization need further improvement, both from the viewpoint of the efficient use of liquid manure and from the viewpoint of the protection of soil and water.

The main health hazards to groundwater resources and surface waters develop from the percolation, washing or flushing of pathogenic germs and biochemical elements of liquid manure. Adequate precautions should be taken around areas demarcated by state authorities as especially vulnerable to water pollutants. Research should be done to monitor and measure the rate and extent of penetration of various components of liquid manure through the surface of the soil profile under different soil and climatic conditions, and thus to establish the maximum doses of liquid manure in single applications, and the total maximum volume to be applied with respect to the soil type, soil species, climate and system of land management being practised.

Suitable application equipment for liquid manure needs to be further developed and internationally tested before being marketed.

Another form of safe waste treatment is composting of the liquid manure after mixing it with water-absorbing materials so as to optimize moisture content. Pulverized municipal solid wastes are considered as suitable for this purpose. It is recommended that such a process be intensively researched and developed as soon as possible because it permits the utilization of two major waste materials. The production of a solid manure pile from liquid manure by mixing it with straw outside the barns is considered inefficient.

Promising prospects are offered by the use of liquid manure for the reclamation of deficient soils and of areas used for disposal of power plant ashes and mining wastes. Research efforts in this field should be continued.

Pig liquid manure can be recommended to be used for the production of substrate for large-scale mushroom growing. The parameters of this method should be refined through research, and recommendations for the practical application of the method should be worked out.

To use dried poultry dung and other processed excrements for refeeding to animals, it is necessary, as soon as possible, to determine the actual nutritive value of the waste, to determine optimum doses for individual categories of animals, and to identify and assess the health implications of waste refeeding, taking into account the new hazards such as toxins, antibiotics, hormonal substances and other biologically active substances present in animal excrements.

Animal excrements should be processed by physical and chemical as well as microbiological methods which result in the conservation of the nitrogenous compounds, and for the production of utilizable energy in the form of methane gas.

Joint treatment of liquid manure with municipal sewage is economical at animal concentrations exceeding 3600 animal equivalent (approximately 23 000–39 000 population equivalent) and only on the condition that the municipality pays for a fair share of the cost.

For a quicker solution of the problems of the utilization of wastes and other by-products from animal production, it is recommended that the feasibility of establishing a network for the prompt exchange of information on results of work being carried out on these problems in various parts of the world be studied immediately and implemented soon.

It is also necessary to conduct a systematic study of the economic effectiveness of different technologies of the treatment and utilization of liquid manure, and to evaluate the best methods from the standpoint of the needs of the national economy, the aspirations of the people and their requirements for nutritious, healthy animal meat, milk and eggs, and with respect to the protection of our natural resources from excessive pollution from animal feedlots.

Appendix

APPENDIX

Glossary of Terms

(ASAE Recommendation ASAE R292.1, reprinted from *Agricultural Engineers Yearbook*, ASAE, St Joseph, Mich., 1975, pp. 515–17)

SECTION 1—PURPOSE AND SCOPE

1.1. The terminology reported herein is intended to establish uniformity in terms used in the field of rural waste management. Terms and definitions were adopted from related fields where applicable.
1.2. Standard procedures for the determination of most of the terms defined herein may be found in *Standard Methods* (American Public Health Association, New York, NY). A source for additional wastewater terms is the Glossary in *Water and Waste Water Control Engineering* (Water Pollution Control Federation, Washington, DC).

SECTION 2—DEFINITIONS

2.1. Activated sludge process. A biological wastewater treatment process in which a mixture of wastewater and activated sludge is agitated and aerated. The activated sludge is subsequently separated from the treated wastewater (mixed liquor) by sedimentation and wasted or returned to the process as needed.
2.2. Adsorption. (1) The adherence of dissolved, colloidal, or finely divided solids on the surfaces of solid bodies with which they are brought into contact. (2) Action causing a change in concentration of gas or solute at the interface of a two-phase system.
2.3. Aerobic bacteria. Bacteria that require free elemental oxygen for their growth. Oxygen in chemical combination will not support aerobic organisms.
2.4. Aerobic decomposition. Reduction of the net energy level of organic matter by aerobic microorganisms.
2.5. Aerobic lagoon, *see* **Lagoon**
2.6. Aeration. (1) The bringing about of intimate contact between air and a liquid by one or more of the following methods: (a) spraying the liquid in the

air, (b) bubbling air through the liquid, (c) agitating the liquid to promote surface absorption of air. (2) The supplying of air to confined spaces under nappes, downstream from gates in conduits, etc., to relieve low pressures and to replenish air entrained and removed from such confined spaces by flowing water. (3) Relief of the effects of cavitation by admitting air to the section affected.

2.7. Aeration tank. A tank in which sludge, wastewater, or other liquid is aerated.

2.8. Aerosol. A system of colloidal particles dispersed in a gas, smoke or fog.

2.9. Agitation. The turbulent remixing of liquid and settled solids.

2.10. Agricultural wastes. Most such wastes are associated with the production of food and fibre on farms, ranges, and forests. These wastes normally include animal manure, crop residues, and dead animals. Agricultural chemicals, fertilizers and pesticides, which find their way into the soil and subsequently into the surface and subsurface water, are classified as agricultural wastes.

2.11. Algae. Primitive plants, one- or many-celled, usually aquatic and capable of synthesizing their foodstuffs by photosynthesis.

2.12. Alkalinity. The capacity of water to neutralize acids, a property imparted by the water's content of carbonates, bicarbonates, hydroxides, and occasionally borates, silicates, and phosphates. It is expressed in milligrammes per litre of equivalent calcium carbonate.

2.13. Anaerobic bacteria. Bacteria not requiring the presence of free or dissolved oxygen for metabolism. Strict anaerobes are hindered or completely blocked by the presence of dissolved oxygen and sometimes by the presence of highly oxidized substances, such as sodium nitrates, nitrites, and perhaps sulphates. Facultative anaerobes can be active in the presence of dissolved oxygen, but do not require it.

2.14. Anaerobic decomposition. Reduction of the net energy level and change in chemical composition of organic matter caused by microorganisms in an anaerobic environment.

2.15. Bacteria. A group of universally distributed, rigid, essentially unicellular microscopic organisms lacking chlorophyll. Bacteria usually appear as spheroid, rod-like or curved entities, but occasionally appear as sheets, chains, or branched filaments. Bacteria are usually regarded as plants.

2.16. Biochemical oxygen demand (BOD). The quantity of oxygen used in the biochemical oxidation of organic matter in a specified time, at a specified temperature, and under specified conditions. A standard test used in assessing wastewater strength.

2.17. Biodegradation (biodegradability). The destruction or mineralization of either natural or synthetic organic materials by the microorganisms populating soils, natural bodies of water, or wastewater treatment systems.

2.18. Biological oxidation. The process whereby living organisms in the presence of oxygen convert the organic matter contained in wastewater into a more stable or a mineral form.

2.19. Biological stabilization. Reduction in the net energy level of organic matter as a result of the metabolic activity of organisms.

2.20. Biological wastewater treatment. Forms of wastewater treatment in which bacterial or biochemical action is intensified to stabilize, oxidize, and nitrify the unstable organic matter present. Intermittent sand filters, contact bed, trickling filters, and activated sludge processes are examples.

2.21. Carbon–nitrogen ratio (C/N). The weight ratio of carbon to nitrogen in a waste material.

2.22. Cesspool. A lined or partially lined underground pit into which raw animal and/or household wastewater is discharged and from which the liquid seeps into the surrounding soil. Sometimes called leaching cesspool.

2.23. Chemical oxidation. Oxidation of organic substances without benefit of living organisms. Examples are by thermal combustion or by oxidizing agents such as chlorine.

2.24. Chemical oxygen demand (COD). A measure of the oxygen-consuming capacity of inorganic and organic matter present in water or wastewater. It is expressed as the amount of oxygen consumed from a chemical oxidant in a specified test. It does not differentiate between stable and unstable organic matter and thus does not necessarily correlate with biochemical oxygen demand. Also known as OC and DOC, oxygen consumed and dichromate oxygen consumed, respectively.

2.25. Chlorination. The application of chlorine to water, sewage, or industrial wastes, generally for the purpose of disinfection, but frequently for accomplishing other biological or chemical results.

2.26. Coagulant. A compound responsible for coagulation. A floc-forming agent.

2.27. Coagulation. In water and wastewater treatment, the destabilization and initial aggregation of colloidal and finely divided suspended matter by the addition of a floc-forming chemical or by biological processes.

2.28. Coliform-group bacteria. A group of bacteria predominantly inhabiting the intestines of man or animal, but also occasionally found elsewhere. It includes all aerobic and facultative anaerobic, Gram-negative, non-spore-forming bacilli that ferment lactose with production of gas. Also included are all bacteria that produce a dark, purplish-green colony with metallic sheen by the membrane-filter technique used for coliform identification. The two groups are not always identical, but they are generally of equal sanitary significance.

2.29. Colloidal matter. Finely divided solids which will not settle but may be removed by coagulation or biochemical action or membrane filtration.

2.30. Composting. Present-day composting is the aerobic, thermophilic decomposition of organic wastes to a relatively stable humus. The resulting humus may contain up to 25 % dead or living organisms and is subject to further, slower decay but should be sufficiently stable not to reheat or cause odour or fly problems. In composting, mixing and aeration are provided to maintain aerobic conditions and permit adequate heat development. The decomposition is done by aerobic organisms, primarily bacteria, actinomycetes and fungi.

2.31. Contamination. Any introduction into water (air or soil) of microorganisms, chemicals, wastes, or wastewater in a concentration that makes the water (air or soil) unfit for its intended use.

2.32. Dehydration. The chemical or physical process whereby water in chemical or physical combination with other matter is removed.

2.33. Denitrification. The reduction of nitrates with nitrogen gas evolved as an end product.

2.34. Detention pond. An earthen basin constructed to store runoff water until such time as the fluids may be recycled on to land.

2.35. Deoxygenation. The depletion of the dissolved oxygen in a liquid. Under natural conditions associated with the biochemical oxidation of organic matter present.

2.36. Digestion. Though aerobic digestion is being used, the term digestion commonly refers to the anaerobic breakdown of organic matter in water solution or suspension into simpler or more biologically stable compounds, or both. Organic matter may be decomposed to soluble organic acids or alcohols and subsequently converted to such gases as methane and carbon dioxide. Complete destruction of organic solid materials by bacterial action alone is never accomplished.

2.37. Disinfection. The art of killing the larger portion of microorganisms in or on a substance with the probability that all pathogenic bacteria are killed by the agent used.

2.38. Dissolved oxygen (DO). The oxygen dissolved in water, wastewater, or other liquid, usually expressed in milligrammes per litre, parts per million, or percent of saturation.

2.39. Effluent. (1) A liquid which flows out of a containing space. (2) Wastewater or other liquid, partially or completely treated, or in its natural state, flowing out of a reservoir, basin, treatment plant, or industrial treatment plant, or part thereof.

2.40. Electroosmosis. An electrokinetic phenomenon in which an interstitial liquid is transported through a porous medium under the influence of an externally applied electromotive force.

2.41. Electrophoresis. The movement of suspended particles through a fluid under the action of an electromotive force applied to electrodes in contact with the suspension.

2.42. Escherichia coli (E. coli). One of the species of bacteria in the coliform group. Its presence is considered indicative of fresh faecal contamination.

2.43. Evaporation rate. The quantity of water, expressed in terms of depth of liquid water, evaporated from a given water surface per unit of time. It is usually expressed in inches per day, month or year.

2.44. Food to microorganism ratio (F/M). The ratio of organic food, BOD, to microorganisms.

2.45. Facultative bacteria. Bacteria which can adapt themselves to growth in the presence, as well as in the absence, of oxygen.

2.46. Facultative decomposition. Reduction of the net energy level of organic matter by microorganisms which are facultative.

2.47. Farm lagoon, *see* **Lagoon.**

2.48. Farm waste, *see* **Agricultural wastes.**

2.49. Fertilizer value. The potential worth of the plant nutrients that are contained in the wastes and could become available to plants when applied on to the soil. A monetary value assigned to a quantity of organic wastes represents the cost of obtaining the same plant nutrients in their commercial form and in the amounts found in the waste. The worth of the waste as a fertilizer can be estimated only for given soil conditions and other pertinent factors such as land availability, time, and handling.

2.50. Filtration. The process of passing a liquid through a filtering medium (which may consist of granular material, such as sand, magnetite, or diatomaceous earth, finely woven cloth, unglazed porcelain, or specially prepared paper) for the removal of suspended or colloidal matter.

2.51. Flocculation. In water and wastewater treatment, the agglomeration of colloidal and finely divided suspended matter after coagulation by gentle stirring by either mechanical or hydraulic means. In biological wastewater treatment where coagulation is not used, agglomeration may be accomplished biologically.

2.52. Gasification. The transformation of soluble and suspended organic materials into gas during waste decomposition.

2.53. Holding unit. A storage unit in which accumulations of manure are collected before subsequent handling or treatment, or both, and ultimate disposal. Water may be added in the pit to promote liquefication.

2.54. Humus. The dark or black carboniferous residue in the soil resulting from the decomposition of vegetable tissues of plants originally growing therein. Residues similar in appearance and behaviour are found in composted manure and well-digested sludges.

2.55. Hydraulic collection and transport system. The collection and transportation or movement of waste material through the use of water.

2.56. Incineration. The rapid oxidation of volatile solids within a specially designed combustion chamber.

2.57. Incubation. Maintenance of viable organisms in or on a nutrient substrate at constant temperature for growth and reproduction.

2.58. Infiltration. The process whereby water enters the soil through the immediate surface.

2.59. Infiltration rate. (1) The rate at which water enters the soil or other porous material under a given condition. (2) The rate at which infiltration takes place, expressed as depth of water per unit time, usually in inches or cm per hour.

2.60. Influent. Water, wastewater, or other liquid flowing into a reservoir, basin, or treatment plant, or any unit thereof.

2.61. Inoculum. Living organisms, or an amount of material containing living organisms (such as bacteria or other microorganisms) which are added to initiate or accelerate a biological process (*e.g.* biological seeding).

2.62. Lagoon. An all-inclusive term commonly given to a water impoundment in which organic wastes are stored or stabilized, or both. Lagoons may be described by the predominant biological characteristics (aerobic, anaerobic, or facultative), by location (indoor, outdoor), by position in a series (first stage, second stage, etc.) and by the organic material accepted (sewage, sludge, manure, or other).

2.63. Leaching. (1) The removal of soluble constituents from soils or other material by water. (2) The removal of salts and alkali from soils by abundant irrigation combined with drainage. (3) The disposal of a liquid through a non-watertight artifical structure, conduit, or porous material by downward or lateral drainage, or both, into the surrounding permeable soil.

2.64. Liquefaction. (1) Act or process of liquefying or of rendering or becoming liquid; reduction to a liquid state. (2) Act or process of converting a solid or a gas to a liquid state by changes in temperature or pressure, or the changing of the organic matter in wastewater from a solid to a soluble state.

2.65. Liquid manure. A suspension of livestock manure in water, in which the concentration of manure solids is low enough so the flow characteristics of the mixture are more like those of Newtonian fluids than plastic fluids.

2.66. Liquor. Water, wastewater, or any combination. Commonly used to designate liquid phase when other phases are present.

2.67. Litter. Vegetative material, such as leaves, twigs, and stems of plants, lying on the surface of the ground in an undecomposed or slightly decomposed state. The bedding material used for poultry.

2.68. Manure. The faecal and urinary defaecations of livestock and poultry. Manure may often contain some spilled feed, bedding or litter.

2.69. Manure flume. Any restricted passageway, open along its full length to the atmosphere, through which liquid moves by gravity.

2.70. Manure stack. A place with an impervious floor and side walls to contain manure and bedding until it may be recycled.

2.71. Manure tank. A storage unit in which accumulations of manure are collected before subsequent handling or treatment, or both, and ultimate disposal. Water may be added in the tank to promote liquefication.

2.72. Milkhouse wastes. The wastewater containing milk residues, detergents which are generated in a milkhouse.

2.73. Mixed liquor. A mixture of activated sludge and organic matter undergoing activated sludge treatment in the aeration tank.

2.74. Odour threshold. The point at which, after successive dilutions with odourless water, the odour of the water sample can just be detected. The threshold odour is expressed quantitatively by the number of times the sample is diluted with odourless water.
2.75. Organic matter. Chemical substances of animal or vegetable origin, or more correctly, of basically carbon structures, comprising compounds consisting of hydrocarbons and their derivatives.
2.76. Oxidation ditch. A modified form of the activated sludge process. An aeration rotor supplies oxygen and circulates the liquid in an oval, racetrack-shaped open channel ditch.
2.77. Oxidation pond. A basin used for retention of wastewater before final disposal, in which biological oxidation of organic material is effected by natural or artificially accelerated transfer of oxygen to the water from air.
2.78. pH. The reciprocal of the logarithm of the hydrogen-ion concentration. The concentration is the weight of hydrogen-ions, in grammes, per litre of solution. Neutral water, for example, has a pH value of 7 and a hydrogen-ion concentration of 10^{-7}.
2.79. Percolation. (1) The flow or trickling of a liquid downward through a contact or filtering medium. The liquid may or may not fill the pores of the medium.
2.80. Percolation rate. The rate of movement of water under hydrostatic pressure through the interstices of the rock or soil, except movement through large openings such as caves.
2.81. Permeability. The property of a material which permits appreciable movement of water through it when saturated and actuated by hydrostatic pressure of the magnitude normally encountered in natural subsurface water.
2.82. Pollution. The presence in a body of water (or soil or air) of material in such quantities that it impairs the water's usefulness or renders it offensive to the senses of sight, taste, or smell. Contamination may accompany pollution. In general, a public-health hazard is created, but, in some instances, only economy or aesthetics are involved as when waste salt brines contaminate surface waters or when foul odours pollute the air.
2.83. Population equivalent (PE). A means of expressing the strength of organic material in wastewater. Domestic wastewater consumes, on an average, 0·17 lb (0·08 kg) of oxygen per capita per day, as measured by the standard BOD test. This figure has been used to measure the strength of organic industrial waste in terms of an equivalent number of persons. For example, if an industry discharges 1000 lb (454 kg) of BOD per day, its waste is equivalent to the domestic wastewater from 6000 persons (1000 ÷ 0·17 = 6000). Caution must be exercised when using population equivalents because of the difficulty in comparing agricultural wastes directly with municipal wastes.

2.84. Putrefaction. Biological decomposition of organic matter with the production of ill-smelling products associated with anaerobic conditions.

2.85. Rural wastes. Wastes produced in rural areas. Most such wastes are associated with the production of food and fibre on farms, ranges and forests. These wastes normally include animal manure, crop residues and dead animals. Residual fertilizers, pesticides, inorganic salts and eroded soil may also be classified as rural wastes when they are in non-urban areas. Domestic solid refuse, human sewage and industrial wastes generated and handled in the rural environment are considered rural wastes.

2.86. Sediment. (1) Any material carried in suspension by water which will ultimately settle to the bottom after the water loses velocity. (2) Fine waterborne matter deposited or accumulated in beds.

2.87. Sedimentation tank. A basin or tank in which water or wastewater containing settleable solids is retained to remove by gravity a part of the suspended matter. Also called sedimentation basin, settling basin, settling tank.

2.88. Seepage. (1) Percolation of water through the lithosphere. Definitive meaning usually is described by an adjective such as influent, effluent (*see* Infiltration). (2) The slow movement of water through small cracks, pores, interstices, of a material into or out of a body surface or subsurface water. (3) The loss of water by infiltration from a canal, reservoir, manure tank or manure stack. It is generally expressed as flow volume per unit time.

2.89. Septage. Septic tank pumpings; the mixed liquid and solid contents pumped from septic tanks and dry wells used for receiving domestic-type sewage.

2.90. Septic tank. A settling tank in which settled solid matter is in immediate contact with the wastewater flowing through the tank and the organic solids are decomposed by anaerobic bacterial action.

2.91. Settleable solids. (1) That matter in wastewater which will not stay in suspension during a preselected settling period, such as one hour, but either settles to the bottom or floats to the top. (2) In the Imhoff cone test, the volume of matter that settles to the bottom of the cone in one hour.

2.92. Settling tank. A basin or tank in which water or wastewater containing settleable solids is retained to remove by gravity a part of the suspended matter. Also called sedimentation basin, sedimentation tank, settling basin.

2.93. Sewage. The spent water of a community. Term now being replaced in technical usage by preferable term wastewater.

2.94. Silt. (1) Soil particles which constitute the physical fraction of a soil between 0·005 mm and 0·05 mm in diameter. (2) Fine particles of soil carried in suspension by flowing water. (3) Deposits of waterborne material in a reservoir, on a delta or on overflowed lands.

2.95. Slotted floors. The floor surface of a building which has open splits, spaces or grooves to allow material to drop below the floor surface.

2.96. Sludge. The accumulated solids separated from liquids, such as water or wastewater, during processing, or deposits on bottoms of streams or other bodies of water. (2) The precipitate resulting from chemical treatment, coagulation, or sedimentation of water or wastewater.

2.97. Sludge volume index (SVI). The ratio of the volume of millilitres of sludge settled fom a 1000 ml sample in 30 min to the concentration of mixed liquor in milligrammes per litre multiplied by 1000.

2.98. Solids content. The residue remaining when the water is evaporated away from a sample of water, sewage, other liquids, or semi-solid masses of material and the residue is then dried at a specified temperature, usually 103 °C.

2.99. Stabilization pond. A type of oxidation pond in which biological oxidation of organic matter is effected by natural or artificially accelerated transfer of oxygen to the water from air.

2.100. Sterilization. The destruction of all living organisms, ordinarily through the agency of heat or of some chemical.

2.101. Supernatant. The liquid standing above a sediment or precipitate.

2.102. Suspended solids. (1) Solids that either float on the surface of, or are in suspension in, water, wastewater, or other liquids, and which are largely removable by laboratory filtering. (2) The quantity of material removed from wastewater in a laboratory test, as prescribed in *Standards Methods for the Examination of Water and Wastewater* and referred to as non-filterable residue.

2.103. Total solids. The sum of dissolved and undissolved constituents in water or wastewater, usually stated in milligrammes per litre.

2.104. Trickling filter. A filter consisting of an artificial bed of coarse material, such as broken stone, clinkers, slate, slats brush, or plastic materials, over which wastewater is distributed or applied in drips, films, or spray from troughs, drippers, moving distributors, or fixed nozzles, and through which it trickles to the underdrains, giving opportunity for the formation of zooleal slimes which clarify and oxidize the wastewater.

2.105. Volatile acids. Fatty acids containing six or less carbon atoms, which are soluble in water and which can be steam-distilled at atmospheric pressure. Volatile acids are commonly reported as equivalent to acetic acid.

2.106. Volatile solids. The quantity of solids in water, wastewater, or other liquids lost in ignition of the dry solids at 600 °C.

2.107. Volatile suspended solids (VSS). That portion of the suspended solids residue driven off as volatile (combustible) gases at a specified temperature and time, usually 600 °C for at least 1 hr.

Index